NANOMATERIALS

NANOMATERIALS

Dr. Brijesh Singh

SHREE PUBLISHERS & DISTRIBUTORS
NEW DELHI-110 002

Edition : 2013

ISBN : 978–81–8329–516–1

Published by :
SHREE PUBLISHERS & DISTRIBUTORS
22/4735 Prakash Deep Building,
Ansari Road, Darya Ganj,
New Delhi–110 002

Printed in India.

Preface

Nanomaterials is a field that takes a materials science-based approach to nanotechnology. It studies materials with morphological features on the nanoscale, and especially those that have special properties stemming from their nanoscale dimensions. Nanoscale is usually defined as smaller than a one tenth of a micrometer in at least one dimension, though this term is sometimes also used for materials smaller than one micrometre.

Nanomaterials are provides a comprehensive impression of the current state of nanomaterials. Employing terminology familiar to materials scientists and engineers, it provides an introduction that delves into the unique nature of nanomaterials. Nanomaterials also offer opportunity for ultra-sensitive biomolecular sensing throughout their local field optical effects, which are several orders of extent higher than the corresponding bulk effects. Optically lively metallic and quantum dot nanomaterials have opened up avenues for newer techniques such as local surface plasmon resonance, surface-enhanced fluorescence, surface-enhanced Raman scattering, fluorescence resonance energy transfer, time-resolved fluorimetry, and others.

Nanomaterials can be metals, ceramics, polymeric materials, or composite materials. Their defining characteristic is a very small feature size in the range of 1-100 nanometres (nm). The unit of nanometre derives its prefix nano from a Greek word meaning dwarf or extremely small. One nanometre spans 3-5 atoms lined up in a row. By comparison, the diameter of a human hair is about 5 orders of magnitude larger than a nanoscale particle. Nanomaterials are not simply another step in miniaturization, but a different arena entirely; the nanoworld lies midway between the scale of atomic and quantum phenomena, and the scale of bulk materials. At the nanomaterial level, some material properties are affected by the laws of atomic physics, rather than behaving as traditional bulk materials do.

There is no strict dividing line between nanoparticles and non-nanoparticles. The size at which materials display different properties to the bulk material is material dependant and can certainly be claimed for many materials much larger in size than 100nm. Definitions certainly become more difficult for materials that are a very long way from being a sphere, such as carbon nanotubes for example. One of the aims for these materials is to grow them into long tubes, certainly not 'nano' in length, but as they have a diameter in the order of 3nm for a single walled tube, they have properties that distinguish them from other allotropes of carbon, and hence can be described as 'nanomaterials'. This sort of nanomaterial has led to the extension of the idea of nanomaterials being considered as such if any one of their structural features are on a scale of less than 100nm, that cause their properties to be different from that of the bulk material.

Nanotechnology refers to engineering materials at the atomic scale: the size of nanometres. Richard Smalley, Robert Curl and Harold Kroto (currently a technical advisor to Luna) were awarded the Nobel prize for their discovery of a spherical crystal form of carbon they named a buckminsterfullerene for its resemblance to the geodesic shape that architect incorporated in many of his designs. The original fullerene sphere, sometimes called a "Buckyball," contained 60 carbon atoms bound by single and double bonds that form a three dimensional geodesic spheroidal crystal. Since that discovery other related forms have included larger spheres. Collectively these materials are called fullerenes. Fullerene structures have a number of unusual properties (chemical, physical and biological) that are significantly different from other organic materials. They are comprised of pure carbon, hence insoluble in water and in many organic solvents. Fullerenes also have an intrinsic thirst for absorbing free radicals, which are molecules that have an uneven number of electrons. Having an uneven number of electrons makes a molecule unstable because the electrons want to mate. The reactive molecule may attach to another molecule to make itself more stable. Sometimes this starts a chain reaction that can harm the cells. This chain reaction of free radicals may be responsible for numerous age-related diseases. Fullerenes are exceptional free radical scavengers, or antioxidants which can intercept free radicals and neutralize them before they cause harm. An area of development at Luna is supplementing natural anti-oxidants with fullerenes to treat conditions where free radicals cause chronic health problems.

The book serves the function of aiding an insightful and thorough discussion of what entails the numerous aspects of Nanomaterials.

The present work is purely a research compilation based on authoritative publications/readings/excerpts/notes/reviews/researches authored by eminent information scientists and IT professionals.

Assistance has been sought from Websites through Internet.

The author, humbly, acknowledges the contributions of all those eminent information scientists and IT experts alongwith their respective publishers and Websites in preparation of this book.

Dr. Brijesh Singh

Contents

1

Introduction

BACKGROUND OF NANOMATERIALS

Nanomaterials are cornerstones of nanoscience and nanotechnology. Nanostructure science and technology is a broad and interdisciplinary area of research and development activity that has been growing explosively worldwide in the past few years. It has the potential for revolutionizing the ways in which materials and products are created and the range and nature of functionalities that can be accessed. It is already having a significant commercial impact, which will assuredly increase in the future.

Nanomaterials can be metals, ceramics, polymeric materials, or composite materials. Their defining characteristic is a very small feature size in the range of 1-100 nanometres (nm). The unit of nanometre derives its prefix nano from a Greek word meaning dwarf or extremely small. One nanometre spans 3-5 atoms lined up in a row. By comparison, the diameter of a human hair is about 5 orders of magnitude larger than a nanoscale particle. Nanomaterials are not simply another step in miniaturization, but a different arena entirely; the nanoworld lies midway between the scale of atomic and quantum phenomena, and the scale of bulk materials. At the nanomaterial level, some material properties are affected by the laws of atomic physics, rather than behaving as traditional bulk materials do.

Although widespread interest in nanomaterials is recent, the concept was raised over 40 years ago. Physicist Richard Feynman delivered a talk in 1959 entitled "There's Plenty of Room at the Bottom", in which he commented that there were no fundamental physical reasons that materials could not be fabricated by manoeuvring individual atoms. Nanomaterials have actually been produced and used by humans for hundreds of years-the beautiful ruby red colour of some glass is due to gold nanoparticles trapped in the glass matrix. The decorative glaze known as luster, found on some medieval pottery, contains metallic spherical nanoparticles dispersed in a complex way in the glaze, which give rise to its special optical properties. The techniques used to produce these materials were considered trade secrets at the time, and are not wholly understood even now.

Development of nanotechnology has been spurred by refinement of tools to see the nanoworld, such as more sophisticated electron microscopy and scanning tunnelling microscopy. By 1990, scientists at IBM had managed to position individual xenon atoms on a nickel surface to spell out the company logo, using scanning tunnelling microscopy probes, as a demonstration of the extraordinary new technology being developed. In the mid-1980s a new class of material-hollow carbon spheres-was discovered. These spheres were called buckyballs or fullerenes, in honour of architect and futurist Buckminster Fuller, who designed a geodesic dome with geometry similar to that found on the molecular level in fullerenes. The C_{60} (60 carbon atoms chemically bonded together in a ball-shaped molecule) buckyballs inspired research that led to fabrication of carbon nanofibres, with diameters under 100 nm. In 1991 S. Iijima of NEC in Japan reported the first observation of carbon nanotubes, which are now produced by a number of companies in commercial quantities. The world market for nanocomposites (one of many types of nanomaterials) grew to millions of pounds by 1999 and is still growing fast.

The variety of nanomaterials is great, and their range of properties and possible applications appear to be enormous, from extraordinarily tiny electronic devices, including miniature batteries, to biomedical uses, and as packaging films, superabsorbants, components of armour, and parts of automobiles. General Motors claims to have the first vehicle to use the materials for exterior automotive applications, in running boards on its mid-size vans. Editors of the journal Science profiled work that resulted in molecular-sized electronic circuits as the most important scientific development in 20012. It is clear that researchers are merely on the threshold of understanding and development, and that a great deal of fundamental work remains to be done.

What makes these nanomaterials so different and so intriguing? Their extremely small feature size is of the same scale as the critical size for physical phenomena-for example, the radius of the tip of a crack in a material may be in the range 1-100 nm. The way a crack grows in a larger-scale, bulk material is likely to be different from crack propagation in a nanomaterial where crack and particle size are comparable. Fundamental electronic, magnetic, optical, chemical, and biological processes are also different at this level. Where proteins are 10-1000 nm in size, and cell walls 1-100 nm thick, their behaviour on encountering a nanomaterial may be quite different from that seen in relation to larger-scale materials. Nanocapsules and nanodevices may present new possibilities for drug delivery, gene therapy, and medical diagnostics.

Surfaces and interfaces are also important in explaining nanomaterial behaviour. In bulk materials, only a relatively small percentage of atoms will be at or near a surface or interface (like a crystal grain boundary). In nanomaterials, the small feature size ensures that many atoms, perhaps half or more in some cases, will be near interfaces. Surface properties such as energy levels, electronic structure, and reactivity can be quite different from interior states, and give rise to quite different material properties.

Let us examine in particular nanocomposites based on polymeric materials, keeping in mind that this is but one small division of nanomaterials. There are several varieties of polymeric nanocomposites, but the most commercially advanced are those that involve dispersion of small amounts of nanoparticles in a polymer matrix. Those most humble of materials, clays, have been found to impart amazing properties. For example, adding such small amounts as 2% by volume of silicate nanoparticles to a polyimide resin increases the strength by 100%. One should keep in mind, of course, that 2% by volume of very small particles is a great many reinforcing particles. Addition of nanoparticles not only improves the mechanical properties, but also has been shown to improve thermal stability, in some cases allowing use of polymer-matrix nanocomposites an additional 100 degrees Centigrade above the normal service conditions. Decrease in material flammability has also been studied, an especially important property for transportation applications where choice of material is influenced by safety concerns. Clay/polymer nanocomposites have been considered as matrix materials for fibre-based composites destined for aerospace components. Aircraft and spacecraft components require lightweight materials with high strength and stiffness, among other qualities. Nanocomposites, with their superior thermal resistance, are also attractive for such applications as housings for electronics.

Others have examined the electrical properties of nanocomposites, with an eye to developing new conductive materials. The use of polymer-based nanocomposites has been expanded to anti-corrosion coatings on metals, and thin-film sensors. Their photoluminescence and other optical properties are being explored. Polymer-matrix nanocomposites can also be used to package films, an application which exploits their superior barrier properties and low permeability. Although some nanomaterials require rather exotic approaches to synthesis and processing, many polymer-matrix nanocomposites can be prepared quite readily. Clay/polymer nanocomposites have been made by subjecting a clay such as montmorillonite to ion exchange or other pretreatment, then mixing the particles with polymer melts. There are also a number of other ways to fabricate the materials, including reactive processes involving in situ polymerization. The low volume fraction of reinforcement particles allows the use of well-established and well-understood processing methods, such as extrusion and injection moulding. Ease of processing and forming may be one explanation for the rapidly expanding applications of the materials. Automotive companies, in particular, have quickly adopted nanocomposites in large scale applications, including structural parts of vehicles.

The most energetic research probably concerns carbon nanotubes. Nanoparticles of carbon-rods, fibres, tubes with single walls or double walls, open or closed ends, and straight or spiral forms-have been synthesized in the past 10 years. There is good reason to devote so much effort to them: carbon nanotubes have been shown to have unique properties, stiffness and strength higher than any other material, for example, as well as extraordinary electronic properties. Carbon nanotubes are reported to be thermally stable in vacuum up to 2800 degrees Centigrade, to have a capacity to carry an electric current a thousand times better than copper wires, and to have twice the thermal conductivity of diamond (which is also a form of carbon). Carbon nanotubes are used as reinforcing particles in nanocomposites, but also have many other potential applications. They could be the basis for a new era of electronic devices smaller and more powerful than any previously envisioned. Nanocomputers based on carbon nanotubes have already been demonstrated.

NANOMATERIALS ESTABLISHMENT

Some nanomaterials occur naturally, but of particular interest are engineered nanomaterials (EN), which are designed for, and already being used in many commercial products and processes. They can be found in such things as sunscreens, cosmetics, sporting goods, stain-resistant clothing, tires, electronics, as well as many other everyday items, and are used in medicine for purposes of diagnosis, imaging and drug delivery. Engineered nanomaterials are resources designed at the molecular (nanometre) level to take advantage of their small size and novel properties which are generally not seen in their conventional, bulk counterparts. The two main reasons why materials at the nano scale can have different properties are increased relative surface area and new quantum effects. Nanomaterials have a much greater surface area to volume ratio than their conventional forms, which can lead to greater chemical reactivity and affect their strength. Also at the nano scale, quantum effects can become much more important in determining the materials properties and characteristics, leading to novel optical, electrical and magnetic behaviours.

Nanomaterials are already in commercial use, with some having been available for several years or decades. The range of commercial products available today is very broad, including stain-resistant and wrinkle-free textiles, cosmetics, sunscreens, electronics, paints and varnishes. Nanocoatings and nanocomposites are finding uses in Book title 1.3 diverse consumer products, such as windows, sports equipment, bicycles and automobiles. There are novel UV-blocking coatings on glass bottles which protect beverages from damage by sunlight, and longer-lasting tennis balls using butylrubber/nano-clay composites. Nanoscale titanium dioxide, for instance, is finding applications in cosmetics, sun-block creams and self-cleaning windows, and nanoscale silica is being used as filler in a range of products, including cosmetics and dental fillings.

ADVANCES IN NANOMATERIALS

The history of nanomaterials began immediately after the big bang when Nanostructures were formed in the early meteorites. Nature later evolved many other Nanostructures like seashells, skeletons etc. Nanoscaled smoke particles were formed during the use of fire by early humans. The scientific story of nanomaterials however began much later. One of the first scientific report is the colloidal gold particles synthesised by Michael Faraday as early as 1857. Nanostructured catalysts have also been investigated for over 70 years. By the early 1940's, precipitated and fumed silica nanoparticles were being manufactured and sold in USA and Germany as substitutes for ultrafine carbon black for rubber reinforcements.

Nanosized amorphous silica particles have found large-scale applications in many everyday consumer products, ranging from non-diary coffee creamer to automobile tires, optical fibres and catalyst supports. In the 1960s and 1970's metallic nanopowders for magnetic recording tapes were developed. In 1976, for the first time, nanocrystals produced by the now popular inert-gas evaporation technique was published by Granqvist and Buhrman. Recently it has been found that the Maya blue paint is a nanostructured hybrid material. The origin of its colour and its resistance to acids and biocorrosion are still not understood but studies of authentic samples from Jaina Island show that the material is made of needle-shaped palygorskite (clay) crystals that form a superlattice with a period of 1.4 nm, with intercalates of amorphous silicate substrate containing inclusions of metal (Mg) nanoparticles. The beautiful tone of the blue colour is obtained only when both these nanoparticles and the superlattice are present, as has been shown by the fabrication of synthetic samples.

NANOMATERIALS ENGINEERING

Today nanophase engineering expands in a rapidly growing number of structural and functional materials, both inorganic and organic, allowing to manipulate mechanical, catalytic, electric, magnetic, optical and electronic functions.

The production of nanophase or cluster-assembled materials is usually based upon the creation of separated small clusters which then are fused into a bulk-like material or on their embedding into compact liquid or solid matrix materials. *e.g.* nanophase silicon, which differs from normal silicon in physical and electronic properties, could be applied to macroscopic semiconductor processes to create new devices. For instance, when ordinary glass is doped with quantized semiconductor "colloids," it becomes a high performance optical medium with potential applications in optical computing.

CLASSIFICATION OF NANOMATERIALS

Nanomaterials have extremely small size which having at least one dimension 100 nm or less. Nanomaterials can be nanoscale in one dimension (*e.g.* surface films), two dimensions (*e.g.* strands or fibres), or three dimensions (*e.g.* particles). They can exist in single, fused, aggregated or agglomerated forms with spherical, tubular, and irregular shapes. Common types of nanomaterials include nanotubes, dendrimers, quantum dots and fullerenes. Nanomaterials have applications in the field of nano technology, and displays different physical chemical characteristics from normal chemicals (*i.e.*, silver nano, carbon nanotube, fullerene, photocatalyst, carbon nano, silica).

Classification

Materials referred to as "nanomaterials" generally fall into two categories: fullerenes, and inorganic nanoparticles.

Fullerenes

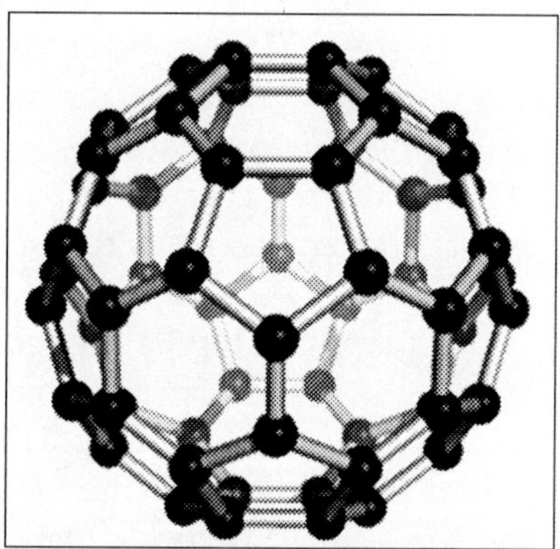

Figure: Rotating View of Buckminsterfullerene C_{60}

The fullerenes are a class of allotropes of carbon which conceptually are graphene sheets rolled into tubes or spheres. These include the carbon nanotubes (or silicon nanotubes) which are of interest both because of their mechanical strength and also because of their electrical properties.

For the past decade, the chemical and physical properties of fullerenes have been a hot topic in the field of research and development, and are likely to continue to be for a long time. In April 2003, fullerenes were under study for potential medicinal use: binding specific antibiotics to the structure of resistant bacteria and even target certain types of cancer cells such as melanoma. The October 2005 issue of Chemistry and Biology contains an article describing the use of fullerenes as light-activated antimicrobial agents. In the field of nanotechnology, heat resistance and superconductivity are among the properties attracting intense research. A common method used to produce fullerenes is to send a large current between two nearby graphite electrodes in an inert atmosphere. The resulting carbon plasma arc between the electrodes cools into sooty residue from which many fullerenes can be isolated. There are many calculations that have been done using ab-initio Quantum Methods applied to fullerenes. By DFT and TDDFT methods one can obtain IR, Raman and UV spectra. Results of such calculations can be compared with experimental results.

Nanoparticles

Nanoparticles or nanocrystals made of metals, semiconductors, or oxides are of particular interest for their mechanical, electrical, magnetic, optical, chemical and other properties. Nanoparticles have been used as quantum dots and as chemical catalysts such as nanomaterial-based catalysts. Nanoparticles are of great scientific interest as they are effectively a bridge between bulk materials and atomic or molecular structures. A bulk material should have constant physical properties regardless of its size, but at the nano-scale this is often not the case. Size-dependent properties are observed such as quantum confinement in semiconductor particles, surface plasmon resonance in some metal particles and superparamagnetism in magnetic materials.

Nanoparticles exhibit a number of special properties relative to bulk material. For example, the bending of bulk copper (wire, ribbon, etc.) occurs with movement of copper atoms/clusters at about the 50 nm scale. Copper nanoparticles smaller than 50 nm are considered super hard materials that do not exhibit the same malleability and ductility as bulk copper. The change in properties is not always desirable. Ferroelectric materials smaller than 10 nm can switch their magnetisation direction using room temperature thermal energy, thus making them useless for memory storage. Suspensions of nanoparticles are possible because the interaction of the particle surface with the solvent is strong enough to overcome differences in density, which usually result in a material either sinking or floating in a liquid. Nanoparticles often have unexpected visual properties because they are small enough to confine their electrons and produce quantum effects. For example gold nanoparticles appear deep red to black in solution. The often very high surface area to volume ratio of nanoparticles provides a tremendous driving force for diffusion, especially at elevated temperatures. Sintering is possible at lower temperatures and over shorter durations than for larger particles. This theoretically does not affect the density of the final product, though flow difficulties and the tendency of nanoparticles to agglomerate do complicate matters. The surface effects of nanoparticles also reduces the incipient melting temperature.

Sol-gel

The sol-gel process is a wet-chemical technique commonly used to synthesise a wide variety of nanomaterials.

CHARACTERIZATION

The first observations and size measurements of nanoparticles were made during the first decade of the 20th century. They are mostly associated with the name of Zsigmondy who made detailed studies of gold sols and other nanomaterials with sizes down to 10 nm and less. He published a book in 1914. He used an ultramicroscope that employs a *dark field* method for seeing particles with sizes much less than light wavelength.

There are traditional techniques developed during 20th century in Interface and Colloid Science for characterizing nanomaterials. These are widely used for *first generation* passive nanomaterials.

These methods include several different techniques for characterizing particle size distribution. This characterization is imperative because many materials that are expected to be nano-sized are actually aggregated in solutions. Some of methods are based on light scattering. Other apply ultrasound, such as ultrasound attenuation spectroscopy for testing concentrated nano-dispersions and microemulsions.

There is also a group of traditional techniques for characterizing surface charge or zeta potential of nanoparticles in solutions. This information is required for proper system stabilzation, preventing its aggregation or flocculation. These methods include microelectrophoresis, electrophoretic light scattering and electroacoustics. The last one, for instance colloid vibration current method is suitable for characterizing concentrated systems.

Safety

Nanomaterials behave differently than other similarly-sized particles. It is therefore necessary to develop specialized approaches to testing and monitoring their effects on human health and on the environment. The OECD Chemicals Committee has established the Working Party on Manufactured Nanomaterials to address this issue and to study the practices of OECD member countries in regards to nanomaterial safety.

While nanomaterials and nanotechnologies are expected to yield numerous health and health care advances, such as more targeted methods of delivering drugs, new cancer therapies, and methods of early detection of diseases, they also may have unwanted effects. Increased rate of absorption is the main concern associated with manufactured nanoparticles.

When materials are made into nanoparticles, their surface area to volume ratio increases. The greater specific surface area (surface area per unit weight) may lead to increased rate of absorption through the skin, lungs, or digestive tract and may cause unwanted effects to the lungs as well as other organs. However, the particles must be absorbed in sufficient quantities in order to pose health risks.

As the use of nanomaterials increases worldwide, concerns for worker and user safety are mounting. To address such concerns, the Swedish

Karolinska Institute conducted a study in which various nanoparticles were introduced to human lung epithelial cells. The results, released in 2008, showed that iron oxide nanoparticles caused little DNA damage and were nontoxic. Zinc oxide nanoparticles were slightly worse. Titanium dioxide caused only DNA damage. Carbon nanotubes caused DNA damage at low levels. Copper oxide was found to be the worst offender, and was the only nanomaterial identified by the researchers as a clear health risk.

HISTORY OF NANOMATERIALS

The history of nanomaterials began immediately after the big bang when Nanostructures were formed in the early meteorites. Nature later evolved many other Nanostructures like seashells, skeletons etc. Nanoscaled smoke particles were formed during the use of fire by early humans.

The scientific story of nanomaterials however began much later. One of the first scientific report is the colloidal gold particles synthesised by Michael Faraday as early as 1857. Nanostructured catalysts have also been investigated for over 70 years. By the early 1940's, precipitated and fumed silica nanoparticles were being manufactured and sold in USA and Germany as substitutes for ultrafine carbon black for rubber reinforcements. Nanosized amorphous silica particles have found large-scale applications in many everyday consumer products, ranging from non-diary coffee creamer to automobile tires, optical fibres and catalyst supports. In the 1960s and 1970s metallic nanopowders for magnetic recording tapes were developed. In 1976, for the first time, nanocrystals produced by the now popular inert-gas evaporation technique was published by Granqvist and Buhrman. Recently it has been found that the Maya blue paint is a nanostructured hybrid material. The origin of its colour and its resistance to acids and biocorrosion are still not understood but studies of authentic samples from Jaina Island show that the material is made of needle-shaped palygorskite (a clay) crystals that form a superlattice with a period of 1.4 nm, with intercalcates of amorphous silicate substrate containing inclusions of metal (Mg) nanoparticles. The beautiful tone of the blue colour is obtained only when both these nanoparticles and the superlattice are present, as has been shown by the fabrication of synthetic samples. Probably most celebrated historical comment on the advancement of nanomaterials was the public speculation by physicist Richard Feynman in 1959 at a meeting of the American Physical Society about the effects of manipulating minuscule bits of condensed matter in his famous talk "There's plenty of room at the bottom".

Today nanophase engineering expands in a rapidly growing number of structural and functional materials, both inorganic and organic, allowing tomanipulate mechanical, catalytic, electric, magnetic, optical and electronic functions. The production of nanophase or cluster-assembled materials, is usually based upon the creation of separated small clusters which then are fused into a bulk-like material or on their embedding into compact liquid or solid matrix materials. E.g. nanophase silicon, which differs from normal silicon in physical and electronic properties, could be applied to macroscopic semiconductor processes to create new devices.

For instance, when ordinary glass is doped with quantized semiconductor "colloids," it becomes a high performance optical medium with potential applications in optical computing.

IMPORTANCE OF NANOMATERIALS

These materials have created a high interest in recent years by virtue of their unusual mechanical, electrical, optical and magnetic properties. Some examples are given below:

- Nanophase ceramics are of particular interest because they are more ductile at elevated temperatures as compared to the coarse-grained ceramics.
- Nanostructured semiconductors are known to show various non-linear optical properties. Semiconductor Q-particles also show quantum confinement effects which may lead to special properties, like the luminescence in silicon powders and silicon germanium quantum dots as infrared optoelectronic devices. Nanostructured semiconductors are used as window layers in solar cells.
- Nanosized metallic powders have been used for the production of gas tight materials, dense parts and porous coatings. Cold welding properties combined with the ductility make them suitable for metal-metal bonding especially in the electronic industry.
- Single nanosized magnetic particles are mono-domains and one expects that also in magnetic nanophase materials the grains correspond with domains, while boundaries on the contrary to disordered walls. Very small particles have special atomic structures with discrete electronic states, which give rise to special properties in addition to the superparamagnetism behaviour. Magnetic nanocomposites have been used for mechanical force transfer (ferrofluids), for high density information storage and magnetic refrigeration.
- Nanostructured metal clusters and colloids of mono- or plurimetallic composition have a special impact in catalytic applications. They may serve as precursors for new type of heterogeneous catalysts (Cortex-catalysts) and have been shown to offer substantial advantages concerning activity, selectivity and lifetime in chemical transformations and electrocatalysis (fuel cells). Enantioselective catalysis was also achieved using chiral modifiers on the surface of nanoscale metal particles.

- Nanostructured metal-oxide thin films are receiving a growing attention for the realization of gas sensors (NOx, CO, CO_2, CH_4 and aromatic hydrocarbons) with enhanced sensitivity and selectivity. Nanostructured metal-oxide (MnO_2) finds application for rechargeable batteries for cars or consumer goods. Nanocrystalline silicon films for highly transparent contacts in thin film solar cell and nano-structured titanium oxide porous films for its high transmission and significant surface area enhancement leading to strong absorption in dye sensitized solar cells.
- Polymer based composites with a high content of inorganic particles leading to a high dielectric constant are interesting materials for photonic band gap structure.

NANOMATERIALS: NOMENCLATURE, NOVELTY, AND NECESSITY

Nanomaterials, it is perhaps best to start with the broad area that is known as nanotechnology. Identification of the concept of nanotechnology has been attributed to Richard Feynman, who presented a speech in 1959 titled "There's Plenty of Room at the Bottom." In his speech, Feynman described manipulating atoms to make materials many decades before it became possible to do so. The term "nanotechnology" was not used until 1974 by Taniguchi at the University of Tokyo, Japan, to refer to the ability to engineer materials precisely at the nanometre level, driven by electronics industry needs. In 1981, the advent of the scanning tunnelling microscope enabled atom clusters to be seen, while in 1991 IBM demonstrated the ability to arrange individual xenon atoms using an atomic force instrument.

NANONOMENCLATURE

Atomic-Force Microscope: A scanning-probe instrument that maps the surface topography by measuring the force acting on a tip as it slides along a surface or moves perpendicular to it.

Bottom-Up Nanofabrication: The building of nanostructures starting with small components such as atoms or molecules. Buckminsterfullerene or Buckyball: A spheroidal fullerene; the first known example of a fullerene.

Fullerene: A form of carbon having a large molecule consisting of an empty cage of 60 or more carbon atoms Nanocomposites: Composite structures whose characteristic dimensions are found at the nanoscale.

Nanocrystalline: Material with average grain size measured in billionths of a meter.

Nanodots: Nanoparticles that consist of homogeneous material, especially those that are almost spherical or cubical in shape.

Nanofabrication: The manufacture or preparation of nanostructures.

Nanoparticles: Aggregates of atoms bridging the continuum between small molecular clusters of a few atoms and dimensions of 0.2–1 nm and solids containing millions of atoms and having the properties of macroscopic bulk material.

Nanorods: Nanostructures that are shaped like long sticks or dowels, with a diameter in the nanoscale and a length very much longer.

Nanoscale: Phenomena that occur on the length scale between 1 and 100 nanometres.

Nanoscience: The study of fundamental principles of molecules and structures with at least one dimension roughly between 1 and 100 nanometres.

Nanostructure/Nanostructured Materials: Structures whose characteristic variation in design length is at the nanoscale.

Nanotechnology: The application of nanoscience in technological devices.

Nanotubes: Almost always carbon nanotubes, referring to the wires of pure carbon that look like rolled sheets of graphite or like carbon soda straws.

Nanowires: Another term for nanorods, especially nanorods that can conduct electricity.

Quantum Dots: Nanostructures of roughly spherical or cubic shape that are small enough to exhibit characteristically quantum behaviour in optical or electrical processes.

Scanning-Probe Instruments: Tools for both measuring and preparing nanostructures on surfaces; they work using the interactions between a scanning-tip structure and the nanostructure on the surface, which can be either manipulated or measured.

Scanning-Tunnelling Microscope: The first of the scanning probe instruments. It works at the scale of the nanostructure and measures electrons tunnelling between a scanning tip and a conducting surface.

Top-Down Nanofabrication: The process of making nanostructures starting with larger structures and taking parts away.

NANOMATERIALS AND NANOTECHNOLOGY

Not unlike the situation in conventional technology, materials are a critical if sometimes underappreciated component in achieving the goals of nanotechnology. A central construct of materials science and engineering is that the structure and processing of a material give rise to its properties, which in turn determine its performance in use. This certainly applies to nanomaterials as well and provides a useful framework for understanding in a familiar context.

Nanomaterials have been important in the materials field for quite a long time. An early example was the incorporation of gold nanoparticles in stained glass in 10 A.D. At sizes in the nanorange, gold can exhibit a range of colours. Also, nano-sized carbon black particles have been used to reinforce tires for

100 years. Perhaps one manifestation of the concept of nanomaterials familiar to traditional metallurgists is precipitation hardening. The accidental discovery of precipitation hardening in 1906 by Wilm in Duralumin alloys opened the way to significant improvements in strength properties for aluminum. Although researchers were unable to image the precipitates with the instruments available at the time, it was subsequently discovered that nanometre sized precipitates were the source of hardening. Through the advent of electron microscopy came a better understanding of the structure of precipitates, including aspects such as size and coherency with the metal matrix. This allowed further improvement in nanoscale structures through composition selection and knowledgeable processing, resulting in further gains in a variety of materials systems. It is interesting to note that the ability to image precipitates was critical to optimizing processing and properties, much as the invention of scanning-probe instruments enabling the imaging of nanostructures has accelerated progress in nanotechnology.

What then are the nanomaterials of today and tomorrow? As a framework for this discussion, the approach of Jones for organizing nanotechnology into three categories can be applied to nanomaterials. The categories are:

- Incremental nanotechnology—improving the properties of materials by controlling their nanoscale structure.
- Evolutionary nanotechnology—taking a step beyond redesigning simple materials at the nanoscale and designing nanoscale devices that do something interesting
- Radical nanotechnology—developing nanoscale machines that would exist at the convergence of nanotechnology, biotechnology, information technology, and cognitive technology.

INCREMENTAL NANOTECHNOLOGY

Incremental nanomaterials are materials that have improved properties at the nanoscale, typically as a result of their greatly increased surface area, but do not typically take advantage of the quantum effects. Nanoparticles are already seeing application, taking advantage primarily of the high surface area of these fine powders. Nanoceramic powders, the most commercially important of which are simple metal oxides, constitute almost 90% of the total market. For example, nano-sized zinc oxide particles are in use in sunscreen. Nanostructured ceramic coatings are adding durability and toughness to hulls of U.S. Navy ships. Metal powders are important, as well. Iron nanoparticles have been used to treat ground-water contaminated with trichloroethylene while aluminum nanoparticles have been developed that, due to their increased surface area, have substantially greater "bang for the buck" as solid rocket propellant.

Incremental nanomaterials also include polymer nanocomposites in which clay nanoparticles are incorporated to increase the hardness and reduce the permeability of the polymer. These have seen application in automotive panels and step assists in vans. Other examples of these types of applications are nanoparticles in tennis balls and carbon nanotubes in tennis racquets. Incremental nanotechnology can also be found in the development of nanostructured materials by techniques such as equal channel angular extrusion and accumulative roll bonding, which fall in the category of severe plastic deformation processes. Articles by Semiatin *et al.* and Zhu discuss these concepts in more depth.

The primary benefit of these processes is to produce ultrafine-grained structures with dimensions in the sub-micrometer range, resulting in increased strengthening along with other enhanced properties. Note again that the benefits obtained in these nanostructured materials come primarily from the increase in surface area (in this case the surface area of grain boundaries) and the impact of the small scale of the structure on dislocation motion rather than quantum effects. Also note that the inherent coupling of the ability to image these structures and their behaviour is inextricably linked to their optimization. For example, recent work has described changes in deformation mode from dislocation-slip to grain boundary sliding as a function of grain size for nanocrystalline nickel using the in-situ microscope at the National Centre for Electron Microscopy at the Lawrence Berkeley Laboratory.

EVOLUTIONARY NANOTECHNOLOGY

Evolutionary nanotechnology takes advantage of the changes that can occur in materials at the nanoscale related both to increased chemical reactivity and the increasing importance of quantum effects. Examples include nanoscale sensors that exploit the large surface area of nanotubes and semiconductor nanostructures such as quantum dots and quantum wells. For example, silica-based nanomaterials, molecular imprinted polymers, and silicon platforms are envisioned for collection, concentration, and detection of chemical weapons and other related compounds in security and defence applications. The biggest steps are being made in evolutionary nanotechnology, with more and more products expected to appear in the market in the next five years.

A material that has generated substantial interest is the class of molecularly perfect carbon structures called fullerenes. The creation of a 60 carbon atom molecule in the form of a geodesic sphere, termed the buckyball by Smalley in 1985, has lead to other fullerenes containing many more atoms of carbon and taking different shapes. One such shape is the carbon nanotube (CNT), which has a range of unique properties. These materials can act as either highly conductive nanowires or as semiconductors, depending on the specific arrangement of the carbon atoms. The perfection of the CNT creates a material with a strength reported to be 30 times that of steel at 1/6 the density. The thermal conductivity of CNTs is 50% higher than that of diamond. These unique properties have already led to the use of CNTs in specialty applications such as high-performance tennis racquets and selected electronics. Longer term, CNTs are seen as key components in fuel cells, electrical transmission lines, and thermoelectric conversion devices.

Carbon nanotubes are by no means the sole materials focus of evolutionary nanotechnology. Other materials are in development that will generate immediate interest and impact on the electronic materials field. As described in an earlier *JOM* paper by Rittner, these include semiconductor nanowires, quantum dots, semiconductor nanocrystals, nanoscale thin films, and organic molecules. It is anticipated that there will be significant impact in the energy area, including high-efficiency solar energy conversion as well as advanced fuel cells and batteries. Medical applications enabled by nanomaterials including diagnosis and treatment, along with markedly different approaches to pharmaceuticals, are also predicted to be on the horizon in the next decade.

RADICAL NANOTECHNOLOGY

Radical nanotechnology is viewed as having an impact much farther into the future. Examples include next-generation military uniforms being developed by the Massachusetts Institute of Technology's Institute for Soldiering Nanotechnology. The uniforms are designed to defend against chemical and biological weapons, provide ballistic protection, monitor health, administer medical aid, and provide communication capabilities. Much farther in the future are concepts in which nanotechnology converges with biology, information technology, and cognition. Novelist Michael Crichton seized upon this convergence as the technological engine for his 2002 novel *Prey*. Based on a hypothetical future in which so-called nanobots develop evolutionary capability and self-replicate, becoming an out-of-control, dangerous force, the book paints a scary picture. Although the book is fiction, it does raise the question of responsibility—a question is being taken sufficiently seriously that there exists a Centre for Responsible Nanotechnology Development. While interesting (for cocktail parties and novels), radical nanotechnology is viewed as being decades away if at all.

NANOFABRICATION

Since nanomaterials have little mass and are dominated by surface area and size effects, the processes and equipment for nanotechnology-based manufacturing are expected to differ significantly from those currently used. One of the key differences is the concept of manufacturing on an atom-by-atom basis, which has been labelled bottom-up processing. Conventional methods, called top-down techniques, start with a block of material, etching or milling it down to the desired shape. The main challenge for top-down manufacture is the creation of increasingly small structures with sufficient accuracy, whereas for bottom-up manufacture, the challenge is to make structures large enough, and of sufficient quality, for use as materials. These two methods have evolved separately and have now reached the point where the best achievable feature size for each technique is approximately the same.

Most of the current nanomaterials with potential for structural use are powder based, produced from processes such as inert-gas condensation, electrodeposition, atomization, and mechanical alloying. They need to be consolidated into shapes with minimal porosity without creating significant structural coarsening and loss of the desirable nanostructural properties. This requires a balance of minimal high-temperature exposure and adequate pressure, and innovative methods of consolidation need to be developed to produce successful materials.

While nanotubes have incredible strength, capturing this strength in a structural component is no small challenge. This challenge is analogous to the challenges faced by early metal matrix composite materials engineers who sought to capture the ultrahigh strength of ceramic whiskers in structural materials. Incorporating these expensive reinforcement materials cost-effectively in large-volume operations was difficult, often resulting in damage to the whiskers and loss of much of their property advantage.

NANOTECHNOLOGY AND THE MATERIALS ENTERPRISE

Billions of dollars are being invested in opportunities centred around materials that exist on the scale of a billionth of a meter. The National Science Foundation has made the bold prediction that the business impact of nanotechnology will be $1 trillion by 2015. Some 800,000 to 2 million new jobs, in a field where the estimates of current number of workers range from 20,000 to 40,000, will be required to reach this level. Pundits claim that within the next decade, nanotech will have huge effects on many practical industries, including manufacturing, health care, energy, agriculture, communication, transportation, and electronics. Just issued, The Nanotech Report 2004 produced by Lux Research highlights the magnitude of financial resources being put into nanotechnology. Specifically, the report notes:

- Governments, corporations, and venture capitalists will spend more than $8.6 billion worldwide on nanotechnology R&D in 2004.
- National and local governments across the world will invest more than $4.6 billion in nanotechnology R&D in 2004. Roughly 35% of this investment is in North America, 35% in Asia, 28% in Europe, and 2% in the rest of the world.
- The U.S. government will spend nearly twice as much on nanotechnology this year as it did on the Human Genome Project (HGP) in its peak year. In 2005, the National Nanotechnology Initiative will surpass the HGP on a cumulative basis. The United States has appropriated more than $3.16 billion for nanotechnology

R&D since 2000 and is proposing $982 million in new funding for 2005.

- Established corporations will spend more than $3.8 billion globally on nanotechnology R&D in 2004.

It is not an exaggeration to say that nanotechnology is the biggest thing to happen to the physical sciences for quite some time. It is estimated that worldwide government funding in nanotechnology increased by seven times from 1997 to 2003, exceeding $3 billion in 2003. In the United States, the 21st Century Nanotechnology Research and Development Act, signed in December 2003, will provide $3.7 billion for four years beginning in 2005. The funding request for $982 million in 2005 is an increase of more than 500% compared to 2001. A key component of this funding is for five user facilities, called the Nanoscale Science Research Centres, which are now under construction at Department of Energy laboratory sites. These research facilities will focus on synthesis, processing, and fabrication of nanoscale materials. They will be co-located with existing user facilities to provide sophisticated characterization and analysis capabilities. Also included is funding for addressing so-called grand challenge areas. Three of these areas, Nanostructured Materials by Design, Manufacturing at the Nanoscale, and Nano-Electronics, -Photonics, and -Magnetics, have a materials focus.

Another indicator of the potential in nanotechnology is indicated by the growth in patent activity. In the United States, for example, the number of patents issued in areas such as atomic force microscopy and quantum dots increased ten-fold from 1994–2003. The U.S. Patent and Trademark Office has been prompted to create a classification system for nanotechnology (as the Japanese Patent Office has already) so that it can track the number, pendency, and assignment of nanotechnology patent applications. While at present there is a classification for fullerene-related patent applications there are no separate classes for areas such as nanotubes or nanowires.

Also harkening the transition from laboratory to commercialization is the presence of organizations such as the Nonbusiness Alliance, which claims to be the first industry association founded to advance the emerging business of nanotechnology and microsystems. The Nonbusiness Alliance's mission is to create a collective voice for the emerging small-tech industry and develop a range of initiatives to support and strengthen the nanotechnology business community.

The business prospects of nanotechnology have caught the attention of Wall Street, where brokerage firm Merrill Lynch has been promoting a nanotech stock index. A nanotechnology company, Nanosys, contemplated an initial public offering of stock but withdrew it, which was viewed in a Wall Street Journal article as "less an indictment of nanotechnology than a sign of how tough it is for any company in an early stage of development right now." The article goes on to note, however, that "nanotechnology remains difficult to define, causing additional uncertainties for investors. Some companies have used the nanotechnology label to hype unrelated products, while many real advances are occurring inside big companies such as Intel, where the developments have only a modest impact on stock prices."

Nanomaterials are one area where real revenues are being realised currently. It is estimated that sales of nanomaterials were $1.5 million in 1999 and grew to $430 million by the end of 2003. With this annual growth rate of 300%, the market size would be projected to be $1.3 billion in 2004 and $5 billion in 2005. A more conservative estimate of $1 billion in revenue in 2007 is provided buy a recent study by the Freedonia Group.

Nanotechnology is already having an impact on the structure of the scientific community. The intersection of physics, chemistry, and materials science at the nanoscale has required a broadening of the educational process, which will be required to meet the need for future nanotechnologists. More than 300 academic programmes in nanotechnology exist today, with at least 200 in the United States and 100 internationally. There are implications for professional societies as well, and it has been noted that the scientific and engineering communities should create new means of interdisciplinary training and communication, reduce the barriers that inhibit individuals from working across disciplines, develop links to a variety of other technical and medical organizations, and address ethical issues related to technological developments.

Nanotechnology is gaining awareness beyond the scientific community. The number of times terms related to "nano" appeared in the popular press has risen from 190 in 1995 to a projection of more than 12,000 in 2004. Recent studies carried out both in the United States and United Kingdom indicated that the public is not as aware yet of nanotechnology. Of those polled in the United States, 71% had heard "little" or "nothing" about nanotechnology while only 29% in the United Kingdom had heard of it. Of that 29%, only 19% were able to offer some sort of definition. These same studies indicated that a majority of those polled in both countries are encouraged by the prospects of nanotechnology to improve life in the future.

As nanotechnology grows into the public consciousness, its potential societal implications are beginning to be seriously addressed. On one level, attention is being directed to the environmental, safety, and health issues that may develop. In addition, the potentially disruptive nature of nanotechnology, particularly if visions of the radical technology are realised, is beginning to be addressed by groups such as the Foresight Institute.

FUTURE FOR THE NANOMATERIALS SECTOR

Three years of negotiations to update the EU Novel Foods regulations fell apart last week when a last-ditch attempt to reach a compromise between the EU Parliament, the European Commission and the EU Council of Ministers failed. While the talks fell down on the use of cloned animals and their descendants, their failure to reach a compromise will have wider ramifications, with the disagreement meaning that there will be no updates on legislation around nanotechnology. What is this likely to mean for the burgeoning sector? Petah Marian reports.

Figure: Scientists are Currently Assesing the Risk Around Nanomaterials

Nanomaterials, products that are less than one millionth of one millimetre, are set to become an important channel for innovation in the food sector. This technology has the potential to offer benefits around product reformulation and functional foods, longer food shelf life, and improved levels of food safety. While the channel is one that is still very nascent, and not currently being used commercially in food products, a UK government report from March last year valued nanotechnology in food and agriculture as being worth US$3.2bn by 2015, up from $265m in 2007.

Current applications are focused around antimicrobial packaging, with nano silver included in food containers to prevent bacteria from sticking to food, as well as the ability to improve the shelf life of products by making packaging less permeable. For example, there are beer bottles that contain nano silica, which Leatherhead microscopy consultant Kathy Groves says, makes it more difficult for oxygen to get through the polymer of the plastic, which would make the beer go off.

A Food Standards Agency (FSA) spokesperson told just-food today that the European Food Services Authority (EFSA) recently provided a "positive opinion" on the use of titanium nitride nanoparticles in the manufacture of polyethylene bottles. It said that this will be reflected in EU legislation from 1 May. Dr. Barry Park from NanoKTN, a knowledge transfer network for the nanotechnology sector, also highlighted the potential for nanotechnology to be used in product reformulation, with nano-sized salt particles set to become available that will allow manufacturers to use less salt, while still maintaining the salty flavour.

He also said that there is work being done in reducing the amount of fat, and sugar in products using nanotechnology, saying that research is underway to reduce the droplet size of fat in emulsions, using water in oil in water emulsions. While nanoparticles are the same as traditional ingredients, only smaller, there are concerns that nanoparticles can penetrate into parts of the body where larger particles can not go.

Park said the FSA is currently investigating what happens when nanomaterials pass through the gut wall and where they end up. "Some of those studies have been done, and there are a lot of studies yet to be completed," he said. "Many of the studies have found that they don't pass through and they don't reside in the gut, which means they don't play any part in issues we should be concerned about," he added.

The current Novel Foods regulation dates back to May 1997 and regulates food and ingredients that have not been significantly used for human consumption in the EU before 15 May 1997. Talks to update the regulation fell down last week when the European Parliament and European Council of Ministers could not come to an agreement on regulations around cloned meat and that of their offspring.

Last year, the European Parliament called for nano-ingredients to be subject to Novel Food regulations. It also called for a moratorium on their use until a specifically designed risk assessment of nanotechnology processes or nano-ingredients can prove them to be safe, expressing concerns that nanotechnology is already being used in food and food packaging. However, the FSA confirmed to just-food that no engineered nanomaterials have been approved for marketing as foods or ingredients in the EU. "We are also not aware of any food additive that is specifically engineered to consist of nano-sized particles," the spokesperson said.

Groves said that some people believe there is no need for new regulations on nanotechnology. "The novel food regulations will address any 'novel' properties of ingredients, say for instance nano," she said. "And I've got lots of sympathy for that, I think that it's true." Park echoed Groves' position, saying that the lack of a change in the regulation gives the industry a sense of certainty, as some companies would have been waiting to understand what the changes will mean.

According to Park, the lack of change means that "clarity [in legislation], at least for the moment" will mean that companies have a "clearer sense of what they can do, based on what they've done in the past". He added that any change may have meant that companies would have "held fire" on their nanotechnology strategies until they understood what the changes in legislation would have meant in practice.

While the industry feels that current Novel Foods regulations cover the sector well, Groves said that it is important to get a clear definition of what is considered "nano", particularly in terms of labelling. She suggested that under many definitions products like icing sugar and boiled eggs would be considered nanomaterials due to the size of the natural 'nano' particles in the products. The new regulations, had they gone ahead, would have meant a legal definition of the term nanomaterials. A European Commission spokesperson told just-food that would have triggered a pre-market approval under Novel Foods Regulation and a mandatory labelling requirement on the ingredients list with the word 'nano' after the name of the ingredient.

Groves said she has been involved in discussions with the FSA on the labelling of nanomaterials and said that her personal position is that it is "not worthwhile". However, she said that in the FSA's consumer study of nanomaterials, which is set to be released next week, found that "many people favour simple labelling". She said the study found that people can understand the need for nanomaterials in packaging, but can not really understand the need for nanomaterials in their food.

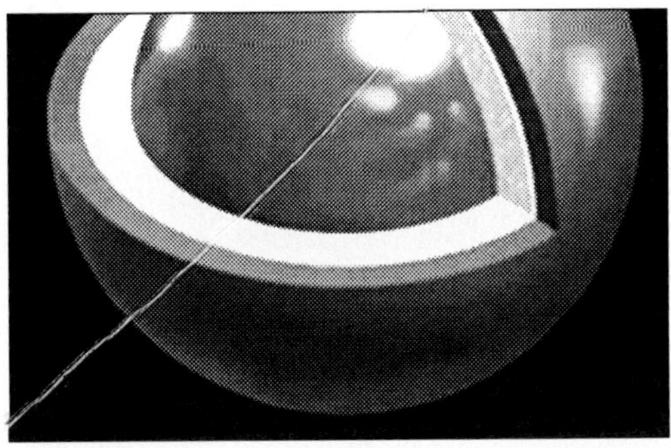

"Even in the beginning, they weren't sure if they wanted technology to tell them if products were fresh or not, but they were asked to keep a diary of the food they bought and the food they threw away and when they came back they were all quite different, and said they were now aware of how much food they do waste and do throw away. So I think it's just about getting people to think about the benefits," says Groves. She says that the argument is that if you put an 'N' or something simple on the product, people can then go away and look it up. However, Groves emphasised that putting an 'N' or a 'nano' on the label is not going to help consumers without more education into the way that science and technology feeds into the food sector. "We do need to educate people more about their food. And they need to take responsibility about learning about it."

While studies into the viability and safety of nanomaterials continue, food companies may face an uphill battle in swinging consumer perception in their favour, with concerns that a lack of education may mean that this technology goes the way of cloned meat and GM. Groves said that the industry needs to be more open, as there is a "climate of fear around the word nanotechnology". A House of Lords report into nanotechnology and food found that food companies' failure to publish details of their research into nanotechnology was "unhelpful" and that the industry's attempts to be "secretive" about its research "is the type of behaviour which may bring about the public reaction it is trying to avert". While it seems as though there is still a some way to go before nanomaterials and technologies make their way into our food, industry may have even further to go before it is able to convince consumers to eat it.

NANOMATERIALS FOR THE FUTURE

A conceptual high-energy-density physics target about 2 millimetres in diameter for use at Livermore's National Ignition Facility. Deuterium gas is encapsulated in high-atomic-number doped foam and deuterium ice and encased in an outer beryllium–copper shell. Less than 20 years ago, nanomaterials—substances on the scale of one-billionth of a meter—were a curiosity. Today, their diverse technological applications seem to be endless. Materials reduced to the nanoscale show very different physical, chemical, and mechanical properties compared to those they exhibit on a macroscale. For example, nanoparticles have electronic structures that lie between that of bulk materials and atomic or molecular structures. Nanomaterials have real potential for impact in applications such as alternative energy conversion and energy harvesting. The development of new nanomaterials will also lead to new mass-produced consumer products in cosmetics, cars, building components, clothes, health, and new smart electronic devices. Novel assembly and manipulation of nanostructured materials in confined geometries with tailored composition and function will have applications in catalysis, sensing, hydrogen storage, advanced nuclear materials, corrosion-resistant coatings, and photonics.

The design of new nanomaterials with novel properties and functionalities requires atomic-level control of the material's structure. The fabrication of functional nanomaterials can be achieved by conventional top-down methods such as lithography or bottom-up approaches such as self-assembly. The latter approach can be very cost-effective for the fabrication of macroscopic three-dimensional nanostructures. However, self-assembly is scientifically very challenging because it requires detailed understanding of materials interactions at the atomic level. The goal of this LDRD project is twofold. The research team is advancing the scientific field of functional nanomaterials by combining the best of directed self-assembly with automated lithography and assembly to develop nanomaterials for fusion-energy applications for advanced laser systems. Second, they are applying this scientific knowledge to fabricate complex laser-fusion target structures by high-throughput techniques such as automated assembly and templating. Success will lay the scientific foundation for efficient production of laser fusion targets with unprecedented complexity.

A successful experiment at fusion-class laser facilities such as the National Ignition Facility creates a unique laboratory environment previously only available in a nuclear test. First among the unique characteristics is an inertially confined burning hydrogen plasma. Second is the bright neutron source that such a plasma will create. The challenge is to create laser targets that take advantage of this environment. To carry out experiments in this environment, researchers must place nanostructured materials with controlled composition and precise geometry in cubic millimetre spaces inside an inertial-confinement fusion capsule. Inertial fusion ignition will enable important laboratory and national missions, as well as new science. A series of such experiments is proposed for the National Ignition Facility, and NNSA's top scientific priority is significant advancement in the understanding of boost physics within the next five years. Thus, among the first proposed experiments is a suite of boost physics experiments. These are designed to investigate the mix in burning plasmas. Complex targets for study of the mix in burning hydrogen plasmas, nuclear physics in high-neutron-brightness environments, and fast-ignition inertial-confinement fusion require precisely placing nanoporous materials with small amounts of additional incorporated elements to alter the material's properties (a process known as doping), inside the target capsule.

Traditionally, target capsules are made by assembling prefabricated and machined components of the structure. This approach limits the complexity of the structures that can be achieved. With self-assembly techniques, complex targets could be built either from the inside out or outside in. In the latter case, the ablator shell with a fill hole would serve as both a template and a container in which materials could be grown or modified. LDRD researchers will use this "chemistry in a capsule" approach to generate complex nanostructures within the ablator shell by in situ self-assembly of functional building blocks. For example, a uniform and conforming film of a porous material lining the inside of the shell could be prepared by spin coating and subsequent curing by light, heat, or chemical initiation. Once formed, the properties of a nanoporous capsule liner could be further fine-tuned by atomic layer deposition. Possible applications include doping with mid-to-high-atomic-number elements as well as surface engineering of the wetting behaviour. In principle, this self-assembly technique could be applied to any nanostructure formation.

In just the first year of this project, the LDRD team has made significant progress in all these areas. They have successfully demonstrated the feasibility of the chemistry in a capsule approach by loading a diamond laser target with the nanomaterial precursor solution and forming nanoporous foam inside the capsule. Team members installed an atomic-layer-deposition system and demonstrated that the technique can be applied to deposit mid-to-high-atomic-number elements on the nanoporous foam liner inside the capsule. They also demonstrated that the capsule liner can be rendered inactive and then reactivated for atomic layer deposition. This discovery opens the door to generating nanoporous foams with three-dimensional doping patterns by area-selective atomic layer deposition. The effects and requirements of this tailored doping on yield and symmetry of foam-filled capsules was assessed by hydrodynamic simulations. X-ray scattering was used to study the effects of wetting organic aerogels with liquid hydrogen on the foam morphology. These experiments demonstrated that even low-density aerogels can survive liquid hydrogen wetting and de-wetting cycles. Finally, an in situ film-stress-monitoring capability was developed to study the microstructure-stress correlation in thick metal films grown by magnetron sputtering. The team was able to control the texture—and thus the stress of thick beryllium films—by applying a bias to the substrate, thereby changing the initial stress state from tensile to compressive. Besides this impressive progress, they also developed novel nanomaterials for energy storage and conversion applications and studied their unique properties.

Future LDRD efforts will include casting of thin films of aerogels in target capsules by growing cross-linkable polymer brushes, for example. Using in situ x-ray techniques, the team will study the effect of aerogel morphology on wetting with liquid hydrogen. They will develop novel atomic-layer-deposition precursor chemistries for doping with high-atomic-number materials. Area-selective doping will be attempted through spatially selective modifications of the substrate's surface chemistry. In addition, new functionalities will be added to their nanomanipulator to measure thickness profiles from three-dimensional targets and to apply picoliter amounts of liquids. Researchers will continue work on stress control in metal deposition and develop predictive multiscale tools.

The successful conclusion of this LDRD project will enable the fabrication of complex targets for experiments to understand boost physics, investigate the dynamics of nuclear excited states, and pursue inertial-confinement fusion fast ignition. It will also create new materials for catalysis, hydrogen storage, and self-healing nuclear reactor materials. A very important long-term benefit of this effort will be the establishment of an outstanding Laboratory team for the design and assembly of tailored nanomaterials and nanostructures.

REFERENCES

- Aviram, A. : *Electronics using Hybrid-molecular and Mono-molecular Devices*, London, 2000.
- Bhatnagar, Vasudev : *Nanostructure Science and Technology*, Campus Books, Delhi, 2009.
- Cerofolini, G. F.: *Nanoscale Devices*, Springer-Verlag, Berlin, 2009.
- Fogelberg, H.: *The Material Culture of Nanotechnology*, Göteborg: Göteborg University, 2003.
- Hosono, H. : *Nanomaterials : From Research to Applications*, Elsevier, A Division of Reed Elsevier India, 2008.

2

Carbon and Inorganic Nanotubes

CARBON

Carbon is the chemical element with symbol C and atomic number 6. As a member of group 14 on the periodic table, it is nonmetallic and tetravalent—making four electrons available to form covalent chemical bonds. There are three naturally occurring isotopes, with ^{12}C and ^{13}C being stable, while ^{14}C is radioactive, decaying with a half-life of about 5,730 years. Carbon is one of the few elements known since antiquity.

There are several allotropes of carbon of which the best known are graphite, diamond, and amorphous carbon. The physical properties of carbon vary widely with the allotropic form. For example, diamond is highly transparent, while graphite is opaque and black. Diamond is among the hardest materials known, while graphite is soft enough to form a streak on paper (hence its name, from the Greek word "to write"). Diamond has a very low electrical conductivity, while graphite is a very good conductor. Under normal conditions, diamond has the highest thermal conductivity of all known materials.

All carbon allotropes are solids under normal conditions with graphite being the most thermodynamically stable form. They are chemically resistant and require high temperature to react even with oxygen. The most common oxidation state of carbon in inorganic compounds is +4, while +2 is found in carbon monoxide and other transition metal carbonyl complexes. The largest sources of inorganic carbon are limestones, dolomites and carbon dioxide, but significant quantities occur in organic deposits of coal, peat, oil and methane clathrates. Carbon forms more compounds than any other element, with almost ten million pure organic compounds described to date, which in turn are a tiny fraction of such compounds that are theoretically possible under standard conditions.

Carbon is the 15th most abundant element in the Earth's crust, and the fourth most abundant element in the universe by mass after hydrogen, helium, and oxygen. It is present in all known life forms, and in the human body carbon is the second most abundant element by mass (about 18.5%) after oxygen. This abundance, together with the unique diversity of organic compounds and their unusual polymer-forming ability at the temperatures commonly encountered on Earth, make this element the chemical basis of all known life.

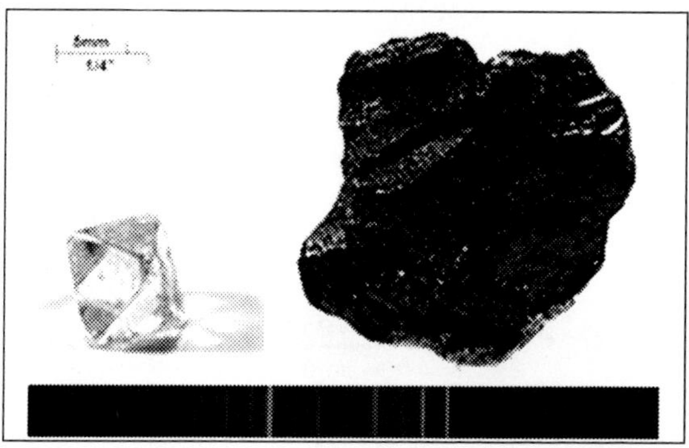

Figure : Spectral Lines of Carbon

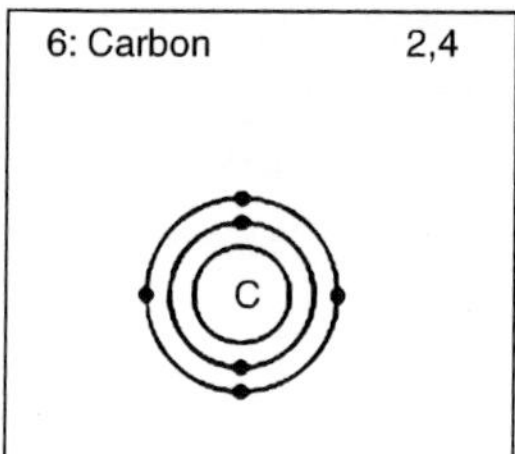

Figure : General Properties

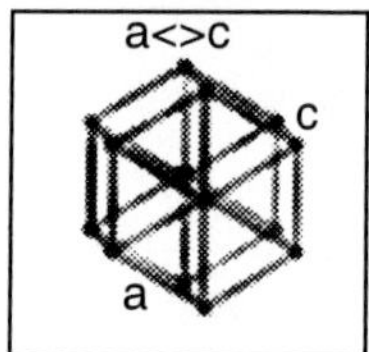

Figure : Crystal Structure

CHARACTERISTICS

The different forms or *allotropes* of carbon include the hardest naturally occurring substance, diamond, and also one of the softest known substances, graphite. Moreover, it has an affinity for bonding with other small atoms, including other carbon atoms, and is capable of forming multiple stable covalent bonds with such atoms. As a result, carbon is known to form almost ten million different compounds; the large majority of all chemical compounds. Carbon also has the highest sublimation point of all elements. At atmospheric pressure it has no melting point as its triple point is at 10.8 ± 0.2 MPa and 4,600 ± 300 K (~4,330 °C or 7,820 °F), so it sublimates at about 3,900 K.

Carbon sublimes in a carbon arc which has a temperature of about 5,800 K (5,530 °C; 9,980 °F). Thus, irrespective of its allotropic form, carbon remains solid at higher temperatures than the highest melting point metals such as tungsten or rhenium. Although thermodynamically prone to oxidation, carbon resists oxidation more effectively than elements such as iron and copper that is weaker reducing agents at room temperature.

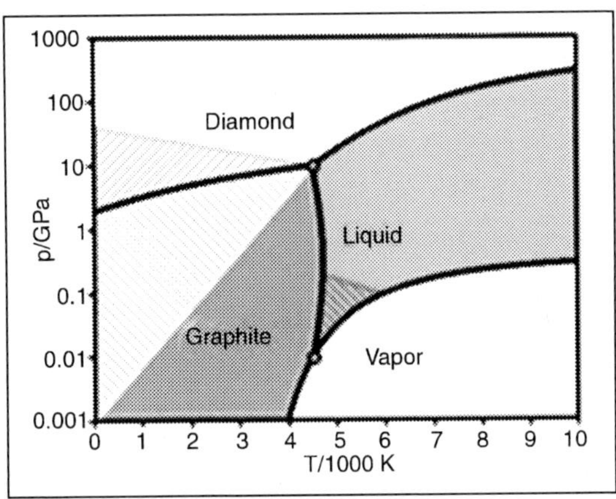

Figure: Theoretically Predicted Phase Diagram of Carbon.

Carbon compounds form the basis of all known life on Earth, and the carbon-nitrogen cycle provides some of the energy produced by the Sun and other stars. Although it forms an extraordinary variety of compounds, most forms of carbon are comparatively unreactive under normal conditions. At standard temperature and pressure, it resists all but the strongest oxidizers. It does not react with sulfuric acid, hydrochloric acid, chlorine or any alkalis. At elevated temperatures carbon reacts with oxygen to form carbon oxides, and will reduce such metal oxides as iron oxide to the metal. This exothermic reaction is used in the iron and steel industry to control the carbon content of steel:

$$Fe_3O_4 + 4\ C_{(s)} \rightarrow 3\ Fe_{(s)} + 4\ CO_{(g)}$$

with sulfur to form carbon disulfide and with steam in the coal-gas reaction:

$$C_{(s)} + H_2O_{(g)} \rightarrow CO_{(g)} + H_{2(g)}.$$

Carbon combines with some metals at high temperatures to form metallic carbides, such as the iron carbide cementite in steel, and tungsten carbide, widely used as an abrasive and for making hard tips for cutting tools.

As of 2009, graphene appears to be the strongest material ever tested. However, the process of separating it from graphite will require some technological development before it is economical enough to be used in industrial processes.

The system of carbon allotropes spans a range of extremes:

Synthetic nanocrystalline diamond is the hardest material known.	Graphite is one of the softest materials known.
Diamond is the ultimate abrasive.	Graphite is a very good lubricant, displaying superlubricity.
Diamond is an excellent electrical insulator.	Graphite is a conductor of electricity.
Diamond is the best known naturally occurring thermal conductor	Some forms of graphite are used for thermal insulation (i.e. firebreaks and heat shields)
Diamond is highly transparent.	Graphite is opaque.
Diamond crystallizes in the cubic system.	Graphite crystallizes in the hexagonal system.
Amorphous carbon is completely isotropic.	Carbon nanotubes are among the most anisotropic materials ever produced.

Allotropes

Atomic carbon is a very short-lived species and, therefore, carbon is stabilized in various multi-atomic structures with different molecular configurations called allotropes. The three relatively well-known allotropes of carbon are amorphous carbon, graphite, and diamond. Once considered exotic, fullerenes are nowadays commonly synthesized and used in research; they include buckyballs, carbon nanotubes, carbon nanobuds and nanofibres. Several other exotic allotropes have also been discovered, such as lonsdaleite, glassy carbon, carbon nanofoam and linear acetylenic carbon (carbyne).

The amorphous form is an assortment of carbon atoms in a non-crystalline, irregular, glassy state, which is essentially graphite but not held in a crystalline macrostructure. It is present as a powder, and is the main constituent of substances such as charcoal, lampblack (soot) and activated carbon. At normal pressures carbon takes the form of graphite, in which each atom is bonded trigonally to three others in a plane composed of fused hexagonal rings, just like those in aromatic hydrocarbons. The resulting network is 2-dimensional, and the resulting flat sheets are stacked and loosely bonded through weak van der Waals forces. This gives graphite its softness and its cleaving properties (the sheets slip easily past one another). Because of the delocalization of one of the outer electrons of each atom to form a π-cloud, graphite conducts electricity, but only in the plane of each covalently bonded sheet. This results in a lower bulk electrical conductivity for carbon than for most metals. The delocalization also accounts for the energetic stability of graphite over diamond at room temperature.

At very high pressures carbon forms the more compact allotrope diamond, having nearly twice the density of graphite. Here, each atom is bonded tetrahedrally to four others, thus making a 3-dimensional network of puckered six-membered rings of atoms. Diamond has the same cubic structure as silicon and germanium and because of the strength of the carbon-carbon bonds, it is the hardest naturally occurring substance in terms of resistance to scratching. Contrary to the popular belief that *"diamonds are forever"*, they are in fact thermodynamically unstable under normal conditions and transform into graphite. However, due to a high activation energy barrier, the transition into graphite is so extremely slow at room temperature as to be unnoticeable. Under some conditions, carbon crystallizes as lonsdaleite. This form has a hexagonal crystal lattice where all atoms are covalently bonded. Therefore, all properties of lonsdaleite are close to those of diamond.

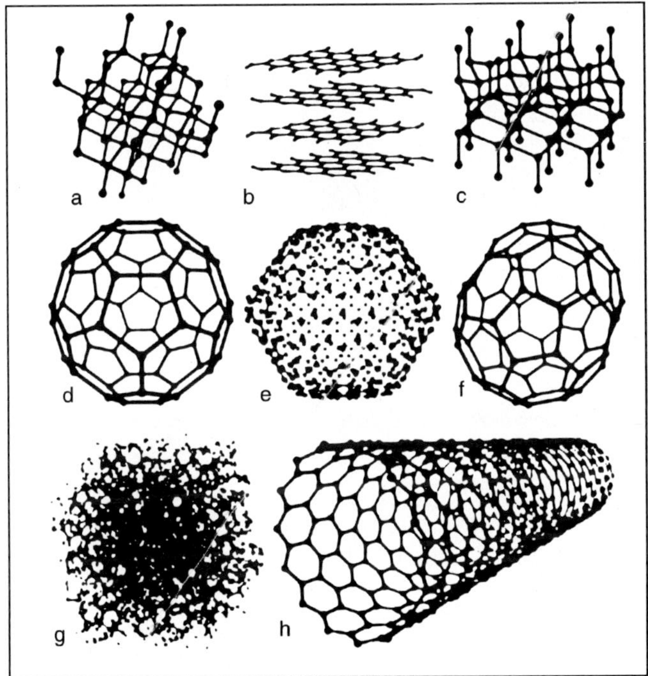

Figure: Some Allotropes of Carbon: (a) Diamond; (b) Graphite; (c) Lonsdaleite; (d–f) Fullerenes (C_{60}, C_{540}, C_{70}); g) Amorphous Carbon; (h) Carbon Nanotube.

Fullerenes have a graphite-like structure, but instead of purely hexagonal packing, they also contain pentagons (or even heptagons) of carbon atoms, which bend the sheet into spheres, ellipses or cylinders. The properties of fullerenes (split into buckyballs, buckytubes and nanobuds) have not yet been fully analysed and represent an intense area of research in nanomaterials. The names *"fullerene"* and *"buckyball"* are given after Richard Buckminster Fuller, popularizer of geodesic domes, which resemble the structure of fullerenes. The buckyballs are fairly large molecules formed completely of carbon bonded trigonally, forming spheroids (the best-known and simplest is the soccerball-shaped C_{60} buckminsterfullerene). Carbon nanotubes are structurally similar to buckyballs, except that each atom is bonded trigonally in a curved sheet that forms a hollow cylinder. Nanobuds were first reported in 2007 and are hybrid bucky tube/buckyball materials (buckyballs are covalently bonded to the outer wall of a nanotube) that combine the properties of both in a single structure.

Of the other discovered allotropes, carbon nanofoam is a ferromagnetic allotrope discovered in 1997. It consists of a low-density cluster-assembly of carbon atoms strung together in a loose three-dimensional web, in which the atoms are bonded trigonally in six- and seven-membered rings. It is among the lightest known solids, with a density of about 2 kg/m^3. Similarly, glassy carbon contains a high proportion of closed porosity, but contrary to normal graphite, the graphitic layers are not stacked like pages in a book, but have a more random arrangement. Linear acetylenic carbon has the chemical structure $-(C:::C)_n-$. Carbon in this modification is linear with *sp* orbital hybridization, and is a polymer with alternating single and triple bonds. This type of carbyne is of considerable interest to nanotechnology as its Young's modulus is forty times that of the hardest known material – diamond.

Occurrence

Carbon is the fourth most abundant chemical element in the universe by mass after hydrogen, helium, and oxygen. Carbon is abundant in the Sun, stars, comets, and in the atmospheres of most planets. Some meteorites contain microscopic diamonds that were formed when the solar system was still a protoplanetary disk. Microscopic diamonds may also be formed by the intense pressure and high temperature at the sites of meteorite impacts.

In combination with oxygen in carbon dioxide, carbon is found in the Earth's atmosphere (approximately 810 gigatonnes of carbon) and dissolved in all water bodies (approximately 36,000 gigatonnes of carbon). Around 1,900 gigatonnes of carbon are present in the biosphere. Hydrocarbons (such as coal, petroleum, and natural gas) contain carbon as well—coal "reserves" (not "resources") amount to around 900 gigatonnes, and oil reserves around 150 gigatonnes.

Proven sources of natural gas are about 175 10^{12} cubic metres (representing about 105 gigatonnes carbon), but it is estimated that there are also about 900 10^{12} cubic metres of "unconventional" gas such as shale gas, representing about 540 gigatonnes of carbon. (In the past, quantities of hydrocarbons were greater. In the period from 1751 to 2008 about 347 gigatonnes of carbon were released as carbon dioxide to the atmosphere from burning of fossil fuels.) Carbon is also locked up as methane and methane hydrates in polar regions. It is estimated that at least 1,400 Gt of carbon is in this form just in (and under) the submarine permafrost of the Siberian Shelf.

Carbon is a major component in very large masses of carbonate rock (limestone, dolomite, marble and so on). Coal is the largest commercial source of mineral carbon, accounting for 4,000 gigatonnes or 80% of fossil carbon fuel. It is also rich in carbon – for example, anthracite contains 92–98%.

Figure: Graphite Ore

Figure: Raw Diamond Crystal.

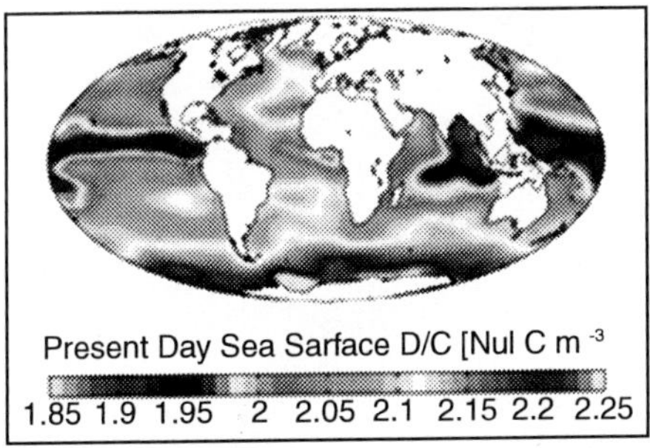

Figure: "Present Day" (1990s) Sea Surface Dissolved Inorganic Carbon Concentration (from the GLODAP Climatology).

As for individual carbon allotropes, graphite is found in large quantities in the United States (mostly in New York and Texas), Russia, Mexico, Greenland, and India. Natural diamonds occur in the rock kimberlite, found in ancient volcanic "necks", or "pipes". Most diamond deposits are in Africa, notably in South Africa, Namibia, Botswana, the Republic of the Congo, and Sierra Leone.

There are also deposits in Arkansas, Canada, the Russian Arctic, Brazil and in Northern and Western Australia. Diamonds are now also being recovered from the ocean floor off the Cape of Good Hope. However, though diamonds are found naturally, about 30% of all industrial diamonds used in the U.S. are now made synthetically.

Carbon-14 is formed in upper layers of the troposphere and the stratosphere, at altitudes of 9–15 km, by a reaction that is precipitated by cosmic rays. Thermal neutrons are produced that collide with the nuclei of nitrogen-14, forming carbon-14 and a proton.

Isotopes

Isotopes of carbon are atomic nuclei that contain six protons plus a number of neutrons (varying from 2 to 16). Carbon has two stable, naturally occurring isotopes. The isotope carbon-12 (^{12}C) forms 98.93% of the carbon on Earth, while carbon-13 (^{13}C) forms the remaining 1.07%. The concentration of ^{12}C is further increased in biological materials because biochemical reactions discriminate against ^{13}C. In 1961 the International Union of Pure and Applied Chemistry (IUPAC) adopted the isotope carbon-12 as the basis for atomic weights. Identification of carbon in NMR experiments is done with the isotope ^{13}C.

Carbon-14 (^{14}C) is a naturally occurring radioisotope which occurs in trace amounts on Earth of up to 1 part per trillion (0.0000000001%), mostly confined to the atmosphere and superficial deposits, particularly of peat and other organic materials. This isotope decays by 0.158 MeV β^- emission. Because of its relatively short half-life of 5730 years, ^{14}C is virtually absent in ancient rocks, but is created in the upper atmosphere (lower stratosphere and upper troposphere) by interaction of nitrogen with cosmic rays. The abundance of ^{14}C in the atmosphere and in living organisms is almost constant, but decreases predictably in their bodies after death. This principle is used in radiocarbon dating, invented in 1949, which has been used extensively to determine the age of carbonaceous materials with ages up to about 40,000 years.

There are 15 known isotopes of carbon and the shortest-lived of these is ^{8}C which decays through proton emission and alpha decay and has a half-life of 1.98739×10^{-21} s. The exotic ^{19}C exhibits a nuclear halo, which means its radius is appreciably larger than would be expected if the nucleus were a sphere of constant density.

Formation in Stars

Formation of the carbon atomic nucleus requires a nearly simultaneous triple collision of alpha particles (helium nuclei) within the core of a giant or supergiant star which is known as the triple-alpha process, as the products of further nuclear fusion reactions of helium with hydrogen or another helium nucleus produce lithium-5 and beryllium-8 respectively, both of which are highly unstable and decay almost instantly back into smaller nuclei. This happens in conditions of temperatures over 100 megakelvin and helium concentration that the rapid expansion and cooling of the early universe prohibited, and therefore no significant carbon was created during the Big Bang. Instead, the interiors of stars in the horizontal branch transform three helium nuclei into carbon by means of this triple-alpha process. In order to be available for formation of life as we know it, this carbon must then later be scattered into space as dust, in supernova explosions, as part of the material which later forms second, third-generation star systems which have planets accreted from such dust. The Solar System is one such third-generation star system. Another of the fusion mechanisms powering stars is the CNO cycle, in which carbon acts as a catalyst to allow the reaction to proceed.

Rotational transitions of various isotopic forms of carbon monoxide (for example, ^{12}CO, ^{13}CO, and $C^{18}O$) are detectable in the submillimetre wavelength range, and are used in the study of newly forming stars in molecular clouds.

Carbon Cycle

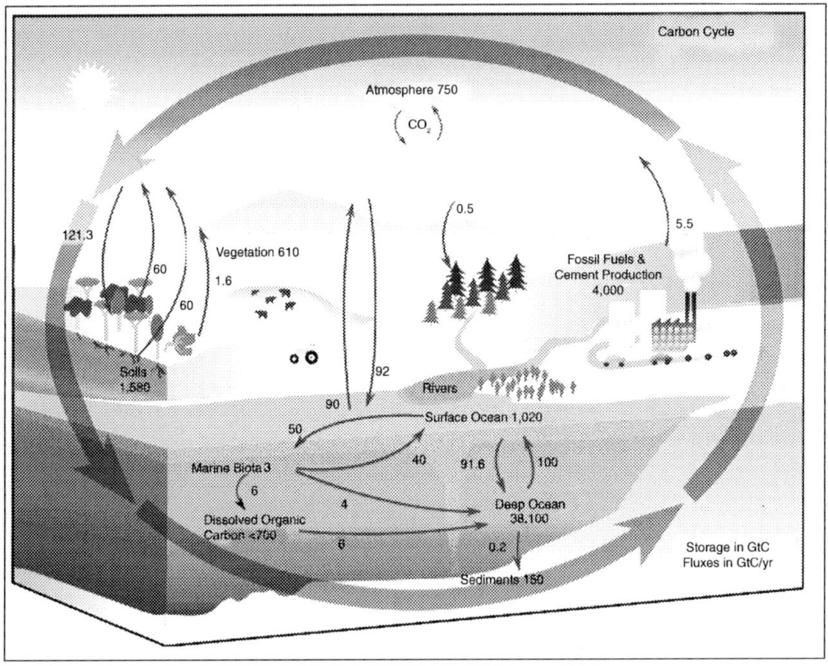

Figure: Diagram of the Carbon Cycle.

Under terrestrial conditions, conversion of one element to another is very rare. Therefore, the amount of carbon on Earth is effectively constant. Thus, processes that use carbon must obtain it somewhere and dispose of it somewhere else. The paths that carbon follows in the environment make up the carbon cycle. For example, plants draw carbon dioxide out of their environment and use it to build biomass, as in carbon respiration or the Calvin cycle, a process of carbon fixation. Some of this biomass is eaten by animals, whereas some carbon is exhaled by animals as carbon dioxide. The carbon cycle is considerably more complicated than this short loop; for example, some carbon dioxide is dissolved in the oceans; dead plant or animal matter may become petroleum or coal, which can burn with the release of carbon, should bacteria not consume it.

TYPES OF CARBON CREDITS

Essentially, the types of carbon credits can be split into two forms, those within the voluntary market and those within the compliance market. Each type of carbon credit adheres to a particular standard or certification.

CERTIFIED EMISSION REDUCTION (CER) UNITS

The most common type of compliance credit is a CER (Certified Emission Reduction unit) which originates from projects in developing counties. Certification and overall approval of these abatement projects and their credits is known as the Clean Development Mechanism (CDM).

EMISSION REDUCTION UNIT (ERU)

Like CER in developing nations, within developed nations, a mechanism known as Joint Implementation or JI, produces compliance credits referred to as Emission Reduction Units or ERUs.

NEW SOUTH WALES GREENHOUSE GAS ABATEMENT CERTIFICATE (NGAC)

The New South Wales Greenhouse Abatement Certificate (NGAC) certification process is comprehensive. It includes Kyoto Protocol measures, but goes beyond these. In summary the NGAC certification process ensures the following:

- That each NGAC represents one tonne of carbon dioxide stored for at least 100 years.
- That the trees have been planted since 1990.
- That the trees weren't planted on old growth forest cleared land (the land must have been clear prior to 1990).
- That should the tree from which your carbon credit came come to any harm within 100 years of your purchase *e.g.*, Fire, disease, logging; that carbon credit will be replaced immediately from another source.

VOLUNTARY CARBON CREDITS

More information on the voluntary market can be found here. The credit types below are just a sample of the most commonly used products in Australia and globally. Many more types exist overseas and if you want more information on these, please contact us.

VOLUNTARY CARBON UNIT (VCU) OR VOLUNTARY CARBON STANDARD (VCS) CREDIT

The VCS Programme provides a robust, global standard for approval of credible voluntary carbon credits. VCS credits or Voluntary Carbon Units (VCU) must be real, the abatement must have occurred, they must be additional by going beyond business-as-usual activities, be measurable, permanent, not temporarily displace emissions, the findings need to be independently verified and unique so they cannot be used more than once to offset emissions.

The VCS is the most widely known and chosen standards in the voluntary market due to it's Kyoto compatibility as well as it's ability to manage a wide range of project types and methodologies.

VERIFIED (OR VOLUNTARY) EMISSIONS REDUCTION (VER) AND GOLD STANDARD VER

The most popular type of carbon credit used to offset emissions around the world voluntarily is a VER, a Verified or Voluntary Emission Reduction unit and there are many different types. Before CDM or JI projects deliver credits used for Compliance purposes such as CERs and ERUs they can produce VERs.

These credits can be verified to a number of specific standards, including the Gold Standard. Not all projects go on to register within the CDM or JI, often due to the size of the project and the inhibitive costs associated with compliance registration, so their choice of one or more of these voluntary standards is made based on it's overall viability and compatibility to them.

RENEWABLE ENERGY CERTIFICATE (REC)

A REC is not a carbon credit that represents one tonne of CO_2e emissions but rather a unit that relates to how much CO_2e is saved by the adoption of renewable energy and how efficiently one mega watt hour (MWh) of electricity can be produced.

This can vary from as little as a 500 kilos of CO_2e, to as much as almost two tonnes from older, less efficient power stations. Like carbon credits, in an attempt to faze out and replace traditional, emission intensive activities, RECs provide financial subsidies for the power sector to help renewable energy projects become more viable around the world.

New technology and innovations to existing technology are rapidly being realised in areas such as; solar Photo Voltaic (PV) cells, wind farms, subterranean geothermal power plants, wave collection technology, hydroelectric, tidal power, renewable biomass and more. Depending on their location, these projects can produce RECs but as they also displace CO_2e they can often be a more viable project if a choice was made in favour of producing carbon credits instead, for example VCUs, VERs or CERs.

THE PHASE DIAGRAM OF CARBON

The stable bonding configuration of carbon at NTP is graphite, with an energy difference between the graphite and the diamond of 0.02 eV per atom. Due to the high energetic barrier between the two phases of carbon, the transition from diamond to the most stable phase of graphite at normal conditions is extremely slow. This transition can also occurs more rapidly, when diamond is exposed to ion bombardment or high temperature for example. Due to the high anisotropy in the graphite structure as compared to that of diamond, the electronic, mechanical and optical properties of these two phases of carbon are very different. Some properties of diamond and graphite crystals are presented. In the column related to graphite, the in-plane properties appears on the left and the transverse one between planes on the right.

Table : Properties of diamond and graphite.

Property	*Graphite*	*Diamond*
Lattice constant (RT) [Å]	2.462 6.708	3.567
Bond length (RT) [Å]	1.421	1.545
Atomic density [cm^{-3}]	1.14×10^{23}	1.77×10^{23}
Thermal conductivity [W/cm-K]	30 0.06	25
Debye temperature [K]	2500 950	1860
Electron mobility [cm^2/V-sec]	$20{\times}10^3$ 100	1800
Hole mobility [cm^2/V-sec]	$15{\times}10^3$ 90	1500
Melting point K	4200	4500
Band gap [eV]	-0.04	5.47

Bridging between these two allotropes of carbon lie a whole variety of carbon materials which include, among others, amorphous sp^2 bonded carbon (such as thermally evaporated carbon), micropolycrystalline sp^2 bonded graphite (such as glassy carbon), nanodiamond films, and amorphous sp^3 bonded carbon (sometimes referred to as amorphous diamond), which is structurally analogous to amorphous Si and is formed during low energy carbon ions deposition.

Nanodiamond films, for example, can been grown by different deposition techniques such as dc assisted plasma chemical vapour deposition (CVD) from a methane-hydrogen mixture and others. The criteria of quality of the nanodiamond films include low-contents of nondiamond phases, nano-sizes crystallites, uniform nanocrystallinity throughout thick films and random grain orientation.

Another polymorphic form of carbon was discovered in 1985. It exists in discrete molecular form, and consists of a hollow spherical cluster of carbon atoms. Each molecule is composed of groups of carbon atoms that are bonded to one another form both hexagons and pentagons geometrical configuration.

The material composed of C^{60} is known as buckminsterfullerene, named in honour of R.

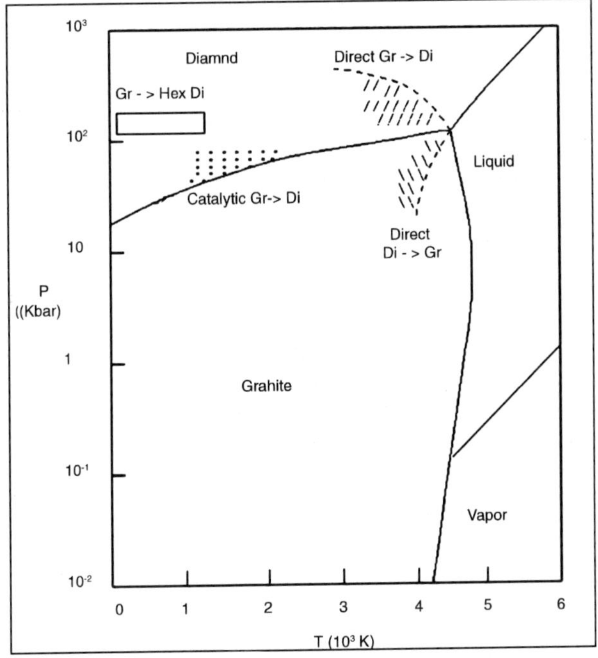

Figure : P, T Phase Diagram of Carbon Reproduced.

Buckminster Fuller, who invented the geodesic dome. In the solid state, the C^{60} units form a crystalline structure and pack together in a face-centred cubic array.

Molecular shapes other than the ball clusters recently have been discovered: these include nanoscale tubular and polyhedral structures. It is anticipated that, with further developments, the fullerenes will become technologically important materials.

USE OF CARBON NANOTUBES

Here, we focus on the recent work on *carbon nanotubes*; their structure, properties and uses in nano-electronic devices. Carbon nanotubes are extremely thin (their diameter is about 10,000 times smaller than a human hair), hollow cylinders made of carbon atoms.

- Nanotube manipulation: We can manipulate the nanotube positions, change their shape, cut them and place them on electrodes.
- Molecular mechanics: We can simulate the mechanical behaviour of nanotubes by calculating the forces acting between nanotubes and other objects such as the substrate.

- Nanotube Field-Effect Transistor: We have successfully used semiconducting single and multi-walled nanotubes as channels of field-effect transistors.
- Nanotube rings: While normally nanotubes are straight, we have devised ways to prepare them in a ring form.
- Nanotube theory: Computation and theory of the electrical and mechanical properties.

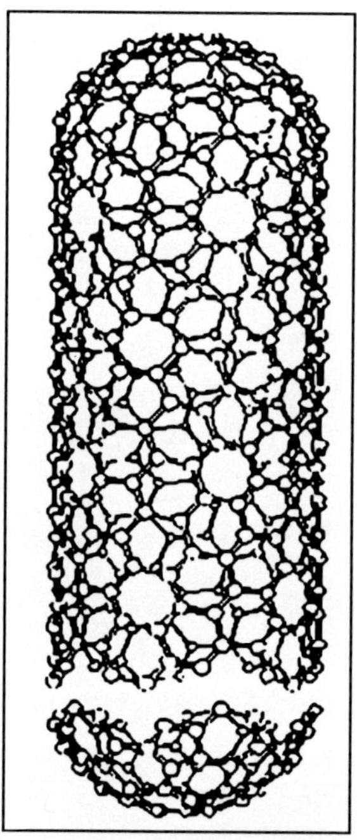

Nanotubes, depending on their structure, can be metals or semi-conductors. They are also extremely strong materials and have good thermal conductivity. The above characteristics have generated strong interest in their possible use in nano-electronic and nano-mechanical devices.

CARBON NANOTUBE SYNTHESIS

What is Carbon Nanotube Synthesis, and What are its Implications for Modern Science? While the mystery and implications for science surrounding carbon nanotubes in modern science have gained them immense popularity with scientists and researchers in myriad fields. The age old question-what is carbon nanotube synthesis? is quickly gaining ground in modern science as well. And its implications may far outweigh and outreach those of just carbon nanotubes, since this process of synthesis may very well open doors for science and for the use of carbon nanotubes that no one ever thought possible. Carbon nanotubes are essentially constructed of a lattice work sheet of graphite that is rolled into a cylindrical shape. These carbon nanotubes are feather light in their construction and much stronger than many other properties used in not only nanotechnology but also in electronics, optics, additional areas of materials science, architecture and even the construction of motor vehicles, space vehicles and other materials useful in a wide array of traditional fields of research and modern living.

Carbon nanotubes are also known to have any unique electrical properties and have even been tested and proven as super efficient conductors of heat, otherwise referred to as thermal conductors. It is only their potential to have some toxic properties that has limited their widespread use and distribution for construction and other purposes.

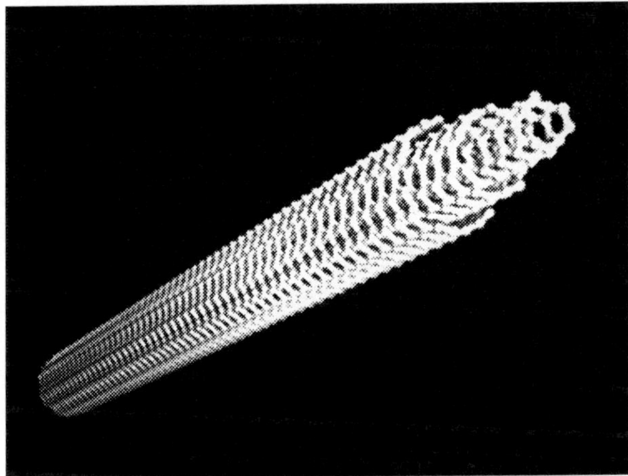

Carbon nanotubes come in a variety of forms, including single walled carbon nanotubes, double walled carbon nanotubes and multi walled carbon nanotubes. Each type of carbon nanotubes has its own set of unique properties that make it useful for different areas of science. For instance, the single walled carbon nanotube has especially strong electric properties that the other varieties lack.

Carbon nanotube synthesis is incredibly important to the research of the potential uses of carbon nanotubes because nanotube synthesis is the method of production used to create carbon nanotubes. The faster the rate of synthesis, or production, the more carbon nanotubes available for researches and scientists to use in testing for the various possible uses of carbon nanotubes in standard modern day living, current science and potential future applications. Although carbon nanotube synthesis can occur naturally through flame synthesis, the most common methods of used in modern science include laser ablation, arc discharge and chemical vapour deposition.

The arc discharge method of nanotube synthesis was initially developed by accident. In 1991, an arc discharge that was meant for producing fullerenes, produced, along with the fullerenes, carbon nanotubes. In 1992, the experiment was reproduced by two researchers at the Fundamental Research Laboratory of the NEC. This process involves high temperatures for discharge which cause the negative electrode in the arc to force sublimate the carbon nanotubes contained within. Additionally, this process has a yield by weight of more than 30 percent and it is able to produce both multi walled (including double walled) and single walled carbon nanotubes.

This method for carbon nanotube synthesis is most popular because it was the first method discovered and used. The carbon nanotubes produced through this method of synthesis often have few defects, if any, and lengths of more than 50 micrometers are able to be produced. But arc discharge is not the only type of nanotube synthesis available to researchers and scientists. Other methods have been developed in order to provide a wide variety of acceptable outcomes—acceptable outcomes being defined as producing usable carbon nanotubes in large quantities with minimal defects.

One of the additional methods of nanotube synthesis is laser ablation method. When scientists use laser ablation synthesis method for carbon nanotubes, both a pulsing laser and an inert gas are paired together in a tightly sealed reactor with high temperatures. The pulsed laser is used to vaporize the graphite as the gas is allowed to slowly seep into the reactor chamber. The graphite vapour condenses as the reactor surfaces of the chamber cool down and nanotubes form out of the vapours.

In the laser ablation method, scientists most often produce single walled carbon nanotubes and the yield is approximately 70 percent by weight. Although this method yields much more by weight than the other two available methods, it is much more expensive to use and many scientists and researchers opt in favour of the less expensive methods. The final popular method of carbon nanotube synthesis is the chemical vapour desposition method. Although this method has been used to synthesize carbon since 1959, the chemical catalytic vapour deposition method of producing carbon nanotubes was not first used until 1993, nearly forty years after its first use for synthesis and two years after the first nanotubes were synthesized using the arc discharge method.

This method was further developed in 2007 by University of Cincinnati researchers and at that time began producing carbon nanotubes at lengths of 18 millimetres, much longer than those produced by the arc discharge method at only fifty micrometers.

In chemical vapour desposition, a substrate made of iron, cobalt, nickel or a combination of any of these metals is heated to precisely 700 degrees Celsius inside a chemical reactor. Process gases such as hydrogen, nitrogen or ammonia along with a carbon gas like methane, ethanol or acetylene, is leaked into the reactor chamber. The meeting of the gases at the substrate cause a reaction that breaks the carbon gases apart and forces the particles to the sides of the chamber, where the carbon nanotubes are formed. This method of nanotube synthesis is the most popular for commercial production because it allows for nanotubes to be grown from and on a specified substrate. The most important factors relating to nanotube synthesis are how fast the nanotubes can be grown, how many can be made at any one time, how much the process to be used will cost and most importantly, how few structural defects will be present in the new carbon nanotubes that have been synthesized. All of these factors determine the future of carbon nanotubes and the development of better method for synthesis will determine the future application of carbon nanotubes in everything that surrounds us.

ALLOTROPES OF CARBON NANOTUBES

Carbon nanotubes (CNTs) are allotropes of carbon with a cylindrical nanostructure. Nanotubes have been constructed with length-to-diameter ratio of up to 132,000,000:1, significantly larger than for any other material. These cylindrical carbon molecules have unusual properties, which are valuable for nanotechnology, electronics, optics and other fields of materials science and technology. In particular, owing to their extraordinary thermal conductivity and mechanical and electrical properties, carbon nanotubes find applications as additives to various structural materials. For instance, nanotubes form only a tiny portion of the material(s) in (primarily carbon fibre) baseball bats, golf clubs, or car parts.

Nanotubes are members of the fullerene structural family, which also includes the spherical buckyballs, and the ends of a nanotube may be capped with a hemisphere of the buckyball structure. Their name is derived from their long, hollow structure with the walls formed by one-atom-thick sheets of carbon, called graphene. These sheets are rolled at specific and discrete ("chiral") angles, and the combination of the rolling angle and radius decides the nanotube properties; for example, whether the individual nanotube shell is a metal or semiconductor. Nanotubes are categorized as single-walled nanotubes (SWNTs) and multi-walled nanotubes (MWNTs). Individual nanotubes naturally align themselves into "ropes" held together by van der Waals forces, more specifically, pi-stacking.

Applied quantum chemistry, specifically, orbital hybridization best describes chemical bonding in nanotubes. The chemical bonding of nanotubes is composed entirely of sp^2 bonds, similar to those of graphite. These bonds, which are stronger than the sp^3 bonds found in alkanes and diamond, provide nanotubes with their unique strength.

CARBON NANOTUBE MATERIAL

A Carbon Nanotube is a tube-shaped material, made of carbon, having a diameter measuring on the nanometre scale. A nanometre is one-billionth of a meter, or about one ten-thousandth of the thickness of a human hair. The graphite layer appears somewhat like a rolled-up chicken wire with a continuous unbroken hexagonal mesh and carbon molecules at the apexes of the hexagons.

Carbon Nanotubes have many structures, differing in length, thickness, and in the type of helicity and number of layers. Although they are formed from essentially the same graphite sheet, their electrical characteristics differ depending on these variations, acting either as metals or as semiconductors.

As a group, Carbon Nanotubes typically have diameters ranging from <1 nm up to 50 nm. Their lengths are typically several microns, but recent advancements have made the nanotubes much longer, and measured in centimeters.

Properties of a Carbon Nanotube

The intrinsic mechanical and transport properties of Carbon Nanotubes make them the ultimate carbon fibres. The following tables compare these properties to other engineering materials.

Overall, Carbon Nanotubes show a unique combination of stiffness, strength, and tenacity compared to other fibre materials which usually lack one or more of these properties. Thermal and electrical conductivity are also very high, and comparable to other conductive materials.

Table : Mechanical Properties of Engineering Fibres

Fibre Material	*Specific Density*	*E (TPa)*	*Strength (GPa)*	*Strain at Break (%)*
Carbon Nanotube	1.3 - 2	1	10 - 60	10
HS Steel	7.8	0.2	4.1	< 10
Carbon Fibre - PAN	1.7 - 2	0.2 - 0.6	1.7 - 5	0.3 - 2.4
Carbon Fibre - Pitch	2 - 2.2	0.4 - 0.96	2.2 - 3.3	0.27 - 0.6
E/S - glass	2.5	0.07/0.08	2.4/4.5	4.8
Kevlar* 49	1.4	0.13	3.6 - 4.1	2.8

Kevlar is a registered trademark of DuPont.

Table : Transport Properties of Conductive Materials

Material	*Thermal Conductivity (W/m.k)*	*Electrical Conductivity*
Carbon Nanotubes	> 3000	106 - 107
Copper	400	6 x 107
Carbon Fibre - Pitch	1000	2 - 8.5 x 106
Carbon Fibre - PAN	8 - 105	6.5 - 14 x 106

Potential Applications for Carbon Nanotubes

Carbon Nanotube Technology can be used for a wide range of new and existing applications:

- Conductive plastics
- Structural composite materials
- Flat-panel displays
- Gas storage

- Antifouling paint
- Micro- and nano-electronics
- Radar-absorbing coating
- Technical textiles
- Ultra-capacitors
- Atomic Force Microscope (AFM) tips
- Batteries with improved lifetime
- Biosensors for harmful gases
- Extra strong fibres.

Nanocyl Produce Carbon Nanotubes

Nanocyl uses the "Catalytic Carbon Vapour Deposition" method for producing Carbon Nanotube Technologies. This proven industrial process is well known for its reliability and scalability. It involves growing nanotubes on substrates, thus enabling uniform, large-scale production of the highest-quality carbon nanotubes worldwide.

TYPES OF CARBON NANOTUBES AND RELATED STRUCTURES

TERMINOLOGY OF CARBON NANOTUBES

There is no consensus on some terms describing carbon nanotubes in scientific literature: both "-wall" and "-walled" are being used in combination with "single", "double", "triple" or "multi", and the letter C is often omitted in the abbreviation; for example, multi-walled carbon nanotube (MWNT).

Single-walled

Figure : Armchair (*n*,*n*)

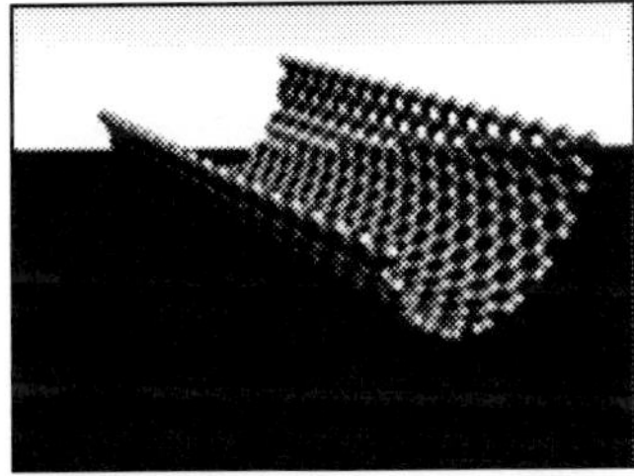

Figure : The Translation Vector is Bent, while the Chiral Vector Stays Straight

Figure : Graphene Nanoribbon

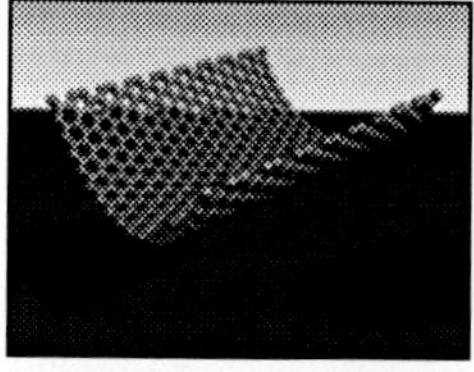

Figure : The Chiral Vector is Bent, while the Translation Vector Stays Straight

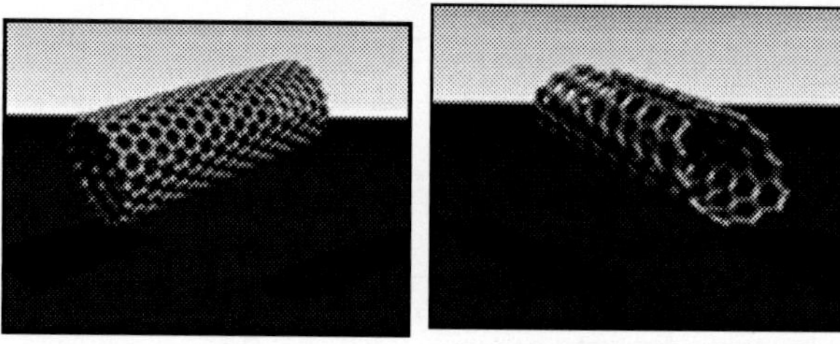

Figure : Zigzag (*n*,0) **Figure** : Chiral (*n*,*m*)

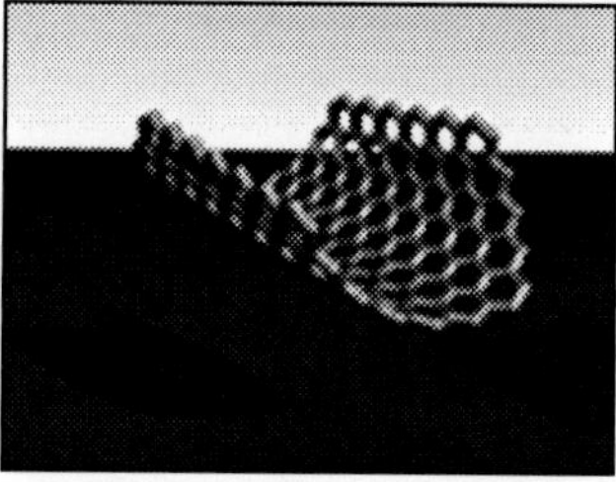

Figure : *n* and *m* can be Counted at the end of the Tube

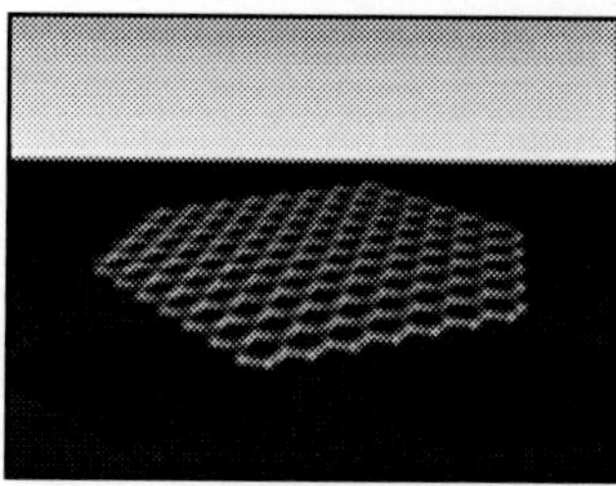

Figure : Graphene Nanoribbon.

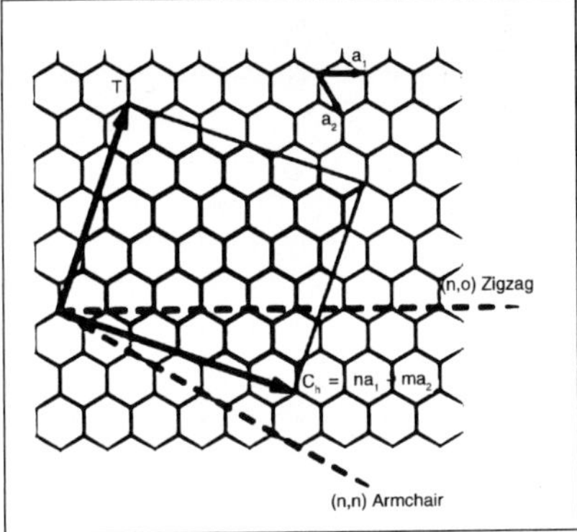

Figure : The *(n,m)* Nanotube Naming Scheme can be thought of as a Vector (C_h) in an Infinite Graphene Sheet that Describes how to "Roll Up" the Graphene Sheet to make the Nanotube. T Denotes the Tube Axis, and a_1 and a_2 are the Unit Vectors of Graphene in Real Space.

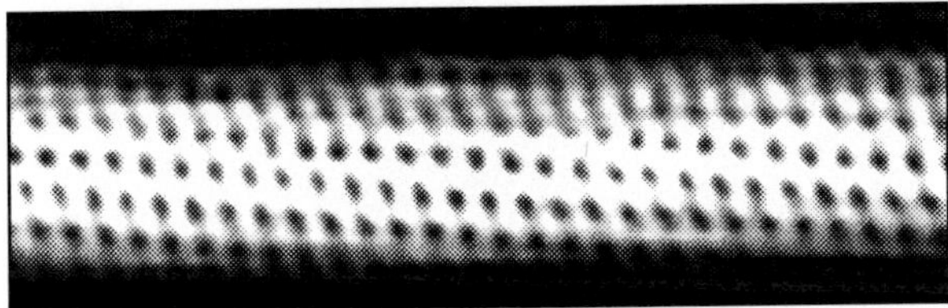

Figure : A Scanning Tunnelling Microscopy Image of Single-walled Carbon Nanotube

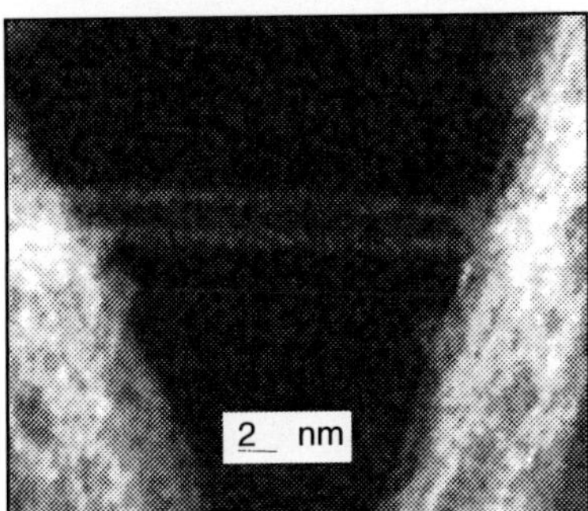

Figure : A Transmission Electron Microscopy Image of a Single-walled Carbon Nanotube

Most single-walled nanotubes (SWNT) have a diameter of close to 1 nanometre, with a tube length that can be many millions of times longer. The structure of a SWNT can be conceptualized by wrapping a one-atom-thick layer of graphite called graphene into a seamless cylinder. The way the graphene sheet is wrapped is represented by a pair of indices *(n,m)*. The integers *n* and *m* denote the number of unit vectors along two directions in the honeycomb crystal lattice of graphene. If $m = 0$, the nanotubes are called zigzag nanotubes, and if $n = m$, the nanotubes are called armchair nanotubes. Otherwise, they are called chiral. The diameter of an ideal nanotube can be calculated from its (n,m) indices as follows

$$d = \frac{a}{\pi}\sqrt{\left(n^2 + nm + m^2\right)}.$$

where $a = 0.246$ nm.

SWNTs are an important variety of carbon nanotube because most of their properties change significantly with the (n,m) values, and this dependence is non-monotonic. In particular, their band gap can vary from zero to about 2 eV and their electrical conductivity can show metallic or semiconducting behaviour. Single-walled nanotubes are likely candidates for miniaturizing electronics.

The most basic building block of these systems is the electric wire, and SWNTs with diameters of an order of a nanometre can be excellent conductors. One useful application of SWNTs is in the development of the first intermolecular field-effect transistors (FET). The first intermolecular logic gate using SWCNT FETs was made in 2001. A logic gate requires both a p-FET and an n-FET. Because SWNTs are p-FETs when exposed to oxygen and n-FETs otherwise, it is possible to protect half of an SWNT from oxygen exposure, while exposing the other half to oxygen. This results in a single SWNT that acts as a NOT logic gate with both p and n-type FETs within the same molecule.

Single-walled nanotubes are dropping precipitously in price, from around \$1500 per gram as of 2000 to retail prices of around \$50 per gram of as-produced 40–60% by weight SWNTs as of March 2010.

Multi-walled

Multi-walled nanotubes (MWNT) consist of multiple rolled layers (concentric tubes) of graphene. There are two models that can be used to describe the structures of multi-walled nanotubes. In the *Russian Doll* model, sheets of graphite are arranged in concentric cylinders, *e.g.*, a (0,8) single-walled nanotube (SWNT) within a larger (0,17) single-walled nanotube. In the *Parchment* model, a single sheet of graphite is rolled in around itself, resembling a scroll of parchment or a rolled newspaper. The interlayer distance in multi-walled nanotubes is close to the distance between graphene layers in graphite, approximately 3.4 Å. The Russian Doll structure is observed more commonly.

Its individual shells can be described as SWNTs, which can be metallic or semiconducting. Because of statistical probability and restrictions on the relative diameters of the individual tubes, one of the shells, and thus the whole MWNT, is usually a zero-gap metal.

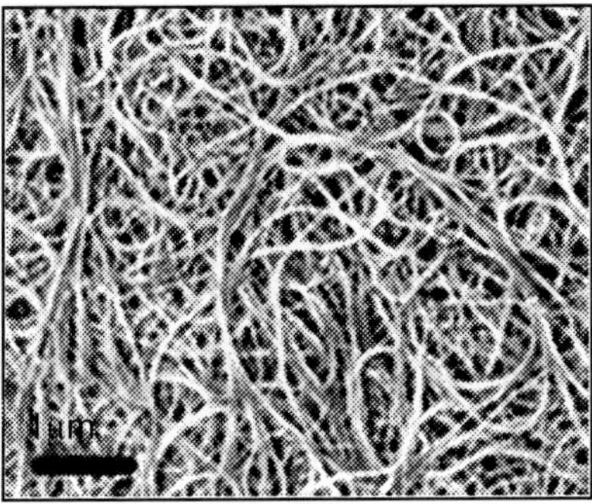

Figure : A Scanning Electron Microscopy Image of Carbon Nanotubes Bundles

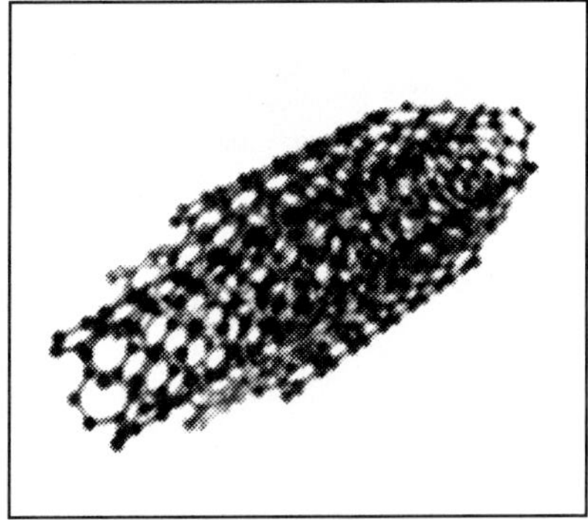

Figure : Triple-walled Armchair Carbon Nanotube

Double-walled carbon nanotubes (DWNT) form a special class of nanotubes because their morphology and properties are similar to those of SWNT but their resistance to chemicals is significantly improved. This is especially important when functionalization is required (this means grafting of chemical functions at the surface of the nanotubes) to add new properties to the CNT. In the case of SWNT, covalent functionalization will break some C=C double bonds, leaving "holes" in the structure on the nanotube and, thus, modifying both its mechanical and electrical properties. In the case of DWNT, only the outer wall is modified. DWNT synthesis on the gram-scale was first proposed in 2003 by the CCVD technique, from the selective reduction of oxide solutions in methane and hydrogen. The telescopic motion ability of inner shells and their unique mechanical properties permit to use multi-walled nanotubes as main movable arms in coming nanomechanical devices. Retraction force that occurs to telescopic motion caused by the Lennard-Jones interaction between shells and its value is about 1.5 nN.

Torus

In theory, a nanotorus is a carbon nanotube bent into a torus (doughnut shape). Nanotori are predicted to have many unique properties, such as magnetic moments 1000 times larger than previously expected for certain specific radii. Properties such as magnetic moment, thermal stability, etc. vary widely depending on radius of the torus and radius of the tube.

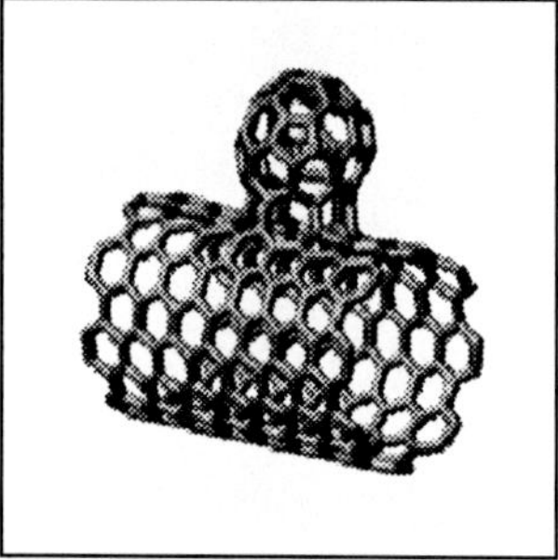

Figure: A Stable Nanobud Structure

Nanobud

Carbon nanobuds are a newly created material combining two previously discovered allotropes of carbon: carbon nanotubes and fullerenes. In this new material, fullerene-like "buds" are covalently bonded to the outer sidewalls of the underlying carbon nanotube. This hybrid material has useful properties of both fullerenes and carbon nanotubes. In particular, they have been found to be exceptionally good field emitters. In composite materials, the attached fullerene molecules may function as molecular anchors preventing slipping of the nanotubes, thus improving the composite's mechanical properties.

Graphenated Carbon Nanotubes (g-CNTs)

Graphenated CNTs are a relatively new hybrid that combines graphitic foliates grown along the sidewalls of multiwalled or bamboo style CNTs. Yu *et al.* reported on "chemically bonded graphene leaves" growing along the sidewalls of CNTs. Stoner *et al.* described these structures as "graphenated CNTs" and reported in their use for enhanced supercapacitor performance. Hsu *et al.* further reported on similar structures formed on carbon fibre paper, also for use in supercapacitor applications. The foliate density can vary as a function of deposition conditions (*e.g.* temperature and time) with their structure ranging from few layers of graphene (< 10) to thicker, more graphite-like. The fundamental advantage of an integrated graphene-CNT structure is the high surface area three-dimensional framework of the CNTs coupled with the high edge density of graphene. Graphene edges provide significantly higher charge density and reactivity than the basal plane, but they are difficult to arrange in a three-dimensional, high volume-density geometry. CNTs are readily aligned in a high density geometry (*i.e.,* a vertically aligned forest) but lack high charge density surfaces—the sidewalls of the CNTs are similar to the basal plane of graphene and exhibit low charge density except where edge defects exist. Depositing a high density of graphene foliates along the length of aligned CNTs can significantly increase the total charge capacity per unit of nominal area as compared to other carbon nanostructures.

Carbon Peapod

Carbon peapod is a novel hybrid carbon material which traps fullerene inside a carbon nanotube. It can possess interesting magnetic properties with heating and irradiating. It can also be applied as an oscillator during theoretical investigations and predictions.

Cup-stacked Carbon Nanotubes

Cup-stacked carbon nanotubes (CSCNTs) differ from other quasi-1D carbon structures, which normally behave as quasi-metallic conductors of electrons. CSCNTs exhibit semiconducting behaviours due to the stacking microstructure of graphene layers.

EXTREME CARBON NANOTUBES

The observation of the *longest* carbon nanotubes (18.5 cm long) was reported in 2009. These nanotubes were grown on Si substrates using an improved chemical vapour deposition (CVD) method and represent electrically uniform arrays of single-walled carbon nanotubes.

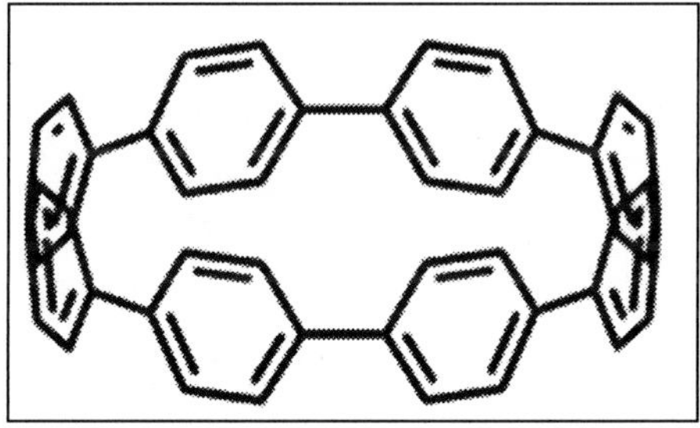

Figure: Cycloparaphenylene

The *shortest* carbon nanotube is the organic compound cycloparaphenylene, which was synthesized in early 2009.

The *thinnest* carbon nanotube is armchair (2,2) CNT with a diameter of 3 Å. This nanotube was grown inside a multi-walled carbon nanotube. Assigning of carbon nanotube type was done by combination of high-resolution transmission electron microscopy (HRTEM), Raman spectroscopy and density functional theory (DFT) calculations.

The *thinnest freestanding* single-walled carbon nanotube is about 4.3 Å in diameter. Researchers suggested that it can be either SWCNT, but exact type of carbon nanotube remains questionable. Carbon nanotubes (all about 4 Å in diameter) were unambiguously identified using more precise aberration-corrected high-resolution transmission electron microscopy. However, they were found inside of double-walled carbon nanotubes.

PROPERTIES

Strength

Carbon nanotubes are the strongest and stiffest materials yet discovered in terms of tensile strength and elastic modulus respectively. This strength results from the covalent sp^2 bonds formed between the individual carbon atoms. In 2000, a multi-walled carbon nanotube was tested to have a tensile strength of 63 gigapascals (GPa). (For illustration, this translates into the ability to endure tension of a weight equivalent to 6422 kg on a cable with cross-section of 1 mm^2.) Further studies, conducted in 2008, revealed that individual CNT shells have strengths of up to ~100 GPa, which is in good agreement with quantum/atomistic models. Since carbon nanotubes have a low density for a solid of 1.3 to 1.4 g/cm^3, its specific strength of up to 48,000 $kN{\cdot}m{\cdot}kg^{-1}$ is the best of known materials, compared to high-carbon steel's 154 $kN{\cdot}m{\cdot}kg^{-1}$.

Under excessive tensile strain, the tubes will undergo plastic deformation, which means the deformation is permanent. This deformation begins at strains of approximately 5% and can increase the maximum strain the tubes undergo before fracture by releasing strain energy.

Although the strength of individual CNT shells is extremely high, weak shear interactions between adjacent shells and tubes leads to significant reductions in the effective strength of multi-walled carbon nanotubes and carbon nanotube bundles down to only a few GPa's. This limitation has been recently addressed by applying high-energy electron irradiation, which crosslinks inner shells and tubes, and effectively increases the strength of these materials to ~60 GPa for multi-walled carbon nanotubes and ~17 GPa for double-walled carbon nanotube bundles.

CNTs are not nearly as strong under compression. Because of their hollow structure and high aspect ratio, they tend to undergo buckling when placed under compressive, torsional, or bending stress.

Comparison of mechanical properties

Material	*Young's modulus (TPa)*	*Tensile strength (GPa)*	*Elongation at break (%)*
SWNT^E	~1 (from 1 to 5)	13–53	16
Armchair SWNT^T	0.94	126.2	23.1
Zigzag SWNT^T	0.94	94.5	15.6–17.5
Chiral SWNT	0.92		
MWNT^E	0.2–0.8–0.95	11–63–150	
Stainless steel^E	0.186–0.214	0.38–1.55	15–50
Kevlar–29&149^E	0.06–0.18	3.6–3.8	~2

^{E}Experimental observation; ^{T}Theoretical prediction.

To axial properties of the nanotube, whereas simple geometrical considerations suggest that carbon nanotubes should be much softer in the radial direction than along the tube axis. Indeed, TEM observation of radial elasticity suggested that even the van der Waals forces can deform two adjacent nanotubes. Nanoindentation experiments, performed by several groups on multiwalled carbon nanotubes and tapping/contact mode atomic force microscope measurement performed on single-walled carbon nanotube, indicated Young's modulus of the order of several GPa confirming that CNTs are indeed rather soft in the radial direction.

Hardness

Standard single-walled carbon nanotubes can withstand a pressure up to 24GPa without deformation. They then undergo a transformation to superhard phase nanotubes. Maximum pressures measured using current experimental techniques are around 55GPa. However, these new superhard phase nanotubes collapse at an even higher, albeit unknown, pressure. The bulk modulus of superhard phase nanotubes is 462 to 546 GPa, even higher than that of diamond(420 GPa for single diamond crystal).

Kinetic Properties

Multi-walled nanotubes are multiple concentric nanotubes precisely nested within one another. These exhibit a striking telescoping property whereby an inner nanotube core may slide, almost without friction, within its outer nanotube shell, thus creating an atomically perfect linear or rotational bearing. This is one of the first true examples of molecular nanotechnology, the precise positioning of atoms to create useful machines. Already, this property has been utilized to create the world's smallest rotational motor. Future applications such as a gigahertz mechanical oscillator are also envisaged.

Electrical Properties

Because of the symmetry and unique electronic structure of graphene, the structure of a nanotube strongly affects its electrical properties. For a given (n,m) nanotube, if $n = m$, the nanotube is metallic; if n " m is a multiple of 3, then the nanotube is semiconducting with a very small band gap, otherwise the nanotube is a moderate semiconductor. Thus all armchair $(n = m)$ nanotubes are metallic, and nanotubes etc. are semiconducting.

However, this rule has exceptions, because curvature effects in small diameter carbon nanotubes can strongly influence electrical properties. Thus, a SWCNT that should be semiconducting in fact is metallic according to the calculations. Likewise, *vice-versa*— zigzag and chiral SWCNTs with small diameters that should be metallic have finite gap (armchair nanotubes remain metallic). In theory, metallic nanotubes can carry an electric current density of 4×10^9 A/cm^2, which is more than 1,000 times greater than those of metals such as copper, where for copper interconnects current densities are limited by electromigration.

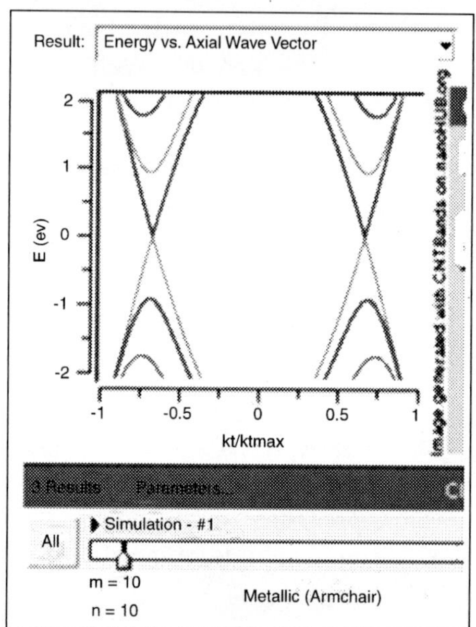

Figure: Band Structures Computed using Tight Binding Approximation for (6,0) CNT (Zigzag, Metallic) (10,2) CNT (Semiconducting) and (10,10) CNT (Armchair, Metallic).

Because of their nanoscale cross-section, electrons propagate only along the tube's axis and electron transport involves quantum effects. As a result, carbon nanotubes are frequently referred to as one-dimensional conductors. The maximum electrical conductance of a single-walled carbon nanotube is $2G_0$, where $G_0 = 2e^2/h$ is the conductance of a single ballistic quantum channel.

There have been reports of intrinsic superconductivity in carbon nanotubes. Many other experiments, however, found no evidence of superconductivity, and the validity of these claims of intrinsic superconductivity remains a subject of debate.

OPTICAL PROPERTIES

Wave Absorption

One of the more recently researched properties of multi-walled carbon nanotubes (MWNTs) is their wave absorption characteristics, specifically microwave absorption. Interest in this research is due to the current military push for radar absorbing materials (RAM) to better the stealth characteristics of aircraft and other military vehicles. There has been some research on filling MWNTs with metals, such as Fe, Ni, Co, etc., to increase the absorption effectiveness of MWNTs in the microwave regime. Thus far, this research has shown improvements in both maximum absorption and bandwidth of adequate absorption. The reason the absorptive properties changed when filled is that the complex permeability (μ_r) and complex permitivity (ε_r), have been shown to vary depending on how the MWNTs are called and what medium they are suspended in. The direct relationship between μ_r, ε_r, and the other system parameters that affect the absorption sample thickness, d, and frequency, f,, where Z_{in} is the normalized input impedance. These characteristics vary by frequency. Because of this, it is convenient to set a baseline reflection loss (R.L.) that is deemed effective and determine the bandwidth within a given frequency that produces the desired reflection loss. A common R.L. to use for this bandwidth determination is "10 dB, which corresponds to a loss of over 90% of the incoming wave. This bandwidth is usually maximized at the same time as the absorption is. This is done by satisfying the impedance matching condition, getting $Z_{in} = 1$. In the work done at Beijing Jiaotong University it was found that Fe filled MWNTs exhibited a maximum reflection loss of –22.73 dB and had a bandwidth of 4.22 GHz for a reflection loss of –10 dB.

$$R.L.(dB) = 20\log_{10}\left[\frac{Z_{in}-1}{Z_{in}+1}\right]$$

$$Z_{in} = \sqrt{\frac{\mu_r}{\varepsilon_r}}\tanh\left[j\left(\frac{2\pi fd}{c}\right)\sqrt{\mu_r\varepsilon_r}\right]$$

Thermal Properties

All nanotubes are expected to be very good thermal conductors along the tube, exhibiting a property known as "ballistic conduction", but good insulators laterally to the tube axis. Measurements show that a SWNT has a room-temperature thermal conductivity along its axis of about 3500 $W{\cdot}m^{-1}{\cdot}K^{-1}$; compare this to copper, a metal well known for its good thermal conductivity, which transmits 385 $W{\cdot}m^{-1}{\cdot}K^{-1}$. A SWNT has a room-temperature thermal conductivity across its axis (in the radial direction) of about 1.52 $W{\cdot}m^{-1}{\cdot}K^{-1}$, which is about as thermally conductive as soil. The temperature stability of carbon nanotubes is estimated to be up to 2800 °C in vacuum and about 750 °C in air.

Defects

As with any material, the existence of a crystallographic defect affects the material properties. Defects can occur in the form of atomic vacancies. High levels of such defects can lower the tensile strength by up to 85%. An important example is the Stone Wales defect, which creates a pentagon and heptagon pair by rearrangement of the bonds. Because of the very small structure of CNTs, the tensile strength of the tube is dependent on its weakest segment in a similar manner to a chain, where the strength of the weakest link becomes the maximum strength of the chain.

Crystallographic defects also affect the tube's electrical properties. A common result is lowered conductivity through the defective region of the tube. A defect in armchair-type tubes (which can conduct electricity) can cause the surrounding region to become semiconducting, and single monoatomic vacancies induce magnetic properties.

Crystallographic defects strongly affect the tube's thermal properties. Such defects lead to phonon scattering, which in turn increases the relaxation rate of the phonons. This reduces the mean free path and reduces the thermal conductivity of nanotube structures. Phonon transport simulations indicate that substitutional defects such as nitrogen or boron will primarily lead to scattering of high-frequency optical phonons. However, larger-scale defects such as Stone Wales defects cause phonon scattering over a wide range of frequencies, leading to a greater reduction in thermal conductivity.

Toxicity

The toxicity of carbon nanotubes has been an important question in nanotechnology. Such research has just begun. The data are still fragmentary and subject to criticism. Preliminary results highlight the difficulties in evaluating the toxicity of this heterogeneous material. Parameters such as structure, size distribution, surface area, surface chemistry, surface charge, and agglomeration state as well as purity of the samples, have considerable impact on the reactivity of carbon nanotubes. However, available data clearly show that, under some conditions, nanotubes can cross membrane barriers, which suggests that, if raw materials reach the organs, they can induce harmful effects such as inflammatory and fibrotic reactions.

A study led by Alexandra Porter from the University of Cambridge shows that CNTs can enter human cells and accumulate in the cytoplasm, causing cell death.

Results of rodent studies collectively show that regardless of the process by which CNTs were synthesized and the types and amounts of metals they contained, CNTs were capable of producing inflammation, epithelioid granulomas (microscopic nodules), fibrosis, and biochemical/toxicological changes in the lungs. Comparative toxicity studies in which mice were given equal weights of test materials showed that SWCNTs were more toxic than quartz, which is considered a serious occupational health hazard when chronically inhaled. As a control, ultrafine carbon black was shown to produce minimal lung responses.

The needle-like fibre shape of CNTs is similar to asbestos fibres. This raises the idea that widespread use of carbon nanotubes may lead to pleural mesothelioma, a cancer of the lining of the lungs or peritoneal mesothelioma, a cancer of the lining of the abdomen (both caused by exposure to asbestos). A recently published pilot study supports this prediction. Scientists exposed the mesothelial lining of the body cavity of mice to long multiwalled carbon nanotubes and observed asbestos-like, length-dependent, pathogenic behaviour that included inflammation and formation of lesions known as granulomas. Authors of the study conclude: This is of considerable importance, because research and business communities continue to invest heavily in carbon nanotubes for a wide range of products under the assumption that they are no more hazardous than graphite. Our results suggest the need for further research and great caution before introducing such products into the market if long-term harm is to be avoided.

According to co-author Dr. Andrew Maynard: This study is exactly the kind of strategic, highly focused research needed to ensure the safe and responsible development of nanotechnology.

It looks at a specific nanoscale material expected to have widespread commercial applications and asks specific questions about a specific health hazard. Even though scientists have been raising concerns about the safety of long, thin carbon nanotubes for over a decade, none of the research needs in the current U.S. federal nanotechnology environment, health and safety risk research strategy address this question.

Although further research is required, the available data suggests that under certain conditions, especially those involving chronic exposure, carbon nanotubes can pose a serious risk to human health.

Synthesis

Figure : Powder of Carbon Nanotubes.

Techniques have been developed to produce nanotubes in sizeable quantities, including arc discharge, laser ablation, high-pressure carbon monoxide (HiPco), and chemical vapour deposition (CVD). Most of these processes take place in vacuum or with process gases. CVD growth of CNTs can occur in vacuum or at atmospheric pressure. Large quantities of nanotubes can be synthesized by these methods; advances in catalysis and continuous growth processes are making CNTs more commercially viable.

Arc Discharge

Nanotubes were observed in 1991 in the carbon soot of graphite electrodes during an arc discharge, by using a current of 100 amps, that was intended to produce fullerenes. However the first macroscopic production of carbon nanotubes was made in 1992 by two researchers at NEC's Fundamental Research Laboratory. The method used was the same as in 1991. During this process, the carbon contained in the negative electrode sublimates because of the high-discharge temperatures. Because nanotubes were initially discovered using this technique, it has been the most widely used method of nanotube synthesis.

The yield for this method is up to 30% by weight and it produces both single- and multi-walled nanotubes with lengths of up to 50 micrometers with few structural defects.

Laser Ablation

In the laser ablation process, a pulsed laser vapourizes a graphite target in a high-temperature reactor while an inert gas is bled into the chamber. Nanotubes develop on the cooler surfaces of the reactor as the vaporized carbon condenses. A water-cooled surface may be included in the system to collect the nanotubes.

This process was developed by Dr. Richard Smalley and co-workers at Rice University, who at the time of the discovery of carbon nanotubes, were blasting metals with a laser to produce various metal molecules. When they heard of the existence of nanotubes they replaced the metals with graphite to create multi-walled carbon nanotubes. Later that year the team used a composite of graphite and metal catalyst particles (the best yield was from a cobalt and nickel mixture) to synthesize single-walled carbon nanotubes.

The laser ablation method yields around 70% and produces primarily single-walled carbon nanotubes with a controllable diameter determined by the reaction temperature. However, it is more expensive than either arc discharge or chemical vapour deposition.

CHEMICAL VAPOUR DEPOSITION (CVD)

The catalytic vapour phase deposition of carbon was reported in 1952 and 1959, but it was not until 1993 that carbon nanotubes were formed by this process. In 2007, researchers at the University of Cincinnati (UC) developed a process to grow aligned carbon nanotube arrays of 18 mm length on a FirstNano ET3000 carbon nanotube growth system.

During CVD, a substrate is prepared with a layer of metal catalyst particles, most commonly nickel, cobalt, iron, or a combination. The metal nanoparticles can also be produced by other ways, including reduction of oxides or oxides solid solutions. The diameters of the nanotubes that are to be grown are related to the size of the metal particles. This can be controlled by patterned (or masked) deposition of the metal, annealing, or by plasma etching of a metal layer. The substrate is heated to approximately 700°C. To initiate the growth of nanotubes, two gases are bled into the reactor: a process gas (such as ammonia, nitrogen or hydrogen) and a carbon-containing gas (such as acetylene, ethylene, ethanol or methane). Nanotubes grow at the sites of the metal catalyst; the carbon-containing gas is broken apart at the surface of the catalyst particle, and the carbon is transported to the edges of the particle, where it forms the nanotubes. This mechanism is still being studied. The catalyst particles can stay at the tips of the growing nanotube during the growth process, or remain at the nanotube base, depending on the adhesion between the catalyst particle and the substrate. Thermal catalytic decomposition of hydrocarbon has become an active area of research and can be a promising route for the bulk production of CNTs. Fluidised bed reactor is the most widely used reactor for CNT preparation. Scale-up of the reactor is the major challenge.

Figure: Nanotubes being grown by Plasma Enhanced Chemical Vapour Deposition

CVD is a common method for the commercial production of carbon nanotubes. For this purpose, the metal nanoparticles are mixed with a catalyst support such as MgO or Al_2O_3 to increase the surface area for higher yield of the catalytic reaction of the carbon feedstock with the metal particles. One issue in this synthesis route is the removal of the catalyst support via an acid treatment, which sometimes could destroy the original structure of the carbon nanotubes. However, alternative catalyst supports that are soluble in water have proven effective for nanotube growth.

If a plasma is generated by the application of a strong electric field during the growth process (plasma enhanced chemical vapour deposition), then the nanotube growth will follow the direction of the electric field. By adjusting the geometry of the reactor it is possible to synthesize vertically aligned carbon nanotubes (*i.e.*, perpendicular to the substrate), a morphology that has been of interest to researchers interested in the electron emission from nanotubes. Without the plasma, the resulting nanotubes are often randomly oriented. Under certain reaction conditions, even in the absence of a plasma, closely spaced nanotubes will maintain a vertical growth direction resulting in a dense array of tubes resembling a carpet or forest.

Of the various means for nanotube synthesis, CVD shows the most promise for industrial-scale deposition, because of its price/unit ratio, and because CVD is capable of growing nanotubes directly on a desired substrate, whereas the nanotubes must be collected in the other growth techniques. The growth sites are controllable by careful deposition of the catalyst. In 2007, a team from Meijo University demonstrated a high-efficiency CVD technique for growing carbon nanotubes from camphor. Researchers at Rice University, until recently led by the late Richard Smalley, have concentrated upon finding methods to produce large, pure amounts of particular types of nanotubes. Their approach grows long fibres from many small seeds cut from a single nanotube; all of the resulting fibres were found to be of the same diameter as the original nanotube and are expected to be of the same type as the original nanotube.

Super-growth CVD

Super-growth CVD (water-assisted chemical vapour deposition) process was developed by Kenji Hata, Sumio Iijima and co-workers at AIST, Japan. In this process, the activity and lifetime of the catalyst are enhanced by addition of water into the CVD reactor. Dense millimetre-tall nanotube "forests", aligned normal to the substrate, were produced. The forests growth rate could be expressed, as

$$H(t) = \beta\tau_0\left(1 - e^{-t/\tau_0}\right).$$

In this equation, β is the initial growth rate and τ_0 is the characteristic catalyst lifetime.

Their specific surface exceeds 1,000 m^2/g (capped) or 2,200 m^2/g (uncapped), surpassing the value of 400–1,000 m^2/g for HiPco samples. The synthesis efficiency is about 100 times higher than for the laser ablation method. The time required to make SWNT forests of the height of 2.5 mm by this method was 10 minutes in 2004. Those SWNT forests can be easily separated from the catalyst, yielding clean SWNT material (purity >99.98%) without further purification. For comparison, the as-grown HiPco CNTs contain about 5–35% of metal impurities; it is therefore purified through dispersion and centrifugation that damages the nanotubes. The super-growth process avoids this problem. Patterned highly organized single-walled nanotube structures were successfully fabricated using the super-growth technique.

The mass density of super-growth CNTs is about 0.037 g/cm^3. It is much lower than that of conventional CNT powders (~1.34 g/cm^3), probably because the latter contain metals and amorphous carbon.

The super-growth method is basically a variation of CVD. Therefore, it is possible to grow material containing SWNT, DWNTs and MWNTs, and to alter their ratios by tuning the growth conditions. Their ratios change by the thinness of the catalyst. Many MWNTs are included so that the diameter of the tube is wide.

The vertically aligned nanotube forests originate from a "zipping effect" when they are immersed in a solvent and dried. The zipping effect is caused by the surface tension of the solvent and the van der Waals forces between the carbon nanotubes. It aligns the nanotubes into a dense material, which can be formed in various shapes, such as sheets and bars, by applying weak compression during the process. Densification increases the Vickers hardness by about 70 times and density is 0.55 g/cm^3. The packed carbon nanotubes are more than 1 mm long and have a carbon purity of 99.9% or higher; they also retain the desirable alignment properties of the nanotubes forest.

Natural, Incidental and Controlled Flame Environments

Fullerenes and carbon nanotubes are not necessarily products of high-tech laboratories; they are commonly formed in such mundane places as ordinary flames, produced by burning methane, ethylene, and benzene, and they have been found in soot from both indoor and outdoor air. However, these naturally occurring varieties can be highly irregular in size and quality because the environment in which they are produced is often highly uncontrolled. Thus, although they can be used in some applications, they can lack in the high degree of uniformity necessary to satisfy the many needs of both research and industry. Recent efforts have focused on producing more uniform carbon nanotubes in controlled flame environments. Such methods have promise for large-scale, low-cost nanotube synthesis, though they must compete with rapidly developing large scale CVD production.

Application-related Issues

Many electronic applications of carbon nanotubes crucially rely on techniques of selectively producing either semiconducting or metallic CNTs, preferably of a certain chirality. Several methods of separating semiconducting and metallic CNTs are known, but most of them are not yet suitable for large-scale technological processes. The most efficient method relies on density-gradient ultracentrifugation, which separates surfactant-wrapped nanotubes by the minute difference in their density. This density difference often translates into difference in the nanotube diameter and (semi)conducting properties. Another method of separation uses a sequence of freezing, thawing, and compression of SWNTs embedded in agarose gel. This process results in a solution containing 70% metallic SWNTs and leaves a gel containing 95% semiconducting SWNTs. The diluted solutions separated by this method show various colours. Moreover, SWNTs can be separated by the column chromatography method. Yield is 95% in semiconductor type SWNT and 90% in metallic type SWNT.

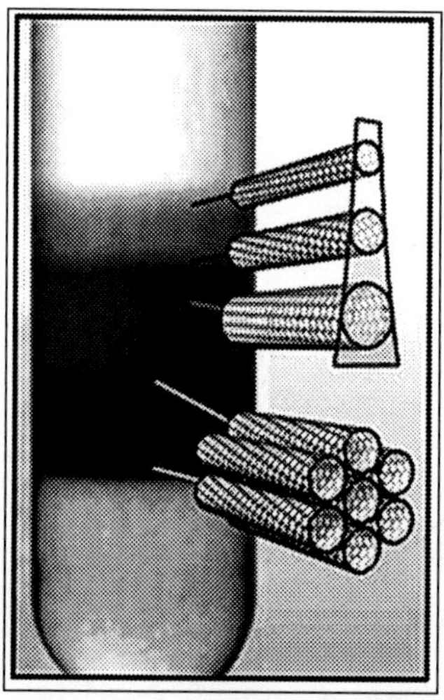

Figure: Centrifuge Tube with a Solution of Carbon Nanotubes, which were Sorted by Diameter using Density-gradient Ultracentrifugation.

In addition to separation of semiconducting and metallic SWNTs, it is possible to sort SWNTs by length, diameter, and chirality. The highest resolution length sorting, with length variation of <10%, has thus far been achieved by size exclusion chromatography (SEC) of DNA-dispersed carbon nanotubes (DNA-SWNT). SWNT diameter separation has been achieved by density-gradient ultracentrifugation (DGU) using surfactant-dispersed SWNTs and by ion-exchange chromatography (IEC) for DNA-SWNT. Purification of individual chiralities has also been demonstrated with IEC of DNA-SWNT: specific short DNA oligomers can be used to isolate individual SWNT chiralities. Thus far, 12 chiralities have been isolated at purities ranging from 70% for (8,3) and (9,5) SWNTs to 90% for (6,5), (7,5) and (10,5)SWNTs. There have been successful efforts to integrate these purified nanotubes into devices, *e.g.* FETs.

An alternative to separation is development of a selective growth of semiconducting or metallic CNTs. Recently, a new CVD recipe that involves a combination of ethanol and methanol gases and quartz substrates resulting in horizontally aligned arrays of 95–98% semiconducting nanotubes was announced. Nanotubes are usually grown on nanoparticles of magnetic metal (Fe, Co), which facilitates production of electronic (spintronic) devices. In particular, control of current through a field-effect transistor by magnetic field has been demonstrated in such a single-tube nanostructure.

Current Applications

Current use and application of nanotubes has mostly been limited to the use of bulk nanotubes, which is a mass of rather unorganized fragments of nanotubes. Bulk nanotube materials may never achieve a tensile strength similar to that of individual tubes, but such composites may, nevertheless, yield strengths sufficient for many applications. Bulk carbon nanotubes have already been used as composite fibres in polymers to improve the mechanical, thermal and electrical properties of the bulk product.

- Easton-Bell Sports, Inc. have been in partnership with Zyvex Performance Materials, using CNT technology in a number of their bicycle components—including flat and riser handlebars, cranks, forks, seatposts, stems and aero bars.
- Zyvex Technologies has also built a 54' maritime vessel, the Piranha Unmanned Surface Vessel, as a technology demonstrator for what is possible using CNT technology. CNTs help improve the structural performance of the vessel, resulting in a lightweight 8,000 lb boat that can carry a payload of 15,000 lb over a range of 2,500 miles.
- Amroy Europe Oy manufactures Hybtonite carbon nanoepoxy resins where carbon nanotubes have been chemically activated to bond to epoxy, resulting in a composite material that is 20% to 30% stronger than other composite materials. It has been used for wind turbines, marine paints and variety of sports gear such as skis, ice hockey sticks, baseball bats, hunting arrows, and surfboards.

Other current applications include:

- Tips for atomic force microscope probes
- In tissue engineering, carbon nanotubes can act as scaffolding for bone growth.

POTENTIAL APPLICATIONS

The strength and flexibility of carbon nanotubes makes them of potential use in controlling other nanoscale structures, which suggests they will have an important role in nanotechnology engineering. The highest tensile strength of an individual multi-walled carbon nanotube has been tested to be is 63 GPa. Carbon nanotubes were found in Damascus steel from the 17th century, possibly helping to account for the legendary strength of the swords made of it.

Structural

Because of the carbon nanotube's superior mechanical properties, many structures have been proposed ranging from everyday items like clothes and sports gear to combat jackets and space elevators. However, the space elevator will require further efforts in refining carbon nanotube technology, as the practical tensile strength of carbon nanotubes can still be greatly improved.

For perspective, outstanding breakthroughs have already been made. Pioneering work led by Ray H. Baughman at the NanoTech Institute has shown that single and multi-walled nanotubes can produce materials with toughness unmatched in the man-made and natural worlds.

Carbon nanotubes are also a promising material as building blocks in bio-mimetic hierarchical composite materials given their exceptional mechanical properties (~1TPa in modulus, and ~100 GPa in strength). Initial attempts to incorporate CNTs into hierarchical structures led to mechanical properties that were significantly lower than these achievable limits. Windle *et al.* have used an *in situ* chemical vapour deposition (CVD) spinning method to produce continuous CNT yarns from CVD grown CNT aerogels. With this technology, they fabricated CNT yarns with strengths as high as ~9 GPa at small gage lengths of ~1 mm, however, defects resulted in a reduction of specific strength to ~1 GPa at 20 mm gage length.

Espinosa *et al.* developed high performance DWNT-polymer composite yarns by twisting and stretching ribbons of randomly oriented bundles of DWNTs thinly coated with polymeric organic compounds. These DWNT-polymer yarns exhibited unusually high energy to failure of ~100 $J \cdot g^{-1}$ (comparable to one of the toughest natural materials – spider silk), and strength as high as ~1.4 GPa. Effort is ongoing to produce CNT composites that incorporate tougher matrix materials, such as Kevlar, to further improve on the mechanical properties towards those of individual CNTs. Because of the high mechanical strength of carbon nanotubes, research is being made into weaving them into clothes to create stab-proof and bulletproof clothing. The nanotubes would effectively stop the bullet from penetrating the body, although the bullet's kinetic energy would likely cause broken bones and internal bleeding.

In Electrical Circuits

Nanotube-based transistors, also known as carbon nanotube field-effect transistors (CNTFETs), have been made that operate at room temperature and that are capable of digital switching using a single electron. However, one major obstacle to realization of nanotubes has been the lack of technology for mass production. In 2001 IBM researchers demonstrated how metallic nanotubes can be destroyed, leaving semiconducting ones behind for use as transistors. Their process is called "constructive destruction," which includes the automatic destruction of defective nanotubes on the wafer. This process, however, only gives control over the electrical properties on a statistical scale.

The potential of carbon nanotubes was demonstrated in 2003 when room-temperature ballistic transistors with ohmic metal contacts and high-k gate dielectric were reported, showing 20–30x higher ON current than state-of-the-art Si MOSFETs. This presented an important advance in the field as CNT was shown to potentially outperform Si. At the time, a major challenge was ohmic metal contact formation. In this regard, palladium, which is a high-work function metal was shown to exhibit Schottky barrier-free contacts to semiconducting nanotubes with diameters >1.7 nm.

The first nanotube integrated memory circuit was made in 2004. One of the main challenges has been regulating the conductivity of nanotubes. Depending on subtle surface features a nanotube may act as a plain conductor or as a semiconductor.

A fully automated method has however been developed to remove non-semiconductor tubes. Another way to make carbon nanotube transistors has been to use random networks of them. By doing so one averages all of their electrical differences and one can produce devices in large scale at the wafer level. This approach was first patented by Nanomix Inc. It was first published in the academic literature by the United States Naval Research Laboratory in 2003 through independent research work. This approach also enabled Nanomix to make the first transistor on a flexible and transparent substrate. Large structures of carbon nanotubes can be used for thermal management of electronic circuits. An approximately 1 mm–thick carbon nanotube layer was used as a special material to fabricate coolers, this materials has very low density, ~20 times lower weight than a similar copper structure, while the cooling properties are similar for the two materials.

Overall, incorporating carbon nanotubes as transistors into logic-gate circuits with densities comparable to modern CMOS technology has not yet been demonstrated.

As Electrical Cables and Wires

Wires for carrying electrical current may be fabricated from pure nanotubes and nanotube-polymer composites. Recently small wires have been fabricated with specific conductivity exceeding copper and aluminum; these cables are the highest conductivity carbon nanotube and also highest conductivity non-metal cables.

As Paper Batteries

A paper battery is a battery engineered to use a paper-thin sheet of cellulose (which is the major constituent of regular paper, among other things) infused with aligned carbon nanotubes. The nanotubes act as electrodes; allowing the storage devices to conduct electricity.

The battery, which functions as both a lithium-ion battery and a supercapacitor, can provide a long, steady power output comparable to a conventional battery, as well as a supercapacitor's quick burst of high energy—and while a conventional battery contains a number of separate components, the paper battery integrates all of the battery components in a single structure, making it more energy efficient.

Solar Cells

One of the promising applications of single-walled carbon nanotubes (SWNTs) is their use in solar panels, due to their strong UV/Vis-NIR absorption characteristics. Research has shown that they can provide a sizeable increase in efficiency, even at their current unoptimized state. Solar cells developed at the New Jersey Institute of Technology use a carbon nanotube complex, formed by a mixture of carbon nanotubes and carbon buckyballs (known as fullerenes) to form snake-like structures.

Buckyballs trap electrons, but they can't make electrons flow. Add sunlight to excite the polymers, and the buckyballs will grab the electrons. Nanotubes, behaving like copper wires, will then be able to make the electrons or current flow.

Additional research has been conducted on creating SWNT hybrid solar panels to increase the efficiency further. These hybrids are created by combining SWNT's with photexcitable electron donors to increase the number of electrons generated. It has been found that the interaction between the photoexcited porphrin and SWNT generates electro-hole pairs at the SWNT surfaces. This phenomenon has been observed experimentally, and contributes practically to an increase in efficiency up to 8.5%.

Hydrogen Storage

In addition to being able to store electrical energy, there has been some research in using carbon nanotubes to store hydrogen to be used as a fuel source. By taking advantage of the capillary effects of the small carbon nanotubes, it is possible to condense gasses in high density inside single-walled nanotubes.

This allows for gasses, most notably hydrogen (H_2), to be stored at high densities without being condensed into a liquid. Potentially, this storage method could be used on vehicles in place of gas fuel tanks for a hydrogen-powered car. A current issue regarding hydrogen-powered vehicles is the onboard storage of the fuel.

Current storage methods involve cooling and condensing the H_2 gas to a liquid state for storage which causes a loss of potential energy (25–45%) when compared to the energy associated with the gaseous state. Storage using SWNTs would allow one to keep the H2 in its gaseous state, thereby increasing the storage efficiency. This method allows for a volume to energy ratio slightly smaller to that of current gas powered vehicles, allowing for a slightly lower but comparable range.

An area of controversy and frequent experimentation regarding the storage of hydrogen by adsorption in carbon nanotubes is the efficiency by which this process occurs. The effectiveness of hydrogen storage is integral to its use as a primary fuel source since hydrogen only contains about one fourth the energy per unit volume as gasoline.

Experimental Capacity

One experiment sought to determine the amount of hydrogen stored in CNTs by utilizing elastic recoil detection analysis (ERDA). CNTs (primarily SWNTs) were synthesized via chemical vapour disposition (CVD) and subjected to a two-stage purification process including air oxidation and acid treatment, then formed into flat, uniform discs and exposed to pure, pressurized hydrogen at various temperatures. When the data was analysed, it was found that the ability of CNTs to store hydrogen decreased as temperature increased. Moreover, the highest hydrogen concentration measured was ~0.18%; significantly lower than commercially viable hydrogen storage needs to be.

In another experiment, CNTs were synthesized via CVD and their structure was characterized using Raman spectroscopy. Utilizing microwave digestion, the samples were exposed to different acid concentrations and different temperatures for various amounts of time in an attempt to find the optimum purification method for SWNTs of the diameter determined earlier.

The purified samples were then exposed to hydrogen gas at various high pressures, and their adsorption by weight percent was plotted. The data showed that hydrogen adsorption levels of up to 3.7% are possible with a very pure sample and under the proper conditions. It is thought that microwave digestion helps improve the hydrogen adsorption capacity of the CNTs by opening up the ends, allowing access to the inner cavities of the nanotubes.

Limitations on Efficient Hydrogen Adsorption

The biggest obstacle to efficient hydrogen storage using CNTs is the purity of the nanotubes. To achieve maximum hydrogen adsorption, there must be minimum graphene, amorphous carbon, and metallic deposits in the nanotube sample. Current methods of CNT synthesis require a purification step. However, even with pure nanotubes, the adsorption capacity is only maximized under high pressures, which are undesirable in commercial fuel tanks.

Ultracapacitors

MIT Laboratory for Electromagnetic and Electronic Systems uses nanotubes to improve ultracapacitors. The activated charcoal used in conventional ultracapacitors has many small hollow spaces of various size, which create together a large surface to store electric charge.

But as charge is quantized into elementary charges, *i.e.* electrons, and each such elementary charge needs a minimum space, a significant fraction of the electrode surface is not available for storage because the hollow spaces are not compatible with the charge's requirements. With a nanotube electrode the spaces may be tailored to size—few too large or too small—and consequently the capacity should be increased considerably.

Radar Absorption

Radars work in the microwave frequency range, which can be absorbed by MWNTs. Applying the MWNTs to the aircraft would cause the radar to be absorbed and therefore seem to have a smaller signature. One such application could be to paint the nanotubes onto the plane. Recently there has been some work done at the University of Michigan regarding carbon nanotubes usefulness as stealth technology on aircraft.

It has been found that in addition to the radar absorbing properties, the nanotubes neither reflect nor scatter visible light, making it essentially invisible at night, much like painting current stealth aircraft black except much more effective.

Current limitations in manufacturing, however, mean that current production of nanotube-coated aircraft is not possible. One theory to overcome these current limitations is to cover small particles with the nanotubes and suspend the nanotube-covered particles in a medium such as paint, which can then be applied to a surface, like a stealth aircraft.

Medical

In the Kanzius cancer therapy, single-walled carbon nanotubes are inserted around cancerous cells, then excited with radio waves, which causes them to heat up and kill the surrounding cells.

Other Applications

Carbon nanotubes have been implemented in nanoelectromechanical systems, including mechanical memory elements (NRAM being developed by Nantero Inc.) and nanoscale electric motors.

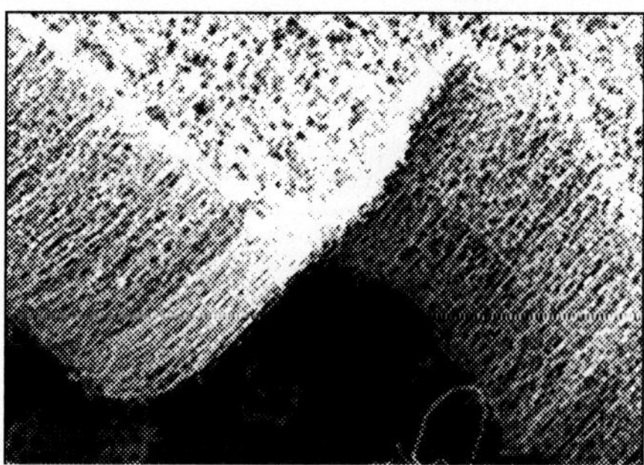

Figure : Aligned Nanotubes are Preferred for Many Applications.

The market a hydrogen sensor that integrated carbon nanotubes on a silicon platform. Since then, Nanomix has been patenting many such sensor applications, such as in the field of carbon dioxide, nitrous oxide, glucose, DNA detection, etc.

Eikos Inc of Franklin, Massachusetts and Unidym Inc. of Silicon Valley, California are developing transparent, electrically conductive films of carbon nanotubes to replace indium tin oxide (ITO). Carbon nanotube films are substantially more mechanically robust than ITO films, making them ideal for high-reliability touchscreens and flexible displays. Printable water-based inks of carbon nanotubes are desired to enable the production of these films to replace ITO. Nanotube films show promise for use in displays for computers, cell phones, PDAs, and ATMs.

A nanoradio, a radio receiver consisting of a single nanotube, was demonstrated in 2007. In 2008 it was shown that a sheet of nanotubes can operate as a loudspeaker if an alternating current is applied. The sound is not produced through vibration but thermoacoustically.

A flywheel made of carbon nanotubes could be spun at extremely high velocity on a floating magnetic axis in a vacuum, and potentially store energy at a density approaching that of conventional fossil fuels. Since energy can be added to and removed from flywheels very efficiently in the form of electricity, this might offer a way of storing electricity, making the electrical grid more efficient and variable power suppliers (like wind turbines) more useful in meeting energy needs. The practicality of this depends heavily upon the cost of making massive, unbroken nanotube structures, and their failure rate under stress. Carbon nanotube springs have the potential to indefinitely store elastic potential energy at ten times the density of lithium-ion batteries with flexible charge and discharge rates and extremely high cycling durability.

Ultra-short SWNTs (US-tubes) have been used as nanoscaled capsules for delivering MRI contrast agents in vivo.

Carbon nanotubes provide a certain potential for metal-free catalysis of inorganic and organic reactions. For instance, oxygen groups attached to the surface of carbon nanotubes have the potential to catalyse oxidative dehydrogenations or selective oxidations. Nitrogen-doped carbon nanotubes may replace platinum catalysts used to reduce oxygen in fuel cells. A forest of vertically aligned nanotubes can reduce oxygen in alkaline solution more effectively than platinum, which has been used in such applications since the 1960s. Here, the nanotubes have the added benefit of not being subject to carbon monoxide poisoning.

Discovery

A 2006 editorial written by Marc Monthioux and Vladimir Kuznetsov in the journal *Carbon* described the interesting and often-misstated origin of the carbon nanotube. A large percentage of academic and popular literature attributes the discovery of hollow, nanometre-size tubes composed of graphitic carbon to Sumio Iijima of NEC in 1991.

In 1952 L.V. Radushkevich and V.M. Lukyanovich published clear images of 50 nanometre diameter tubes made of carbon in the Soviet *Journal of Physical Chemistry*. This discovery was largely unnoticed, as the article was published in the Russian language, and Western scientists' access to Soviet press was limited during the Cold War. It is likely that carbon nanotubes were produced before this date, but the invention of the transmission electron microscope (TEM) allowed direct visualization of these structures.

Carbon nanotubes have been produced and observed under a variety of conditions prior to 1991. A paper by Oberlin, Endo, and Koyama published in 1976 clearly showed hollow carbon fibres with nanometre-scale diameters using a vapour-growth technique. Additionally, the authors show a TEM image of a nanotube consisting of a single wall of graphene. Later, Endo has referred to this image as a single-walled nanotube. In 1979, John Abrahamson presented evidence of carbon nanotubes at the 14th Biennial Conference of Carbon at Pennsylvania State University. The conference paper described carbon nanotubes as carbon fibres that were produced on carbon anodes during arc discharge. A characterization of these fibres was given as well as hypotheses for their growth in a nitrogen atmosphere at low pressures.

In 1981, a group of Soviet scientists published the results of chemical and structural characterization of carbon nanoparticles produced by a thermocatalytical disproportionation of carbon monoxide. Using TEM images and XRD patterns, the authors suggested that their "carbon multi-layer tubular crystals" were formed by rolling graphene layers into cylinders. They speculated that by rolling graphene layers into a cylinder, many different arrangements of graphene hexagonal nets are possible. They suggested two possibilities of such arrangements: circular arrangement (armchair nanotube) and a spiral, helical arrangement (chiral tube).

In 1987, Howard G. Tennett of Hyperion Catalysis was issued a U.S. patent for the production of "cylindrical discrete carbon fibrils" with a "constant diameter between about 3.5 and about 70 nanometres..., length 10^2 times the diameter, and an outer region of multiple essentially continuous layers of ordered carbon atoms and a distinct inner core...."

Iijima's discovery of multi-walled carbon nanotubes in the insoluble material of arc-burned graphite rods in 1991 and Mintmire, Dunlap, and White's independent prediction that if single-walled carbon nanotubes could be made, then they would exhibit remarkable conducting properties helped create the initial buzz that is now associated with carbon nanotubes. Nanotube research accelerated greatly following the independent discoveries by Bethune at IBM and Iijima at NEC of *single-walled* carbon nanotubes and methods to specifically produce them by adding transition-metal catalysts to the carbon in an arc discharge. The arc discharge technique was well-known to produce the famed Buckminster fullerene on a preparative scale, and these results appeared to extend the run of accidental discoveries relating to fullerenes. The original observation of fullerenes in mass spectrometry was not anticipated, and the first mass-production technique by Krätschmer and

Huffman was used for several years before realizing that it produced fullerenes. The discovery of nanotubes remains a contentious issue. Many believe that Iijima's report in 1991 is of particular importance because it brought carbon nanotubes into the awareness of the scientific community as a whole.

CARBON NANOTUBE SCIENCE AND TECHNOLOGY

Carbon nanotubes are molecular-scale tubes of graphitic carbon with outstanding properties. They are among the stiffest and strongest fibres known, and have remarkable electronic properties and many other unique characteristics. For these reasons they have attracted huge academic and industrial interest, with thousands of papers on nanotubes being published every year. Commercial applications have been rather slow to develop, however, primarily because of the high production costs of the best quality nanotubes.

HISTORY OF CARBON NANOTUBES

The current huge interest in carbon nanotubes is a direct consequence of the synthesis of buckminsterfullerene, C_{60}, and other fullerenes, in 1985. The discovery that carbon could form stable, ordered structures other than graphite and diamond stimulated researchers worldwide to search for other new forms of carbon. The search was given new impetus when it was shown in 1990 that C_{60} could be produced in a simple arc-evaporation apparatus readily available in all laboratories.

It was using such an evaporator that the Japanese scientist Sumio Iijima discovered fullerene-related carbon nanotubes in 1991. The tubes contained at least two layers, often many more, and ranged in outer diameter from about 3 nm to 30 nm. They were invariably closed at both ends.

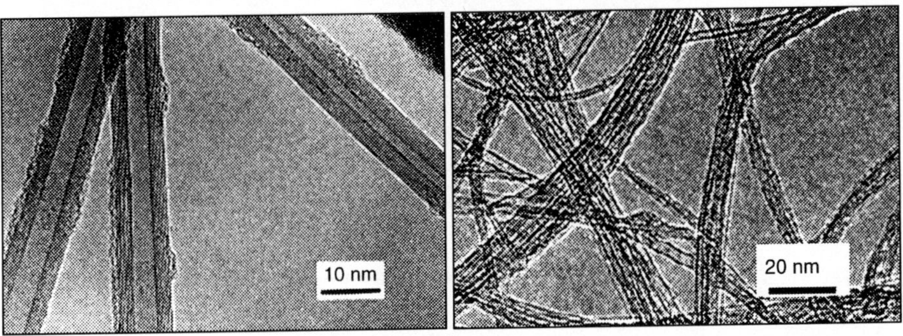

A transmission electron micrograph of some multiwalled nanotubes. In 1993, a new class of carbon nanotube was discovered, with just a single layer. These single-walled nanotubes are generally narrower than the multiwalled tubes, with diameters typically in the range 1-2 nm, and tend to be curved rather than straight. The image on the right shows some typical single-walled tubes It was soon established that these new fibres had a range of exceptional properties, and this sparked off an explosion of research into carbon nanotubes. It is important to note, however, that nanoscale tubes of carbon, produced catalytically, had been known for many years before Iijima's discovery. The main reason why these early tubes did not excite wide interest is that they were structurally rather imperfect, so did not have particularly interesting properties. Recent research has focused on improving the quality of catalytically-produced nanotubes.

Structure

The bonding in carbon nanotubes is sp^2, with each atom joined to three neighbours, as in graphite. The tubes can therefore be considered as rolled-up graphene sheets (graphene is an individual graphite layer). There are three distinct ways in which a graphene sheet can be rolled into a tube, as shown in the diagram below.

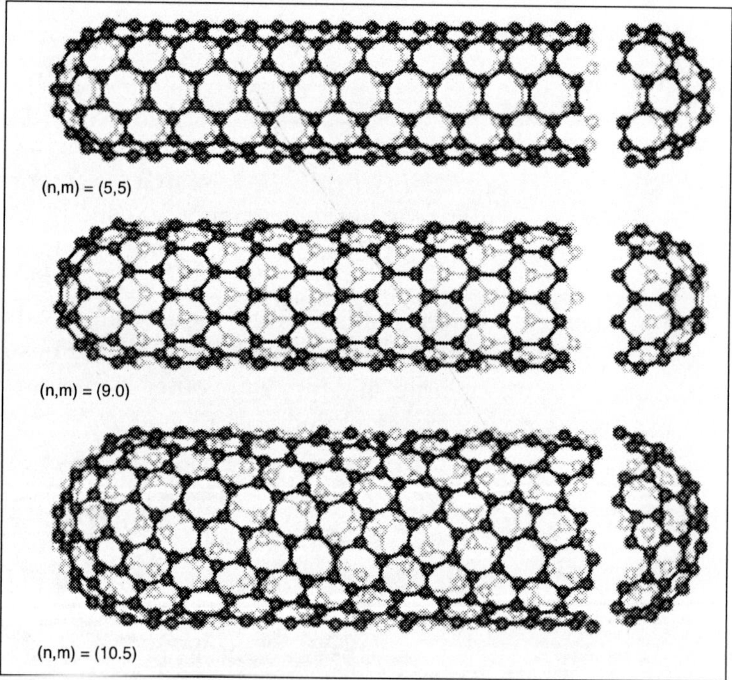

The first two of these, known as "armchair" (top left) and "zig-zag" (middle left) have a high degree of symmetry. The terms "armchair" and "zig-zag" refer to the arrangement of hexagons around the circumference. The third class of tube, which in practice is the most common, is known as chiral, meaning that it can exist in two mirror-related forms. An example of a chiral nanotube is shown at the bottom left.

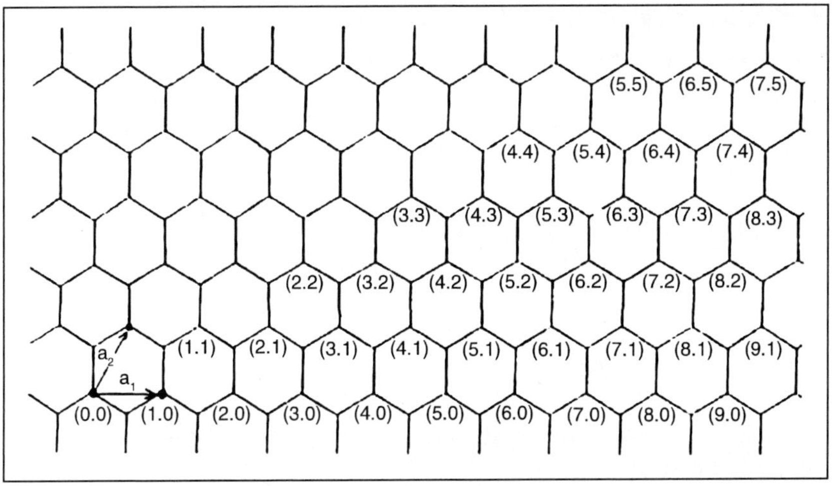

The structure of a nanotube can be specified by a vector, (n,m), which defines how the graphene sheet is rolled up. To produce a nanotube with the indices say, the sheet is rolled up so that the atom labelled is superimposed on the one labelled. That m = 0 for all zig-zag tubes, while n = m for all armchair tubes.

Synthesis

The arc-evaporation method, which produces the best quality nanotubes, involves passing a current of about 50 amps between two graphite electrodes in an atmosphere of helium. This causes the graphite to vaporise, some of it condensing on the walls of the reaction vessel and some of it on the cathode. It is the deposit on the cathode which contains the carbon nanotubes. Single-walled nanotubes are produced when Co and Ni or some other metal is added to the anode.

It has been known since the 1950s, if not earlier, that carbon nanotubes can also be made by passing a carbon-containing gas, such as a hydrocarbon, over a catalyst. The catalyst consists of nano-sized particles of metal, usually Fe, Co or Ni.

These particles catalyse the breakdown of the gaseous molecules into carbon, and a tube then begins to grow with a metal particle at the tip. It was shown in 1996 that single-walled nanotubes can also be produced catalytically. The perfection of carbon nanotubes produced in this way has generally been poorer than those made by arc-evaporation, but great improvements in the technique have been made in recent years. The big advantage of catalytic synthesis over arc-evaporation is that it can be scaled up for volume production. The third important method for making carbon nanotubes involves using a powerful laser to vaporise a metal-graphite target. This can be used to produce single-walled tubes with high yield.

Properties

The strength of the sp^2 carbon-carbon bonds gives carbon nanotubes amazing mechanical properties. The stiffness of a material is measured in terms of its Young's modulus, the rate of change of stress with applied strain. The Young's modulus of the best nanotubes can be as high as 1000 GPa which is approximately 5x higher than steel. The tensile strength, or breaking strain of nanotubes can be up to 63 GPa, around 50x higher than steel. These properties, coupled with the lightness of carbon nanotubes, gives them great potential in applications such as aerospace. It has even been suggested that nanotubes could be used in the "space elevator", an Earth-to-space cable first proposed by Arthur C. Clarke. The electronic properties of carbon nanotubes are also extraordinary. Especially notable is the fact that nanotubes can be metallic or semiconducting depending on their structure. Thus, some nanotubes have conductivities higher than that of copper, while others behave more like silicon. There is great interest in the possibility of constructing nanoscale electronic devices from nanotubes, and some progress is being made in this area. However, in order to construct a useful device we would need to arrange many thousands of nanotubes in a defined pattern, and we do not yet have the degree of control necessary to achieve this. There are several areas of technology where carbon nanotubes are already being used. These include flat-panel displays, scanning probe microscopes and sensing devices. The unique properties of carbon nanotubes will undoubtedly lead to many more applications.

Nanohorns

Single-walled carbon cones with morphologies similar to those of nanotube caps were first prepared by Peter Harris, Edmạn Tsang and colleagues in 1994. They were *not* discovered by NEC scientists, as stated in a press release. They were produced by high temperature heat treatments of fullerene soot. Sumio Iijima's group subsequently showed that they could also be produced by laser ablation of graphite, and gave them the name "nanohorns". This group has demonstrated that nanohorns have remarkable adsorptive and catalytic properties, and that they can be used as components of a new generation of fuel cells.

NANOTUBES AND THEIR APPLICATIONS

The properties of nanotubes have caused researchers and companies to consider using them in several fields. For example, because carbon nanotubes have the highest strength to weight ratio of any known material, researchers at NASA are combining carbon nanotubes with other materials into composites as shown in the photo below that can be used to build lightweight spacecraft. Another property of nanotubes is that they can easily penetrate membraneous such as cell walls. In fact, nanotubes long, narrow shape make them look like miniature needles, so it makes sense that they can function like a needle at the cellular level. Medical researchers are using this property by attaching molecules that are attracted to cancer cells to nanotubes to deliver drugs directly to diseased cells.

Another interesting property of carbon nanotubes is that their electrical resistance changes significantly when other molecules attach themselves to the carbon atoms. Companies are using this property to develop sensors that can detect chemical vapours such as carbon monoxide or biological molecules.

These are just a few of the potential uses of carbon nanotubes The following survey of carbon nanotube applications introduces these and many other uses. A survey of carbon nanotube applications under development:

Researchers and companies are working to use carbon nanotubes in various fields.

Nanotubes bound to an antibody that is produced by chickens have been shown to be useful in lab tests to destroy breast cancer tumours. The antibody carrying nanotubes are attracted to proteins produced by a one type of breast cancer cell. Then the nanotubes absorb light from an infrared laser, incinerating the nanotubes and the tumor they are attached to.

Researchers are developing materials, such as a carbon nanotube based composite developed by NASA that bends with when a voltage is applied, that will need only an electrical voltage to change the shape (morphing) of aircraft wings and other structures. This video from NASA gives you an idea of what a future morphing aircraft might look like.

Longer lasting concrete, researchers have found that carbon nanotubes can fill the voids that occur in conventionial concrete. Because it's these voids that allow water to penetrate into concrete, resulting in the formation of cracks; including nanotubes in the mix stops the cracks from forming.

Carbon nanotubes that perform as a catalyst, without using expensive platinum which most catalysts are based upon. Researchers have found that incorporating nitrogen and iron atoms into the carbon lattice of nanotubes result in nanotubes with catalytic properties.

Lightweight windmill blades made with an epoxy containing carbon nanotubes. The strength and low weight provided by the use of nanotube filled epoxy allows longer windmill blades to be used. This increases the amount of electricity generated by each windmill.

Using carbon nanotubes as the pores in reverse osmosis membranes. This can decrease the power needed to run reverse osmosis desalination plants because water molecules pass through carbon nanotubes more easily than through other types of nanopores. This is because the inside walls of the nanotubes are much smoother than the walls of other types of nanopores. Aircraft using carbon nanotubes to increase strength and flexibility in highly stressed components. Using carbon nanotubes to make a current collecting layer for the cathode in a battery. The battery can be formed on many different types of surfaces by spray painting layers that make up the battery.

Nanotube electodes in thermocells that generate electricity from waste heat. Inexpensive nanotube based sensor that detects bacteria in drinking water. Antibodies sensitive to the particular bacteria are bound to the nanotubes, which are then deposited onto a paper strip. When the bacteria is present it attaches to the antibodies, changing the spacing between the nanotubes and the resistance of the paper strip containing the nanotubes.

Combining carbon nanotubes, bucky-balls and polymers to produce inexpensive solar cells that can be formed by simply painting a surface. A lightweight, low power anti-icing system using carbon nanotubes in a layer coated onto aircraft wing surfaces.

THE NEW INORGANIC SEMICONDUCTOR OF CARBON NANOTUBES

Carbon nanotubes, also known as "bucky tubes" could be an alternative to organic semiconductors. Carbon nanotubes can be considered as extended carbon "bucky balls"- hollow cylinders of structured carbon atoms (tubular fullerene) which can be capped at the two ends with a hemisphere of a bucky ball. Multi wall and single wall nanotubes are available, with single wall nanotubes exhibiting a regular, near perfect structure when carefully produced. These have a diameter of 0.7-2nm with length being thousands of times the diameter.

Developed first by the Japanese and now globally, nanotubes exhibit unique thermal and electrical conductive capabilities. They are even being billed as a wire strong enough to connect a satellite to the Earth. The electrical conductivity of single walled nanotubes is similar to that of metals. This is because its unique structure means electrons can freely pass along the material. This is currently the first and only known molecule that can electronically act as a true metallic conductor. By spacing and controlling the size of the tubes they can behave as semiconductors, similar to silicon.

Single walled nanotubes are also suitable additives for conductive plastics. For applications such as antistatic and EMI shielding coatings, carbon black is currently the most common filler because it is relatively inexpensive. However, a lot of it is required to provide the necessary conductivity. At 30-50%, it also weakens the bulk properties of the material. The cylindrical shape of the nanotube means the aspect ratio (length-width) enables less amount of nanotubes to be used to achieve the same desired conductivity-less than 1% in some cases.

However, being a relatively new technology it is expensive. Some of the purest forms of carbon nanotubes cost $500 per gram and some have likened the cost of nanotube production plants to that of silicon foundries. However, if cheap enough in future it could outperform some organic semiconductors and enable faster circuits.

PURIFICATION TECHNIQUES FOR CARBON NANOTUBES

PURIFICATION OF CARBON NANOTUBES

Purification of carbon nanotubes (CNTs) generally refers to the separation of CNTs from other entities, such as carbon nanoparticles, amorphous carbon, residual catalyst, and other unwanted species. The classic chemical techniques for purification have been tried, but they have not been found to be effective in removing the undesirable impurities. Three basic methods have been used with varying degrees of success, namely:

- Gas-phase
- Liquid-phase and
- Intercalation methods.

Centrifugal Concentration

Generally, a centrifugal separation is necessary to concentrate the single walled nanotubes in low-yield soot before the micro filtration operation, since the nanoparticles easily contaminate membrane filters. The advantage of this method is that unwanted nanoparticles and amorphous carbon are removed simultaneously and the CNTs are not chemically modified. However 2-3 mol nitric acid is useful for chemically removing impurities.

Sonication to Cut Carbon Nanotubes into Smaller Segments

It is now possible to cut CNTs into smaller segments, by extended sonication in concentrated acid mixtures. The resulting CNTs form a colloidal suspension in solvents. They can be deposited on substrates, or further manipulated in solution, and can have many different functional groups attached to the ends and sides of the CNTs.

Gas Phase

The first successful technique for purification of nanotubes was developed by Thomas Ebbesen and coworkers. Following the demonstration that nanotubes could be selectively attached by oxidizing gases these workers realised that nanoparticles, with their defect rich structures might be oxidised more readily than the relatively perfect nanotubes. They found that a significant relative enrichment of nanotubes could be achieved this way, but only at the expense of losing the majority of the original sample. A new gas-phase method has been developed at the NASA Glenn Research Centre to purify gram-scale quantities of single-wall CNTs. This method, a modification of a gas-phase purification technique previously reported by Smalley and others, uses a combination of high-temperature oxidations and repeated extractions with nitric and hydrochloric acid. This improved procedure significantly reduces the amount of impurities such as residual catalyst, and non-nanotube forms of carbon) within the CNTs, increasing their stability significantly.

Liquid Phase

The current liquid-phase purification procedure follows certain essential steps:

- Preliminary filtration- to get rid of large graphite particles;
- Dissolution in organic solvents and concentrated acids to remove fullerenes and catalyst particles.
- Centrifugal separation
- Microfiltration- and
- Chromatography to either separate multi walled nanotubes and unwanted nanoparticles or single walled nanotubes and the amorphous carbon impurities.

It is important to keep the CNTs well-separated in solution, so the CNTs are typically dispersed using a surfactant prior to the last stage of separation.

Intercalation

An alternative approach to purifying multi walled nanotubes was introduced in 1994 by a Japanese research group. This technique made use of the fact that nanoparticles and other graphitic contaminants have relatively "open" structures and can therefore be more readily intercalated with a variety of materials that can close nanotubes. By intercalating with copper chloride, and then reducing this to metallic copper, the research group was able to preferentially oxidize the nanoparticles away, using copper as an oxidation catalyst. Since 1994, this has become a popular method for purification of nanotubes.

"The first stage is to immerse the crude cathodic deposit in a molten copper chloride and potassium chloride mixture at 400°C and leave it for one week. The product of this treatment, which contains intercalated nanoparticles and graphitic fragments, is then washed in ion exchanged water to remove excess copper chloride and potassium chloride. In order to reduce the intercalated copper chloride-potassium chloride metal, the washed product is slowly heated to 500°C in a mixture of Helium and hydrogen and held at this temperature for 1 hour.

Finally, the material is oxidized in flowing air at a rate of 10°C/min to a temperature of 555°C. Samples of cathodic soot which have been treated this way consist almost entirely of nanotubes. A disadvantage of this method is that some amount of nanotubes are inevitably lost in the oxidation stage, and the final material may be contaminated with residues of intercalates. A similar purification technique, which involves intercalation with bromine followed by oxidation, has also been described.

INORGANIC NANOTUBE

Figure: Scanning Electron Microscopy Image of WS_2 Nanotube Bundles.

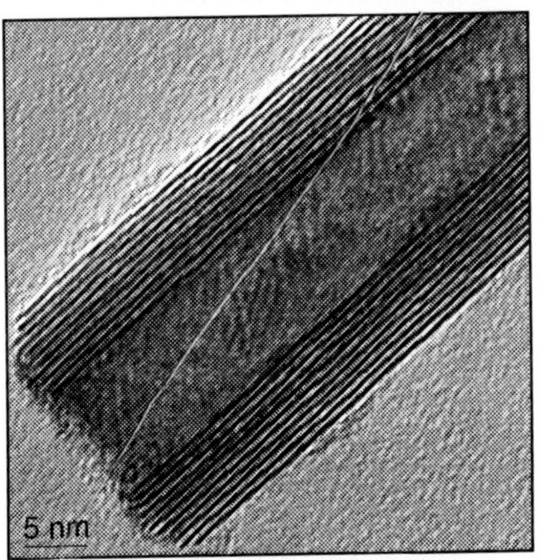

Figure: Transmission Electron Microscopy Image of an Individual WS_2 Multi-wall Nanotube.

An inorganic nanotube is a cylindrical molecule often composed of metal oxides, and morphologically similar to a carbon nanotube. Inorganic nanotubes have been observed to occur naturally in some mineral deposits.

Although Linus Pauling mentioned the possibility of curved layers in minerals as early as 1930, synthetic inorganic nanotubes did not appear until Reshef Tenne *et al.* reported the synthesis of nanotubes composed of tungsten disulfide (WS_2) in 1992. In the intervening years, nanotubes have been synthesised of many other inorganic materials, such as vanadium oxide and manganese oxide, and are being researched for such applications as redox catalysts and cathode materials for batteries.

HISTORY AND OCCURRENCE

Inorganic nanotubes are morphologically similar to carbon nanotubes and are observed in some mineral deposits of natural origin. Synthetic structures of this type were first reported by the group of Reshef Tenne in 1992.

Materials

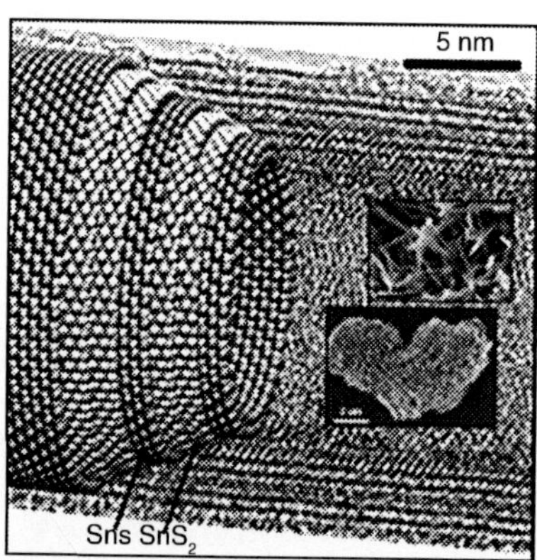

Figure: Illustration of SnS_2/SnS Nanotube Superstructures, Made on Basis of High-resolution Transmission Electron Microscopy Images.

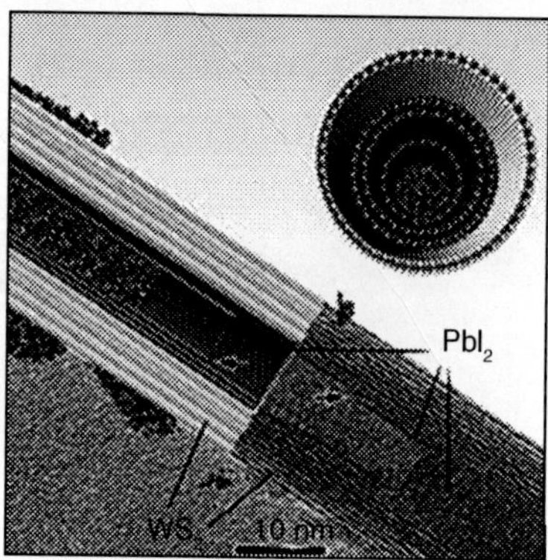

Figure: Illustration of PbI_2/WS_2 Core-shell Nanostructure.

Typical inorganic nanotube materials are 2D layered solids such as tungsten (IV) sulfide (WS_2), molybdenum disulfide (MoS_2) and tin (IV) sulfide (SnS_2). WS_2 and SnS_2/tin (II) sulfide (SnS) nanotubes have been synthesized in macroscopic amounts.

However, traditional ceramics like titanium dioxide (TiO_2) and zinc oxide (ZnO) also form inorganic nanotubes. More recent nanotube and nanowire materials are transition metal/chalcogen/halogenides (TMCH), described by the formula $TM_6C_yH_z$, where TM is transition metal (molybdenum, tungsten, tantalum, niobium), C is chalcogen (sulfur, selenium, tellurium), H is halogen (iodine), and the composition is given by $8.2<(y+z)<10$. TMCH tubes can have a subnanometre-diameter, lengths tunable from hundreds of nanometres to tens of microns and show excellent dispersiveness owing to extremely weak mechanical coupling between the tubes.

In recent year, Chinese scientists announced the creation in the laboratory of copper and bismuth nanotubes.

PROPERTIES AND POTENTIAL APPLICATIONS

Inorganic nanotubes are an alternative material to better-explored carbon nanotubes, showing advantages such as easy synthetic access and high crystallinity, good uniformity and dispersion, predefined electrical conductivity depending on the composition of the starting material and needle-like morphology, good adhesion to a number of polymers and high impact-resistance.

They are therefore promising candidates as fillers for polymer composites with enhanced thermal, mechanical, and electrical properties. Target applications for this kind of composites are materials for heat management, electrostatic dissipaters, wear protection materials, photovoltaic elements, etc. Inorganic nanotubes are heavier than carbon nanotubes and not as strong under tensile stress, but they are particularly strong under compression, leading to potential applications in impact-resistant applications such as bulletproof vests.

The mechanical strength of cellulose fibres can be increased by an order of magnitude by adding only 0.1 wt% of TMCH nanotubes, and measurements of electrical conductivity of polycaprolactone doped with TMCH nanotubes revealed a percolative behaviour with an extremely low percolation threshold. The addition of WS_2 nanotubes to epoxy resin improved adhesion, fracture toughness and strain energy release rate.

The wear of the nanotubes-reinforced epoxy was eight times lower than that of pure epoxy. WS_2 nanotubes were also embedded into a poly(methyl methacrylate) (PMMA) nanofibre matrix via electrospinning. The nanotubes were well dispersed and aligned along fibre axis. The enhanced stiffness and toughness of PMMA fibre meshes by means of inorganic nanotubes addition may have potential applications as impact-absorbing materials.

Optical properties of semiconductor quantum dot-inorganic nanotube hybrids reveal efficient resonant energy transfer from the quantum dot to the inorganic nanotubes upon photoexcitation.

Nanodevices based on one-dimensional nanomaterials are thought for next-generation electronic and photoelectronic systems having small size, faster transport speed, higher efficiency and less energy consumption. A high-speed photodetector for visible and near-infrared light based on individual WS_2 nanotubes has been prepared in laboratory. Inorganic nanotubes are hollow and can be filled with another material, to preserve or guide it to a desired location or generate new properties in the filler material which is confined within a nanometre-scale diameter. To this goal, inorganic nanotube hybrids were made by filling WS_2 nanotubes with molten lead, antimony or bismuth iodide salt by a capillary wetting process, resulting in $PbI_2@WS_2$, $SbI_3@WS_2$ or $BiI_3@WS_2$ core-shell nanotubes.

REFERENCES

- Allen, Robin: *An Introduction to Micro and Nanotechnology,* New York: Oxford University Press, 2002.
- Basnet, J.: *Unique Features and Properties of Nanostructured Materials,* Delhi: Indian J Community Med, 2008.
- Costlow, Laura and April Peter: *Nanomaterials and Nanostructures : Integral Components to Nanomachines, Nanorobots and Nanofactories Making Wonder Products for New Century,* Dominant, Delhi, 2007.
- Drexler, K. Eric: *Nanosystems: Molecular Machinery, Manufacturing, and Computation,* New York: Wiley, 1992.
- Michio Kaku: *Quantum Field Theory: A Modern Introduction,* Oxford University Press, 1993.
- Nanda, S.: *Properties of Individual Nanoparticles,* India: Indian Journal of Preventive and Medicine, 2006.

3

Nanoparticles

DEFINITION OF NANOPARTICLES

There is no accepted international definition of a nanoparticle, but one given in the new PAS71 document developed in the UK is: "A particle having one or more dimensions of the order of 100nm or less". There is a note associated with this definition: "Novel properties that differentiate nanoparticles from the bulk material typically develop at a critical length scale of under 100nm". The "novel properties" mentioned are entirely dependent on the fact that at the nano-scale, the physics of nanoparticles mean that their properties are different from the properties of the bulk material. This makes the size of particles or the scale of its features the most important attribute of nanoparticles.

DIFFERENT ABOUT A NANOPARTICLE

There is no strict dividing line between nanoparticles and non-nanoparticles. The size at which materials display different properties to the bulk material is material dependant and can certainly be claimed for many materials much larger in size than 100nm. Definitions certainly become more difficult for materials that are a very long way from being a sphere, such as carbon nanotubes for example. One of the aims for these materials is to grow them into long tubes, certainly not 'nano' in length, but as they have a diameter in the order of 3nm for a single walled tube, they have properties that distinguish them from other allotropes of carbon, and hence can be described as 'nanomaterials'.

This sort of nanomaterial has led to the extension of the idea of nanomaterials being considered as such if any one of their structural features are on a scale of less than 100nm, that cause their properties to be different from that of the bulk material.

MANUFACTURING METHODS FOR NANOPARTICLES

Many of these nanomaterials are made directly as dry powders, and it is a common myth that these powders will stay in the same state when stored. In fact, they will rapidly aggregate through a solid bridging mechanism in as little as a few seconds. Whether these aggregates are detrimental will depend entirely on the application of the nanomaterial. If the nanoparticles need to be kept separate, then they must be prepared and stored in a liquid medium designed to facilitate sufficient interparticle repulsion forces to prevent aggregation.

NANOPARTICLES PROPERTIES

In nanotechnology, a particle is defined as a small object that behaves as a whole unit in terms of its transport and properties. It is further classified according to size: in terms of diameter, fine particles cover a range between 100 and 2500 nanometres, while ultrafine particles, on the other hand, are sized between 1 and 100 nanometres.

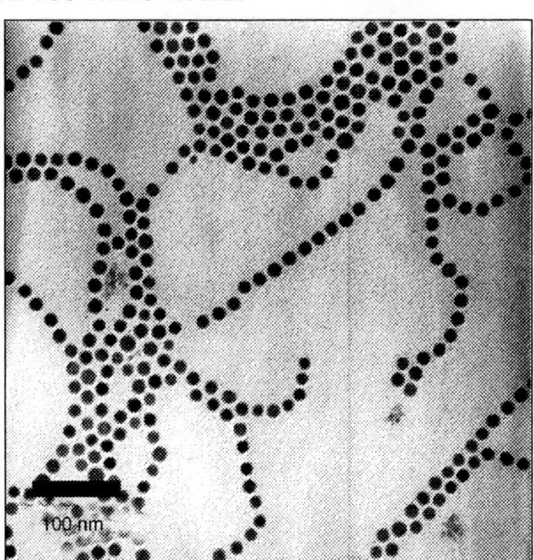

Figure: Colorized Transmission Electron Micrograph Showing Chains of Cobalt Nanoparticles.

Similar to ultrafine particles, nanoparticles are sized between 1 and 100 nanometres. Nanoparticles may or may not exhibit size-related properties that differ significantly from those observed in fine particles or bulk materials. Although the size of most molecules would fit into the above outline, individual molecules are usually not referred to as nanoparticles. Nanoclusters have at least one dimension between 1 and 10 nanometres and a narrow size distribution. Nanopowders are agglomerates of ultrafine particles, nanoparticles, or nanoclusters.

Nanometre-sized single crystals, or single-domain ultrafine particles, are often referred to as nanocrystals. Nanoparticle research is currently an area of intense scientific interest due to a wide variety of potential applications in biomedical, optical and electronic fields. The National Nanotechnology Initiative has led to generous public funding for nanoparticle research in the United States.

Although nanoparticles are generally considered an invention of modern science, they actually have a very long history. Nanoparticles were used by artisans as far back as the 9th century in Mesopotamia for generating a glittering effect on the surface of pots. Even these days, pottery from the Middle Ages and Renaissance often retain a distinct gold or copper coloured metallic glitter. This so called luster is caused by a metallic film that was applied to the transparent surface of a glazing. The luster can still be visible if the film has resisted atmospheric oxidation and other weathering.

The luster originated within the film itself, which contained silver and copper nanoparticles dispersed homogeneously in the glassy matrix of the ceramic glaze. These nanoparticles were created by the artisans by adding copper and silver salts and oxides together with vinegar, ochre and clay, on the surface of previously-glazed pottery. The object was then placed into a kiln and heated to about 600°C in a reducing atmosphere.

In the heat the glaze would soften, causing the copper and silver ions to migrate into the outer layers of the glaze. There the reducing atmosphere reduced the ions back to metals, which then came together forming the nanoparticles that give the colour and optical effects. Luster technique showed that ancient craftsmen had a rather sophisticated empirical knowledge of materials. The technique originated in the Islamic world. As Muslims were not allowed to use gold in artistic representations, they had to find a way to create a similar effect without using real gold. The solution they found was using luster. Michael Faraday provided the first description, in scientific terms, of the optical properties of nanometre-scale metals in his classic 1857 paper. In a subsequent paper, the author (Turner) points out that: "It is well known that when thin leaves of gold or silver are mounted upon glass and heated to a temperature which is well below a red heat (~500 °C), a remarkable change of properties takes place, whereby the continuity of the metallic film is destroyed. The result is that white light is now freely transmitted, reflection is correspondingly diminished, while the electrical resistivity is enormously increased."

NANOPARTICLE UNIFORMITY

The chemical processing and synthesis of high performance technological components for the private, industrial and military sectors requires the use of high purity ceramics, polymers, glass-ceramics and material composites.

In condensed bodies formed from fine powders, the irregular particle sizes and shapes in a typical powder often lead to non-uniform packing morphologies that result in packing density variations in the powder compact.

Uncontrolled agglomeration of powders due to attractive van der Waals forces can also give rise to in microstructural inhomogeneities. Differential stresses that develop as a result of non-uniform drying shrinkage are directly related to the rate at which the solvent can be removed, and thus highly dependent upon the distribution of porosity. Such stresses have been associated with a plastic-to-brittle transition in consolidated bodies, and can yield to crack propagation in the unfired body if not relieved.

In addition, any fluctuations in packing density in the compact as it is prepared for the kiln are often amplified during the sintering process, yielding inhomogeneous densification. Some pores and other structural defects associated with density variations have been shown to play a detrimental role in the sintering process by growing and thus limiting end-point densities. Differential stresses arising from inhomogeneous densification have also been shown to result in the propagation of internal cracks, thus becoming the strength-controlling flaws.

It would therefore appear desirable to process a material in such a way that it is physically uniform with regard to the distribution of components and porosity, rather than using particle size distributions which will maximize the green density. The containment of a uniformly dispersed assembly of strongly interacting particles in suspension requires total control over interparticle forces. Monodisperse nanoparticles and colloids provide this potential.

Monodisperse powders of colloidal silica, for example, may therefore be stabilized sufficiently to ensure a high degree of order in the colloidal crystal or polycrystalline colloidal solid which results from aggregation. The degree of order appears to be limited by the time and space allowed for longer-range correlations to be established. Such defective polycrystalline colloidal structures would appear to be the basic elements of submicrometer colloidal materials science, and, therefore, provide the first step in developing a more rigorous understanding of the mechanisms involved in microstructural evolution in high performance materials and components.

PHYSICAL PROPERTIES OF NANOPARTICLES

Nanoparticles are of great scientific interest as they are effectively a bridge between bulk materials and atomic or molecular structures. A bulk material should have constant physical properties regardless of its size, but at the nano-scale size-dependent properties are often observed. Thus, the properties of materials change as their size approaches the nanoscale and as the percentage of atoms at the surface of a material becomes significant.

For bulk materials larger than one micrometer (or micron), the percentage of atoms at the surface is insignificant in relation to the number of atoms in the bulk of the material. The interesting and sometimes unexpected properties of nanoparticles are therefore largely due to the large surface area of the material, which dominates the contributions made by the small bulk of the material.

For example, nanoparticles of usually yellow gold and gray silicon are red in colour; gold nanoparticles melt at much lower temperatures (~300 °C for 2.5 nm size) than the gold slabs (1064 °C); and absorption of solar radiation in photovoltaic cells is much higher in materials composed of nanoparticles than it is in thin films of continuous sheets of material – the smaller the particles, the greater the solar absorption. Other size-dependent property changes include quantum confinement in semiconductor particles, surface plasmon resonance in some metal particles and superparamagnetism in magnetic materials. Ironically, the changes in physical properties are not always desirable.

Ferroelectric materials smaller than 10 nm can switch their magnetisation direction using room temperature thermal energy, thus making them unsuitable for memory storage. Suspensions of nanoparticles are possible since the interaction of the particle surface with the solvent is strong enough to overcome density differences, which otherwise usually result in a material either sinking or floating in a liquid.

Nanoparticles also often possess unexpected optical properties as they are small enough to confine their electrons and produce quantum effects. For example gold nanoparticles appear deep red to black in solution. The high surface area to volume ratio of nanoparticles provides a tremendous driving force for diffusion, especially at elevated temperatures.

Sintering can take place at lower temperatures, over shorter time scales than for larger particles. This theoretically does not affect the density of the final product, though flow difficulties and the tendency of nanoparticles to agglomerate complicates matters. Moreover, nanoparticles have been found to impart some extra properties to various day to day products. For example the presence of titanium dioxide nanoparticles imparts what we call the self-cleaning effect, and the size being nanorange, the particles can not be observed. Zinc oxide particles have been found to have superior UV blocking properties compared to its bulk substitute. This is one of the reasons why it is often used in the preparation of sunscreen lotions., and is completely photostable.

Clay nanoparticles when incorporated into polymer matrices increase reinforcement, leading to stronger plastics, verifiable by a higher glass transition temperature and other mechanical property tests. These nanoparticles are hard, and impart their properties to the polymer (plastic). Nanoparticles have also been attached to textile fibres in order to create smart and functional clothing.

Metal, dielectric, and semiconductor nanoparticles have been formed, as well as hybrid structures (*e.g.*, core-shell nanoparticles). Nanoparticles made of semiconducting material may also be labelled quantum dots if they are small enough (typically sub 10 nm) that quantization of electronic energy levels occurs. Such nanoscale particles are used in biomedical applications as drug carriers or imaging agents.

Semi-solid and soft nanoparticles have been manufactured. A prototype nanoparticle of semi-solid nature is the liposome. Various types of liposome nanoparticles are currently used clinically as delivery systems for anticancer drugs and vaccines.

SYNTHESIS OF NANOPARTICLES

There are several methods for creating nanoparticles, including both attrition and pyrolysis. In attrition, macro or micro scale particles are ground in a ball mill, a planetary ball mill, or other size reducing mechanism. The resulting particles are air classified to recover nanoparticles. In pyrolysis, a vaporous precursor (liquid or gas) is forced through an orifice at high pressure and burned. The resulting solid (a version of soot) is air classified to recover oxide particles from by-product gases. Pyrolysis often results in aggregates and agglomerates rather than singleton primary particles.

A thermal plasma can also deliver the energy necessary to cause evaporation of small micrometer size particles. The thermal plasma temperatures are in the order of 10,000 K, so that solid powder easily evaporates. Nanoparticles are formed upon cooling while exiting the plasma region. The main types of the thermal plasma torches used to produce nanoparticles are dc plasma jet, dc arc plasma and radio frequency (RF) induction plasmas. In the arc plasma reactors, the energy necessary for evaporation and reaction is provided by an electric arc which is formed between the anode and the cathode. For example, silica sand can be vaporised with an arc plasma at atmospheric pressure. The resulting mixture of plasma gas and silica vapour can be rapidly cooled by quenching with oxygen, thus ensuring the quality of the fumed silica produced. In RF induction plasma torches, energy coupling to the plasma is accomplished through the electromagnetic field generated by the induction coil.

The plasma gas does not come in contact with electrodes, thus eliminating possible sources of contamination and allowing the operation of such plasma torches with a wide range of gases including inert, reducing, oxidizing and other corrosive atmospheres. The working frequency is typically between 200 kHz and 40 MHz. Laboratory units run at power levels in the order of 30–50 kW while the large scale industrial units have been tested at power levels up to 1 MW. As the residence time of the injected feed droplets in the plasma is very short it is important that the droplet sizes are small enough in order to obtain complete evaporation. The RF plasma method has been used to synthesize different nanoparticle materials, for example synthesis of various ceramic nanoparticles such as oxides, carbours/carbides and nitrides of Ti and Si. Inert-gas condensation is frequently used to make nanoparticles from metals with low melting points. The metal is vaporised in a vacuum chamber and then supercooled with an inert gas stream. The supercooled metal vapour condenses into nanometre-sized particles, which can be entrained in the inert gas stream and deposited on a substrate or studied in situ.

Sol-gel

The sol-gel process is a wet-chemical technique (also known as chemical solution deposition) widely used recently in the fields of materials science and ceramic engineering. Such methods are used primarily for the fabrication of materials (typically a metal oxide) starting from a chemical solution (''sol'', short for solution) which acts as the precursor for an integrated network (or ''gel'') of either discrete particles or network polymers.

Typical precursors are metal alkoxides and metal chlorides, which undergo hydrolysis and polycondensation reactions to form either a network "elastic solid" or a colloidal suspension (or dispersion) – a system composed of discrete (often amorphous) submicrometer particles dispersed to various degrees in a host fluid. Formation of a metal oxide involves connecting the metal centres with oxo (M-O-M) or hydroxo (M-OH-M) bridges, therefore generating metal-oxo or metal-hydroxo polymers in solution. Thus, the sol evolves towards the formation of a gel-like diphasic system containing both a liquid phase and solid phase whose morphologies range from discrete particles to continuous polymer networks.

In the case of the colloid, the volume fraction of particles (or particle density) may be so low that a significant amount of fluid may need to be removed initially for the gel-like properties to be recognized. This can be accomplished in any number of ways. The most simple method is to allow time for sedimentation to occur, and then pour off the remaining liquid. Centrifugation can also be used to accelerate the process of phase separation. Removal of the remaining liquid (solvent) phase requires a drying process, which is typically accompanied by a significant amount of shrinkage and densification. The rate at which the solvent can be removed is ultimately determined by the distribution of porosity in the gel.

The ultimate microstructure of the final component will clearly be strongly influenced by changes implemented during this phase of processing. Afterwards, a thermal treatment, or firing process, is often necessary in order to favour further polycondensation and enhance mechanical properties and structural stability via final sintering, densification and grain growth. One of the distinct advantages of using this methodology as opposed to the more traditional processing techniques is that densification is often achieved at a much lower temperature. The precursor sol can be either deposited on a substrate to form a film (*e.g.* by dip-coating or spin-coating), cast into a suitable container with the desired shape (*e.g.* to obtain a monolithic ceramics, glasses, fibres, membranes, aerogels), or used to synthesize powders (*e.g.* microspheres, nanospheres). The sol-gel approach is a cheap and low-temperature technique that allows for the fine control of the product's chemical composition. Even small quantities of dopants, such as organic dyes and rare earth metals, can be introduced in the sol and end up uniformly dispersed in the final product. It can be used in ceramics processing and manufacturing as an investment casting material, or as a means of producing very thin films of metal oxides for various purposes. Sol-gel derived materials have diverse applications in optics, electronics, energy, space, (bio)sensors, medicine (*e.g.* controlled drug release) and separation (*e.g.* chromatography) technology.

NANOPARTICLE COLLOIDS

The term colloid is used primarily to describe a broad range of solid-liquid (and/or liquid-liquid) mixtures, all of which contain distinct solid (and/or liquid) particles which are dispersed to various degrees in a liquid medium. The term is specific to the size of the individual particles, which are larger than atomic dimensions but small enough to exhibit Brownian motion. If the particles are large enough, then their dynamic behaviour in any given period of time in suspension would be governed by forces of gravity and sedimentation. But if they are small enough to be colloids, then their irregular motion in suspension can be attributed to the collective bombardment of a myriad of thermally agitated molecules in the liquid suspending medium, as described originally by Albert Einstein in his dissertation. Einstein proved the existence of water molecules by concluding that this erratic particle behaviour could adequately be described using the theory of Brownian motion, with sedimentation being a possible long-term result. This critical size range (or particle diameter) typically ranges from nanometres (10^{-9} m) to micrometers (10^{-6} m).

MORPHOLOGY OF NANOPARTICLES

Scientists have taken to naming their particles after the real world shapes that they might represent. Nanospheres, nanoreefs, nanoboxes and more have appeared in the literature. These morphologies sometimes arise spontaneously as an effect of a templating or directing agent present in the synthesis such as miscellar emulsions or anodized alumina pores, or from the innate crystallographic growth patterns of the materials themselves. Some of these morphologies may serve a purpose, such as long carbon nanotubes being used to bridge an electrical junction, or just a scientific curiosity like the stars shown at right. Amorphous particles usually adopt a spherical shape (due to their microstructural isotropy) – whereas the shape of anisotropic microcrystlaline whiskers corresponds to their particular crystal habit. At the small end of the size range, nanoparticles are often referred to as clusters. Spheres, rods, fibres, and cups are just a few of the shapes that have been grown. The study of fine particles is called micromeritics.

CHARACTERIZATION OF NANOPARTICLES

Nanoparticle characterization is necessary to establish understanding and control of nanoparticle synthesis and applications. Characterization is done by using a variety of different techniques, mainly drawn from materials science. Common techniques are electron microscopy (TEM, SEM), atomic force microscopy (AFM), dynamic light scattering (DLS), x-ray photoelectron spectroscopy (XPS), powder X-ray diffraction (XRD), Fourier transform infrared spectroscopy (FTIR), matrix-assisted laser desorption/ionization time-of-flight mass spectrometry (MALDI-TOF), ultraviolet-visible spectroscopy, dual polarisation interferometry and nuclear magnetic resonance (NMR). Whilst the theory has been known for over a century, the technology for Nanoparticle tracking analysis (NTA) allows direct tracking of the Brownian motion and this method therefore allows the sizing of individual nanoparticles in solution.

SAFETY OF NANOPARTICLES

Nanoparticles present possible dangers, both medically and environmentally. Most of these are due to the high surface to volume ratio, which can make the particles very reactive or catalytic. They are also able to pass through cell membranes in organisms, and their interactions with biological systems are relatively unknown. However, free nanoparticles in the environment quickly tend to agglomerate and thus leave the nano-regime, and nature itself presents many nanoparticles to which organisms on earth may have evolved immunity (such as salt particulates from ocean aerosols, terpenes from plants, or dust from volcanic eruptions). A more complete analysis is provided in the article on nanotechnology. According to the ''San Francisco Chronicle'', "Animal studies have shown that some nanoparticles can penetrate cells and tissues, move through the body and brain and cause biochemical damage they also have shown to cause a risk factor in men for testicular cancer. But whether cosmetics and sunscreens containing nanomaterials pose health risks remains largely unknown, pending completion of long-range studies recently begun by the FDA and other agencies." Diesel nanoparticles have been found to damage the cardiovascular system in a mouse model.

PLATINUM NANOPARTICLES

Platinum nanoparticles are usually in the form of a suspension or colloid of sub-micrometre-sized particles of platinum in a fluid, usually water. A colloid is technically defined as a stable dispersion of particles in a fluid medium (liquid or gas). Platinum nanoparticles can be made with sizes between about 2 and 20 nanometres (nm), depending on reaction conditions. Trillions of platinum nanoparticles are suspended in the brownish red or black coloured colloidal solution. Nanoparticles come in wide variety of shapes including spheres, rods, cubes and tetrahedra.

Due to the antioxidant properties of the platinum nanoparticles, they are the subject of substantial research with potential applications in a wide variety of areas, including nanotechnology, catalysis, medicine and the synthesis of novel materials with unique properties.

SYNTHESIS

Platinum nanoparticles are made by reduction of hexachloroplatinate. A solution is rapidly stirred while a reducing agent such as hydrogen gas is added, causing platinum ions to be reduced to neutral platinum atoms. As more and more of these platinum atoms form, the solution quickly becomes supersaturated and platinum starts to precipitate in the form of sub-nanometre particles. As the reaction proceeds, platinum atoms are added to these nuclei, which grow in size. If the solution is stirred vigorously enough, the particles will be fairly uniform in size and shape. To prevent the particles from aggregating, some sort of stabilizing agent or stabilizer that sticks to the nanoparticle surface is usually added, such as sodium polyacrylate. They can be functionalized with various organic ligands to create organic-inorganic hybrid materials with advanced functionality.

BIOLOGICAL EFFECTS

Research by Yusei Miyamoto at University of Tokyo, Japan, resulted in the use of platinum nanoparticles of the size 2-3 nm to increase the lifespan of the roundworm *Caenorhabditis elegans*. Nanoparticles may present possible safety issues both medically and environmentally. Most of these issues usually arise due to the high surface to volume ratio, which can make the particles of some metals very reactive or catalytic. In particular, inhaled nanoparticles can pose health risks, and may cause inflammation and disease in the lung. They are able to pass through cell membranes in organisms and their interactions with biological systems are relatively unknown.

THE GOLD NANOPARTICLE FOR NANOTECHNOLOGY

Nanopartz™ is recognized as both an innovator and a quality supplier of gold nanoparticle based products including Accurate™ spherical gold nanoparticles, gold nanorods, micron sized gold (Microgold™), gold nanowires, and the newly released constructs: Ramanprobes™ and Plasmonic Fluorophores™. Nanopartz™ also offers platinum and palladium coated gold nanoparticles. Products developed at Nanopartz™ constitute an essential core technology for the life sciences in diagnostics, imaging, and therapeutics-particularly cancer. In other fields, Nanopartz™ products are well suited for sensors, solar cells, liquid crystals, non-linear optics, polarizers, negative refractive index materials, standards, catalytics, and as robust subcomponents in electronics. Nanopartz™ offers products for the researcher as well as the materials manufacturer.

Over the past four years, Nanopartz™ nanomaterials have appeared in over 200 publications, 2 covers, and 18 patents. Numerous companies have employed Nanopartz™ innovative nanomaterials into their new technologies and improved old technologies. Partnering with Nanopartz™ not only ensures innovation, our audited ISO 9000 system ensures repeated success.

REFERENCES

- Ahmad, Lone: *Nanoscience: Nanotechnologies and Nanophysics,* Jammu Kashmir: South Kashmir Press, 2006.
- Bhatia, A.S. : *Nano Electronics,* Deep and Deep, Delhi, 2011.
- Gupta, Sushmita: *Understanding of Nano Science and Technology,* Global Vision, Delhi, 2006.
- Ranganathan, S.: *Handbook of Nanoparticles,* Chennai: T.T.Ranganathan Clinical research Foundation, 2008.
- Sinha, D.N.: *Nanotubes and Nanowires,* Patna: Indian Academy of Sciences, 2005.
- Varghese, Thomas and Balakrishna, K.M. : *Nanotechnology : An Introduction To Synthesis Properties and Applications of Nanomaterials,* Atlantic, Delhi, 2012.

4

Fullerenes and Molecular Mimics

INTRODUCTION TO FULLERENES

Carbon, the common element in organic compounds, is known to exist in two allotropic forms, *viz,* diamond and graphite. In 1985, a third form of carbon called fullerenes was discovered. The group of scientists led by Smalley, Kroto and Curl attempted to simulate the conditions under which carbon nucleates in the atmosphere of red giant stars. In the experiment, a surface of solid graphite was vaporized by irradiation with the laser into plasma containing atoms and free ions. The free atoms and ions were chilled down due to the collision with the helium atoms. Through the collision, clusters containing various numbers of carbon atoms were formed. The clusters were examined in a mass spectrometer, and it was found that clusters that had 60 and 70 carbon atoms dominated, and that most clusters had 60 carbon atoms. In the beginning, the scientists had problem in producing sufficient amount of fullerenes. They had succeeded to prepare only less than 10-15 g. But after 5 years, Kratschmer and Huffmann, and then Kroto *et al* developed new high yielding preparative methods for fullerenes.

These scientists named the newly found molecule after the architect Richard Buckminster Fuller, who created the dome in 1967 with the same shape as the carbon cluster. Fullerenes generated so much interest and excitement among research scientists that the three scientists who discovered fullerenes received Nobel Prize in Chemistry in 1996. Fullerenes were later found to exist naturally in interstellar dust as well as in geological formations on Earth, though only in the ppm-range. Some of these places are Shunga/ Russia, New Zealand and Sudbury/Canada.

STRUCTURE AND PROPERTIES

Fullerenes are large carbon cage molecules considered to be three-dimensional analogues of benzene. The most abundant form of fullerenes is Buckminster fullerene (C_{60}) with 60 carbon atoms arranged in a spherical structure. The shape of the molecule, known as truncated icosahedron, resembles that of a soccer ball, which contains 12 pentagons and 20 hexagons. Fullerenes fulfil the EULER's theorem, *i.e.*, if a polyeder is to build a closed structure from pentagons and hexagons; it has to contain exact 12 pentagons. Following this rule, the smallest stable fullerene is C_{60}, which has no two pentagons side by side, making it the most stable structure.

An important property of C_{60} molecule is its high symmetry. There are 120 symmetrical operations, like rotation around the axis and reflection in a plane, which map the molecule onto itself. This makes C_{60} the most symmetrical molecule. There are two types of bond lengths in the fullerene: C_5 - C_5 single bonds in the pentagons and C_5 - C_6 double bonds in the hexagons; the first is 1.45 ± 0.015 Å and the other one is 1.40 ± 0.015 Å. Each carbon atom forms bond to three other adjacent atoms with sp hybridization. The set of orbitals is arranged at 120-degrees angles and is centred in the xy-plane. Hence these delocalized pi electrons stabilize the spheroid structure by resonance.

A C_{60} molecule, also known as Buckyball or Buckminsterfullerene, is about 7 Å in diameter. C_{60} molecules condense to form a solid of weakly bound molecules. This crystalline state is called fullerites. This solid is cubic, weakly bound with a lattice constant a=14.71 Å and electrically insulating. It occurs as yellow powder, which turns pink when dissolved in toluene. On exposure to strong UV light, the Buckyballs polymerize, forming bonds between adjacent balls.

In the polymerized state, C_{60} no longer dissolves in toluene. NMR studies of C_{60} benzene solvates show free rotation at room temperature. At about -13°, the balls spin freely in their crystalline positions. At lower temperature, their movements begin to limit to certain orientations. Eventually, below -183°, the balls become completely struck. Chemically the molecule is quite stable; breaking the balls requires temperature of over 1000°. By heating fullerenes up to 1500° in absence of air, they transform to graphite.

Other than C_{60}, fullerenes can contain between 30 to 980 carbon atoms forming different structures with different properties and field of applications. As hexagons are added or removed from the basic soccer ball structure, the molecule begins to lose its roundness. C_{70}, which has 25 hexagons, is shaped more like a rugby ball. Giant fullerenes take on a pentagonal shape. Smaller fullerenes look like asteroids. With the loss of roundness comes a loss of stability, C_{60} being the most stable of all. Other than C_{60} and C_{70}, C_{76}, C_{78}, C_{84}, C_{86} have been isolated and studied in detail. Fullerenes have low density relative to diamond (1.65 g/cc compared to 3.51 g/cc). Fullerenes are unstable in water. Good solvents for fullerenes are carbon disulphide, o-dichlorobenzene, toluene and xylene. Thin layers of fullerenes are coloured from yellow to yellowish-green. The colour of fullerene solutions is attributed to the pi-pi electron transition.

SPECIES OF FULLERENES

Some time after discovery of fullerenes, many chemical modifications of fullerenes were discovered. Some of the important fullerene species are as follows.

Alkali-doped Fullerenes

As fullerene molecule is highly electronegative, it readily forms compounds with electron donating atoms, the most common examples being alkali metals. This reaction leads to production of an interesting class of compounds known as alkali-doped fullerides, wherein alkali metal atoms fill in the space between Buckyballs and donate valence electron to the neighbouring C_{60} molecule. If alkali atoms are potassium or rubidium, the compounds are superconductors, and they conduct electric current without any resistance at temperatures below 20-40 K, *e.g.*, $K_3 C_{60}$, $Rb_3 C_{60}$.

Endohedral Fullerenes

Since fullerenes are hollow with a closed shell of carbon atoms, it is possible to enclose another atom inside. This class of fullerene derivatives is known as Endohedral fullerenes. When the atom trapped inside is a metal, they are known as metallofullerenes. Even though C_{60} is the most common fullerene, few endohedral materials have been synthesized using C_{60} as cage molecule, the reason being small size of the molecule. Most of endohedral materials are made out of C_{82}, C_{84} or even higher fullerenes. The atoms that form stable endohedral compounds include lanthanum, yttrium, scandium, and some of the noble gases. As it is very difficult to open up carbon cage molecules to enclose a foreign atom inside, endohedral material must be synthesized while formation of the cage itself. (The accepted notation for endohedral material is to use the @ symbol to show that the first material is inside the second, *e.g.*, $La@C_{82}$ and $Sc_2 @C_{84}$.)

Exohedral Fullerenes

The most important and most versatile of all species of fullerenes is Exohedral fullerenes or fullerene derivatives, which are molecules formed by a chemical reaction between fullerenes and other chemical groups. Fullerene derivatives are also known as functionalized fullerenes. As fullerenes possess the conjugated p-system of electron, two main types of primary chemical transformations are possible on fullerene surface: addition reactions and redox reactions, which lead to covalent exohedral adducts and salts respectively. As fullerenes are insoluble in water, numerous derivatives of fullerenes have been synthesized with improved solubility profile.

Heterofullerenes

Heterofullerenes represent another fundamental group of modified fullerenes. They represent heteroanalogues of C_{60} and higher fullerenes; one or more carbon atoms of the cage are substituted by hetero-atoms, *e.g.*, trivalent nitrogen or boron atom. The simplest derivative of nitrogen fullerene is aza (60) fullerene C_{59} N. and its dimer $(C_{59}\ N)_2$.

Another significant spin-off product of fullerene research is carbon nanotubes based on carbon or other elements. These systems consist of graphitic sheets seamlessly wrapped into cylinders. The diameter of nanotubes range from 0.4 to > 3 nm for Single-Walled NanoTubes (SWNT) to 1.4 to at least 100 nm for Multi-Walled NanoTubes (MWNT). Also nanotubes are quite stiff and exceptionally strong, having high tensile strength. The calculated Young's Modulus for an individual nanotube is » 0.64TPa. This unique combination of diameter and strength is potentially useful in everything from tennis rackets to space travel.

BIOLOGICAL APPLICATIONS OF FULLERENES

Fullerenes are inert, hollow and indefinitely modifiable. When administered orally in the water-soluble form, they are not absorbed; while on *i.e.* injection, they get rapidly distributed to various body tissues. They are excreted unchanged by kidney. Acute toxicity of water-soluble fullerenes was found to be quite low. All these interesting properties offer possibilities of utilizing fullerenes in biology and medicinal chemistry and promise a bright future for fullerenes as medicinal agents. However, this possibility faces a significant problem, *i.e.*, natural repulsion of fullerenes to water. To overcome this limitation, a number of methodologies are being developed. These include synthesis of fullerene derivatives having modified solubility profile, encapsulation of C_{60} in cyclodextrins or in calixarenes or water suspension preparations. Many derivatives of fullerenes have been synthesized. There have been many patents written on fullerenes, and fullerene patent database is growing rapidly. Fullerenes and their derivatives are being intensively investigated as they exhibited promising preliminary activities in the following medical streams.

Diagnostic Applications

Endohedral metallofullerenes are the fullerenes with metal ion trapped inside fullerene cage. These have shown potential applications in diagnostics. As for example, water solublised forms like $M@C_{82}\ (OH)_{30}$ are being used as Magnetic Resonance Imaging contrast agents (M = Gd ^{3}sub+), X-ray contrast agents (M = Ho^{3}+)[37] and radiopharmaceuticals (M= ^{166}sub Ho ^{3}+ and ^{170}Tm ^{2}sub+). One of these derivatives, *i.e.*, 166sub Ho ^{3}sub + $@C_{82}\ (OH)_{30}$ has been extensively studied as radioactive tracer for imaging of diseased organs and for killing cancerous tumours. The radioactive metal is trapped inside the carbon shell, which is very stable and resistant to metabolism by body. The metallofullerenes have been found to be nontoxic, and they stay in the body for approximately one hour, allowing imaging of circulatory system.

Anti-HIV Activity

The enzyme protease specific to HIV-1 has been shown to be a viable target for antiviral therapy. The active site of this enzyme can be roughly described as an open-ended cylinder lined almost exclusively by hydrophobic amino acids except for two catalytic aspartic acids. Wudl *et al* hypothesized that since C_{60} molecule has approximately same radius as the cylinder that describes the active site of HIV-P, an opportunity exists for a strong hydrophobic interaction between C_{60} derivatives and the active site surfaces. Inhibition of HIV-P in presence of C_{60} was demonstrated through molecular modelling studies and experimental observations. Virus inactivation assays confirmed the activity of fullerene derivatives against HIV-1 and HIV-2. A water-soluble derivative was found to be active against HIV-1 and HIV-2 (EC_{50} approximately 6 mM) in acute and chronic infected human lymphocytes and also noncytotoxic up to 100 mM in peripheral mononuclear cells and H9, Vero and CEM cells.

As the binding constant found experimentally was not significant in terms of affinity, structural optimization of C_{60} derivatives for HIV-P interaction was investigated. On this basis, some authors proposed ideal inhibitors, in which two ammonium groups at distance of 5.5 Å or 5.1 Å are directly linked to C_{60}. Molecular modelling studies on these derivatives showed that they could fit well inside the HIV-P cavity and can interact with carboxylic residues of aspartatic acid.

DNA Photocleavage

A water-soluble fullerene carboxylic derivative was found to be cytotoxic when exposed to visible light. In an initial experiment, C_{60} carboxylic acid was incubated with cultured cells for three days. No biological activity was observed. However on exposure to low power visible light, the compound killed the cells. It was later shown that cytotoxicity of C_{60} derivatives was mediated by its ability to cleave DNA. Hence other potential application of fullerenes is related to the easy photoexcitation of fullerenes on the basis of which fullerene-based photodynamic compounds are being developed for the treatment of cancer.

In this field, many fullerene conjugates with different units possessing biological affinity to nucleic acids or proteins are being investigated for anticancer activity. In particular, conjugate between C_{60} and specific agents that interact with nucleic acids, such as acridine or complementary oligonucleotide, have been synthesized with the objective of increasing both cytotoxicity and sequence selectivity. However, DNA cleavage was found to occur at guanine residue without any significant sequence selectivity. A C_{60} - PEG conjugate irradiated by light strongly induced tumour necrosis without any damage to the overlying normal tissue, with complete cure achieved by a C_{60} -PEG dose of 0.424 mg/kg and irradiation power of 10^{11}sub J/m^2sub, making it an excellent candidate for photodynamic tumour therapy. Cytotoxicity of dendritic C_{60} monoadduct and malonic acid C_{60} trisadduct was recently investigated on Jurkat cells, and on exposure to UV light, the cell number was found to reduce to approximately 19% within two weeks.

Free Radical Scavenger

Free radicals have been linked to onset and progression of foetal neurogenerative diseases like amytrophic lateral sclerosis, commonly known as Lou Gehrig's disease. Fullerene compounds with their unique cage structure, combined with high number of conjugated double bonds in the core, interact with biomolecules and have avid reactivity with free radicals. Buckminster fullerenes, for example, are capable of adding multiple radicals per molecule; the addition of as many as 34 methyl radicals to a single C_{60} sphere has been reported. Hence they are regarded as "free radical sponge". However to improve native solubility profile of fullerenes, several fullerene derivatives have been prepared, and these are reported to possess protective effect in various systems. A water-soluble C_{60} derivative, fullerenol, has been shown to scavenge free radicals *in vitro* and *in vivo*. Fullerenol has been shown to block hydrogen peroxide-induced inhibition of population spikes in rat hippocampus. Carboxy-fullerene, another water-soluble derivative of fullerene, has been shown to prevent apoptotic injury evoked by *N*-methyl *D*-aspartate (NMDA) without interfering with NMDA-induced Ca^{2+} sub influx in cortical cell cultures and has been found to have effective neuroprotective antioxidant activity *in vitro* and *in vivo* (hundreds of times more powerful than Vitamin E). As a free radical scavenger, the protective effect of fullerene derivatives has been demonstrated in various systems, including reduced injury on ischemia reperfusion intestine, protecting cell types from undergoing apoptosis, reduced free radical level in organ perfusate, and having neuroprotective effect.

Additional factors that fullerene derivatives have shown promising results are in pretoxicity studies and the fact that water as wall as lipid soluble derivatives are possible greatly increases their usefulness as antioxidant in health and personal care products, *e.g.*, skin creams, burn creams and nutritional supplements.

Antimicrobial Activity

Fullerenes were anticipated to possess antibacterial activity based on the hypothesis that C_{60} could produce membrane disruption by insertion into phospholipids bilayers. The resulting membrane disorder, causing altered permeability, leads to the release of metabolites and cell death. As for example, monomethoxy triethylene glycol (mTEG) substituted fulleropyrollidines showed complete inhibition of *Mycobacterium avium* at dose 260 mg/ml and *M. tuberculosis* at dose @ 50 mg/ml. The plausible explanation for the antitubercular activity has been proposed that presence of carbon cage destabilizes the tubercular cell wall by intercalation in the hydrophobic part. The affinity of carbon cage for the hydrophobic constituents of *Mycobacterium* cell wall, *i.e.*, ceramide has been further proved by administration of intraperitoneal injection of mTEG fullerene derivative to mice. The derivative was found to be deposited on Tyson's gland, which is again a tissue rich in ceramide.

Antimicrobial activity of another fullerene derivative, *i.e.*, carboxy fullerenes has been intensively investigated. Carboxy fullerenes were found to inhibit *E. coli*-induced meningitis by reducing the damage caused by infiltrating neutrophils on the blood brain barrier but not by direct inhibition of *E. coli*. Moreover, it was found that the carboxy fullerenes inhibited the growth of *Streptococcus pyrogens* strain A-20 (MIC 5-50 mg/l). Since these derivatives exhibited varying effects on Gram+ve and Gram-ve bacteria, further investigations led to the observations that carboxy fullerenes can insert into cell wall of Gram+ve cocci, disrupt the cell wall structure and cause bacterial death.

The cell wall of Gram-ve organisms, having an outer membrane consisting of lipoproteins, lipopolysachharides and phospholipids, is not susceptible to access by fullerenes. This suggests that carboxy fullerenes could be considered as new antimicrobial agents against Gram+ve cocci.

Osteoporosis

To date, biphosphonate compounds and fluoride anions are the drugs used in the treatment of osteoporosis and other bone disorders. However, the former are not absorbed orally and latter are fairly toxic. Advantage is being taken of the preferential localization of fullerene derivatives in bones. Hence polyfluoro biphosphonated fullerene derivatives are being developed as bimodal drug for osteoporosis therapy.

However, the use of fullerenes as biological or pharmacological agents requires the measurement of dosage and serum levels by sensitive and simple immunological procedures. This in turn requires that specific antibodies to fullerenes should be available. The requirement led to the question of the ability of these carbon-made spherical structures to induce the production of specific antibodies.

However, it was determined that immunization of mice with a water-soluble C_{60} fullerene derivative conjugated to bovine thyroglobulin yielded a population of fullerene-specific antibodies of the IgG isotype, showing that the immune system was diverse enough to recognize and process fullerenes as protein conjugates. A monoclonal fullerene-specific antibody has been isolated and the sequences of its light and heavy chains have been reported. These findings further substantiate the practical utility of fullerenes as therapeutic agents, diagnostic agents and as imaging vectors.

CHEMICAL PROPERTIES OF FULLERENES

Fullerene or C_{60} is soccer-ball-shaped or I_h with 12 pentagons and 20 hexagons. According to Euler's theorem these 12 pentagons are required for closure of the carbon network consisting of *n* hexagons and C_{60} is the first stable fullerene because it is the smallest possible to obey this rule. In this structure none of the pentagons make contact with each other. Both C_{60} and its relative C_{70} obey this so-called isolated pentagon rule (IPR). The next homologue C_{84} has 24 IPR isomers of which several are isolated and another 51,568 non-IPR isomers. Non-IPR fullerenes have thus far only been isolated as endohedral fullerenes such as $Tb_3N@C_{84}$ with two fused pentagons at the apex of an egg-shaped cage or as fullerenes with exohedral stabilization such as $C_{50}Cl_{10}$ and reportedly $C_{60}H_8$.

Because of the molecule's spherical shape the carbon atoms are highly pyramidalized, which has far-reaching consequences for reactivity. It is estimated that strain energy constitutes 80% of the heat of formation. The conjugated carbon atoms respond to deviation from planarity by orbital rehybridization of the sp^2 orbitals and pi orbitals to a $sp^{2.27}$ orbital with a gain in p-character. The p lobes extend further outside the surface than they do into the interior of the sphere and this is one of the reasons a fullerene is electronegative. The other reason is that the empty low-lying pi^* orbitals also have high s character.

The double bonds in fullerene are not all the same. Two groups can be identified: 30 so-called double bonds connect two hexagons and 60 bonds connect a hexagon and a pentagon. Of the two the bonds are shorter with more double-bond character and therefore a hexagon is often represented as a cyclohexatriene and a pentagon as a pentalene or radialene. In other words, although the carbon atoms in fullerene are all conjugated the superstructure is not a super aromatic compound. The X-ray diffraction bond length values are 135.5 pm for the bond and 146.7 pm for the bond.

C_{60} fullerene has 60 pi electrons but a closed shell configuration requires 72 electrons. The fullerene is able to acquire the missing electrons by reaction with potassium to form first the K_6C_{6-60} salt and then the $K_{12}C_{12-60}$ In this compound the bond length alternation observed in the parent molecule has vanished.

FULLERENE REACTIONS

Fullerenes tend to react as electrophiles. An additional driving force is relief of strain when double bonds become saturated. Key in this type of reaction is the level of functionalization *i.e.* monoaddition or multiple additions and in case of multiple additions their topological relationships (new substituents huddled together or evenly spaced). In conformity with IUPAC rules, the terms methanofullerene are used to indicate the ring-closed (cyclopropane) fullerene derivatives, and fulleroid to ring-open (methanoannulene) structures.

Nucleophilic Addition

Fullerenes react as electrophiles with a host of nucleophiles in nucleophilic additions. The intermediary formed carbanion is captured by another electrophile. Examples of nucleophiles are Grignard reagents and organolithium reagents. For example the reaction of C_{60} with methylmagnesium chloride stops quantitatively at the penta-adduct with the methyl groups centred around a cyclopentadienyl anion which is subsequently protonated. Another nucleophilic reaction is the Bingel reaction. Fullerene reacts with chlorobenzene and aluminium chloride in a Friedel-Crafts alkylation type reaction. In this hydroarylation the reaction product is the 1,2-addition adduct (Ar-CC-H).

Pericyclic Reactions

The bonds of fullerenes react as dienes or dienophiles in cycloadditions for instance Diels-Alder reactions. 4-membered rings can be obtained by [2+2]cycloadditions for instance with benzyne. An example of a 1,3-dipolar cycloaddition to a 5-membered ring is the Prato reaction. Fullerenes react with carbenes to methanofullerenes.

Hydrogenation

Fullerenes are easily hydrogenated by several methods. Examples of hydrofullerenes are $C_{60}H_{18}$ and $C_{60}H_{36}$. However, completely hydrogenated $C_{60}H_{60}$ is only hypothetical because of large strain. Highly hydrogenated fullerenes are not stable, prolonged hydrogenation of fullerenes by direct reaction with hydrogen gas at high temperature conditions results in collapse of cage structure with formation of polycyclic aromatic hydrocarbons.

Oxidation

Although more difficult than reduction, oxidation of fullerene is possible for instance with oxygen and osmium tetraoxide.

Hydroxylation

Fullerenes can be hydroxylated to fullerenols or fullerols. Water solubility depends on the total number of hydroxyl groups that can be attached. One method is fullerene reaction in diluted sulfuric acid and potassium nitrate to $C_{60}(OH)_{15}$. Another method is reaction in diluted sodium hydroxide catalysed by TBAH adding 24 to 26 hydroxyl groups. Hydroxylation has also been reported using solvent-free NaOH/hydrogen peroxide. $C_{60}(OH)_8$ was prepared using a multistep procedure starting from a mixed peroxide fullerene. The maximum number of hydroxyl groups that can be attached (hydrogen peroxide method) stands at 36-40.

Electrophilic Addition

Fullerenes react in electrophilic additions as well. The reaction with bromine can add up to 24 bromine atoms to the sphere. The record holder for fluorine addition is $C_{60}F_{48}$. According to in silico predictions the as yet elusive $C_{60}F_{60}$ may have some of the fluorine atoms in endo positions (pointing inwards) and may resemble a tube more than it does a sphere.

Retro Additions

Protocols have been investigated for removing substituents via retroadditions after they have served their purpose. Examples are the retro-Bingel reaction and the retro-Prato reaction.

Fullerenes as Ligands

Fullerene is a ligand in organometallic chemistry. The double bond is electron-deficient and usually forms metallic bonds with $\eta = 2$ hapticity. Bonding modes such as $\eta = 5$ or $\eta = 6$ can be induced by modification of the coordination sphere.

- C_{60} fullerene reacts with tungsten hexacarbonyl $W(CO)_6$ to the $(\eta^2\text{-}C_{60})W(CO)_5$ complex in a hexane solution in direct sunlight.

MULTISTEP FULLERENE SYNTHESIS

Although the procedure for the synthesis of the C_{60} fullerene is well established (generation of a large current between two nearby graphite electrodes in an inert atmosphere) a 2002 study described an organic synthesis of the compound starting from simple organic compounds.

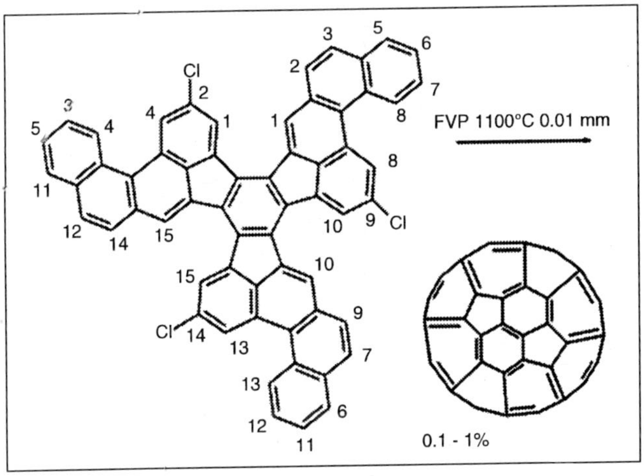

FVP 1100°C 0.01 mm

0.1 - 1%

In the final step a large polycyclic aromatic hydrocarbon consisting of 13 hexagons and three pentagons is submitted to flash vacuum pyrolysis at 1100°C and 0.01 Torr. The three carbon chlorine bonds serve as free radical incubators and the ball is stitched up in a no-doubt complex series of radical reactions. The chemical yield is low: 0.1 to 1%. A small percentage of fullerenes is formed in any process which involves burning of hydrocarbons, *e.g.* in candle burning. The yield through a combustion method is often above 1%. The method proposed above does not provide any advantage for synthesis of fullerenes compared to the usual combustion method, and therefore, the organic synthesis of fullerenes remains a challenge for chemistry.

A similar exercise aimed at construction of a C78 cage in 2008 (but leaving out the precursor's halogens) did not result in a sufficient yield but at least the introduction of Stone Wales defects could be ruled out.

Multistep Nanoribbon Synthesis

In the field of graphene nanoribbons a bottom-up approach has also been investigated

Open-cage Fullerenes

A part of fullerene research is devoted to so-called open-cage fullerenes whereby one or more bonds are removed chemically exposing an orifice. In this way it is possible to insert into it small molecules such as hydrogen, helium or lithium. The first such open-cage fullerene was reported in 1995. In endohedral hydrogen fullerenes the opening, hydrogen insertion and closing back up has already been demonstrated.

Heterofullerenes

In heterofullerenes at least one carbon atom is replaced by another element. Based on spectroscopy, substitutions have been reported with boron (borafullerenes), nitrogen (azafullerenes), oxygen, arsenic, germanium, phosphorus, silicon, iron, copper, nickel, rhodium and iridium. Reports on isolated heterofullerenes are limited to those based on nitrogen and oxygen.

Fullerene Dimers

The C_{60} fullerene dimerizes in a formal [2+2] cycloaddition to a C_{120} bucky dumbbell in the solid state by mechanochemistry (high-speed vibration milling) with potassium cyanide as a catalyst. The trimer has also been reported using 4-aminopyridine as catalyst (4% yield) and observed with Scanning Tunnelling Microscopy as a monolayer

Nanotube Chemistry

Carbon nanotubes, also part of the fullerene family, can be described as graphene sheets rolled into a cylindrical tube. Unlike the spherical fullerenes made up of hexagons and pentagons, nanotubes only have hexagons present but in terms of reactivity both systems have much in common. Due to electrostatic forces nanotubes have a nasty tendency to cluster together into bundles and many potential applications require an exfoliation process. One way to do this is by chemical surface modification. A useful tool for the analysis of derivatised nanotubes is Raman spectroscopy which shows a G-band (G for graphite) for the native nanotubes at 1580 cm^{-1} and a D-band (D for defect) at 1280 cm^{-1} when the graphite lattice is disrupted with conversion of sp^2 to sp^3 hybridized carbon. The ratio of both peaks I_D/I_G is taken as a measure of functionalization. Other tools are UV spectroscopy where pristine nanotubes show distinct Van Hove singularities where functionalized tubes do not, and simple TGA analysis.

In one type of chemical modification, aniline is oxidized to a diazonium intermediate. After expulsion of nitrogen, it forms a covalent bond as an aryl radical:

H20, 80°C, 12 Hrs.

Also known are protocols for cycloadditions such as Diels-Alder reactions, 1,3-dipolar cycloadditions of azomethine ylides and azide–alkyne cycloaddition reactions. One example is a DA reaction assisted by chromium hexacarbonyl and high pressure. The I_D/I_G ratio for reaction with Danishefsky's diene is 2.6. Nanotubes can also be alkylated with alkyl halides using lithium or sodium metal and liquid ammonia (Birch reduction conditions). The initial nanotube salt can function as a polymerization initiator and can react with peroxides to form alkoxy functionalized nanotubes

Fullerene Purification

Fullerene purification is the process of obtaining a fullerene compound free of contamination. In fullerene production mixtures of C_{60}, C_{70} and higher homologues are always formed. Fullerene purification is key to fullerene science and determines fullerene prices and the success of practical applications of fullerenes. The first available purification method for C_{60} fullerene was by HPLC from which small amounts could be generated at large expense.A practical laboratory-scale method for purification of soot enriched in C_{60} and C_{70} starts with extraction in toluene followed by filtration with a paper filter. The solvent is evaporated and the residue (the toluene-soluble soot fraction) redissolved in toluene and subjected to column chromatography. C_{60} elutes first with a purple colour and C_{70} is next displaying a reddish-brown colour.

In nanotube processing the established purification method for removing amorphous carbon and metals is by competitive oxidation (often a sulfuric acid/nitric acid mixture). It is assumed that this oxidation creates oxygen containing groups (hydroxyl, carbonyl, carboxyl) on the nanotube surface which electrostatically stabilize them in water and which can later be utilized in chemical functionalization. One report reveals that the oxygen containing groups in actuality combine with carbon contaminations absorbed to the nanotube wall that can be removed by a simple base wash. Cleaned nanotubes are reported to have reduced D/G ratio indicative of less functionalization, and the absence of oxygen is also apparent from IR spectroscopy and X-ray photoelectron spectroscopy.

Experimental Purification Strategies

A recent kilogram-scale fullerene purification strategy was demonstrated by Nagata *et al.* In this method C_{60} was separated from a mixture of C_{60}, C_{70} and higher fullerene compounds by first adding the amidine compound DBU to a solution of the mixture in 1,2,3-trimethylbenzene. DBU as it turns out only reacts to C_{70} fullerenes and higher which reaction products separate out and can be removed by filtration. C_{60} fullerenes do not have any affinity for DBU and are subsequently isolated. Other diamine compounds like DABCO do not share this selectivity.

C_{60} but not C_{70} forms a 1:2 inclusion compound with cyclodextrin (CD). A separation method for both fullerenes based on this principle is made possible by anchoring cyclodextrin to colloidal gold particles through a sulfur-sulfur bridge. The Au/CD compound is very stable and soluble in water and selectively extracts C_{60} from the insoluble mixture after refluxing for several days. The C_{70} fullerene component is then removed by simple filtration. C_{60} is driven out from the Au/CD compound by adding adamantol which has a higher affinity for the cyclodextrin cavity. Au/CD is completely recycled when adamantol in turn is driven out by adding ethanol and ethanol removed by evaporation. 50 mg of Au/CD captures 5 mg of C_{60} fullerene.

FULLERENE AROMACITY

Researchers have been able to increase the reactivity of fullerenes by attaching active groups to their surfaces. Buckminsterfullerene does not exhibit "superaromaticity": that is, the electrons in the hexagonal rings do not delocalize over the whole molecule.

A spherical fullerene of n carbon atoms has n pi-bonding electrons, free to delocalize. These should try to delocalize over the whole molecule. The quantum mechanics of such an arrangement should be like one shell only of the well-known quantum mechanical structure of a single atom, with a stable filled shell for n = 2, 8, 18, 32, 50, 72, 98, 128, etc.; *i.e.* twice a perfect square number; but this series does not include 60. This 2(N + 1)2 rule (with N integer) for spherical aromaticity is the three-dimensional analogue of Hückel's rule. The 10+ cation would satisfy this rule, and should be aromatic. This has been shown to be the case using quantum chemical modelling, which showed the existence of strong diamagnetic sphere currents in the cation. As a result, C_{60} in water tends to pick up two more electrons and become an anion. The result of C_{60} trying to form a loose metallic bond.

FULLERENE BORON

A new type of buckyball using boron atoms instead of the usual carbon has been predicted and described in 2007. The B80 structure, with each atom forming 5 or 6 bonds, is predicted to be more stable than the C_{60} buckyball. One reason for this given by the researchers is that the B-80 is actually more like the original geodesic dome structure popularized by Buckminster Fuller which uses triangles rather than hexagons. However, this work has been subject to much criticism by quantum chemists as it was concluded that the predicted Ih symmetric structure was vibrationally unstable and the resulting cage undergoes a spontaneous symmetry break yielding a puckered cage with rare Th symmetry (symmetry of a volleyball). The number of six atom rings in this molecule is 20 and number of five member rings is 12. There is an additional atom in the centre of each six member ring, bonded to each atom surrounding it.

FULLERENE BUCKMINSTER

Buckminsterfullerene is the smallest fullerene molecule in which no two pentagons share an edge (which can be destabilizing, as in pentalene). It is also the most common in terms of natural occurrence, as it can often be found in soot. The structure of C_{60} is a truncated (T = 3) icosahedron, which resembles a soccer ball of the type made of twenty hexagons and twelve pentagons, with a carbon atom at the vertices of each polygon and a bond along each polygon edge. The van der Waals diameter of a C_{60} molecule is about 1.1 nanometres (nm). The nucleus to nucleus diameter of a C_{60} molecule is about 0.71 nm.

The C_{60} molecule has two bond lengths. The 6:6 ring bonds (between two hexagons) can be considered "double bonds" and are shorter than the 6:5 bonds (between a hexagon and a pentagon). Its average bond length is 1.4 angstroms. Silicon buckyballs have been created around metal ions.

FULLERENE NAMING

The icosahedral $C_{60}H_{60}$ cage was mentioned in 1965 as a possible topological structure. The existence of C_{60} was predicted by Eiji Osawa of Toyohashi University of Technology in 1970. He noticed that the structure of a corannulene molecule was a subset of a soccer-ball shape, and he hypothesised that a full ball shape could also exist. His idea was reported in Japanese scientific journals, but did not reach Europe or the Americas.

Also in 1970, R. W.Henson (then of the Atomic Energy Research Establishment) proposed the structure and made a model of C_{60}.

Unfortunately, the evidence for this new form of carbon was very weak and was not accepted, even by his colleagues. Using mass spectrometry, discrete peaks appeared corresponding to molecules with the exact mass of sixty or seventy or more carbon atoms. In 1985, Harold Kroto (then of the University of Sussex), James R. Heath, Sean O'Brien, Robert Curl and Richard Smalley, from Rice University, discovered C_{60}, and shortly thereafter came to discover the fullerenes. Kroto, Curl, and Smalley were awarded the 1996 Nobel Prize in Chemistry for their roles in the discovery of this class of molecules. C_{60} and other fullerenes were later noticed occurring outside the laboratory (*e.g.*, in normal candle soot). By 1991, it was relatively easy to produce gram-sized samples of fullerene powder using the techniques of Donald Huffman and Wolfgang Krätschmer. Fullerene purification remains a challenge to chemists and to a large extent determines fullerene prices. So-called endohedral fullerenes have ions or small molecules incorporated inside the cage atoms. Fullerene is an unusual reactant in many organic reactions such as the Bingel reaction discovered in 1993. Carbon nanotubes were recognized in 1991. Minute quantities of the fullerenes, in the form of C_{60}, C_{70}, C_{76}, and C_{84} molecules, are produced in nature, hidden in soot and formed by lightning discharges in the atmosphere. In 1992, fullerenes were found in a family of minerals known as Shungites in Karelia, Russia.

In 2010, fullerenes (C_{60}) have been discovered in a cloud of cosmic dust surrounding a distant star 6500 light years away. Using NASA's Spitzer infrared telescope the scientists spotted the molecules' unmistakable infrared signature. Sir Harry Kroto, who shared the 1996 Nobel Prize in Chemistry for the discovery of buckyballs commented: "This most exciting breakthrough provides convincing evidence that the buckyball has, as I long suspected, existed since time immemorial in the dark recesses of our galaxy."

Naming Buckminsterfullerene (C_{60}) was named after Richard Buckminster Fuller, a noted architectural modeler who popularized the geodesic dome. Since buckminsterfullerenes have a shape similar to that sort of dome, the name was thought appropriate. As the discovery of the fullerene family came after buckminsterfullerene, the shortened name 'fullerene' is used to refer to the family of fullerenes. The suffix "ene" indicates that each C atom is covalently bonded to three others (instead of the maximum of four), a situation that classically would correspond to the existence of bonds involving two pairs of electrons ("double bonds").

FUNCTIONALIZATION OF GOLD NANOPARTICLES

Nanotechnology is expected to revolutionize medicine. A variety of medical processes occur at nanometre length scales. Among the approaches for exploiting developments in nanotechnology in medicine, nanoparticles offer some unique advantages as sensing, delivery, and image-enhancement agents. Several varieties of nanoparticles are available : polymeric nanoparticles, dendrimers, metal nanoparticles, quantum dots, liposomes, micelles, and other types of nano-assemblies. All of these nanostructures can play a major role in medicine, and especially in disease diagnosis and therapy.

Rapid advances in molecular biology and genetic engineering provide an unprecedented opportunity for delivery of drugs and genes to intracellular targets. In cancer, for instance, the effectiveness of many anticancer drugs and genes is limited due to the inability to reach the target site in sufficient concentrations and exert the pharmacological effect. Current intracellular delivery systems are classified as being either viral or non-viral in origin. Viruses are efficient in delivery; however, they suffer from poor safety profiles. Non-viral delivery systems, albeit not as efficient as viruses, have the promise of safety and reproducibility in manufacturing. To enhance delivery of drugs and genes to intracellular targets using non-viral delivery systems, it is necessary to have a detailed understanding of the transport process and identify ways of overcoming the cellular barriers.

Nanoparticles based on gold chemistry have attracted significant research and practical attention recently. They are versatile agents with a variety of biomedical applications including use in highly sensitive diagnostic assays, thermal ablation, and radiotherapy enhancement, as well as for drug and gene delivery. For instance, antibody-modified gold nanoparticles, when used for detection of prostate-specific antigen, had an almost million-fold higher sensitivity than conventional ELISA-based assays. Near-infrared-radiation-absorbing gold – silica nanoshells have been prepared and evaluated for thermal ablation of tumours after systemic administration.

For biomedical applications, surface functionalization of gold nanoparticles is essential to apply them to specific disease areas and allow them to selectively interact with cells or biomolecules. Surface conjugation of antibodies and other targeting moieties is usually achieved by adsorption of the ligand to the gold surface. Surface adsorption, however, can denature the proteins or, in some cases, limit the interactions of the ligand with the target on the cell surface due to steric hinderance. Additionally, for systemic applications, long-circulating nanoparticles are desired for passive targeting to tumours and inflammatory sites. Poly(ethylene glycol) (PEG) modification of nanoparticles affords a long-circulating property by evading macrophage-mediated uptake and removal from the systemic circulation. Surface modification of gold nanoparticles through the PEG spacer would, therefore, allow the modified nanoparticles to remain in the systemic circulation for the prolonged period and provide flexibility to the attached ligand for efficient interaction with its target.

The synthesis and characterization of hetero-bifunctional PEG with a thiol group on one terminus and a reactive functional group on the other for conjugation to a biologically relevant targeting moiety. Using thiol-PEG-coumarin, a model fluorescent dye, we describe the "proof-of-concept" for surface functionalization of gold nanoparticles, cellular cytotoxicity evaluations, and fluorescence confocal analysis of cell uptake and nanoparticle distribution.

EXPERIMENTAL PROCEDURES

Synthesis and Characterization of Coumarin-PEG-thiol

A versatile hetero-bifunctional PEG compound containing end-groups of thiol and alcohol derivatized as the coumarin carbamate was efficiently. The key step for the PEG synthesis involves the desymmetrization of the PEG-diol compound. This was achieved by monotosylation of the PEG-diol using the method of Bouzide and Sauve.

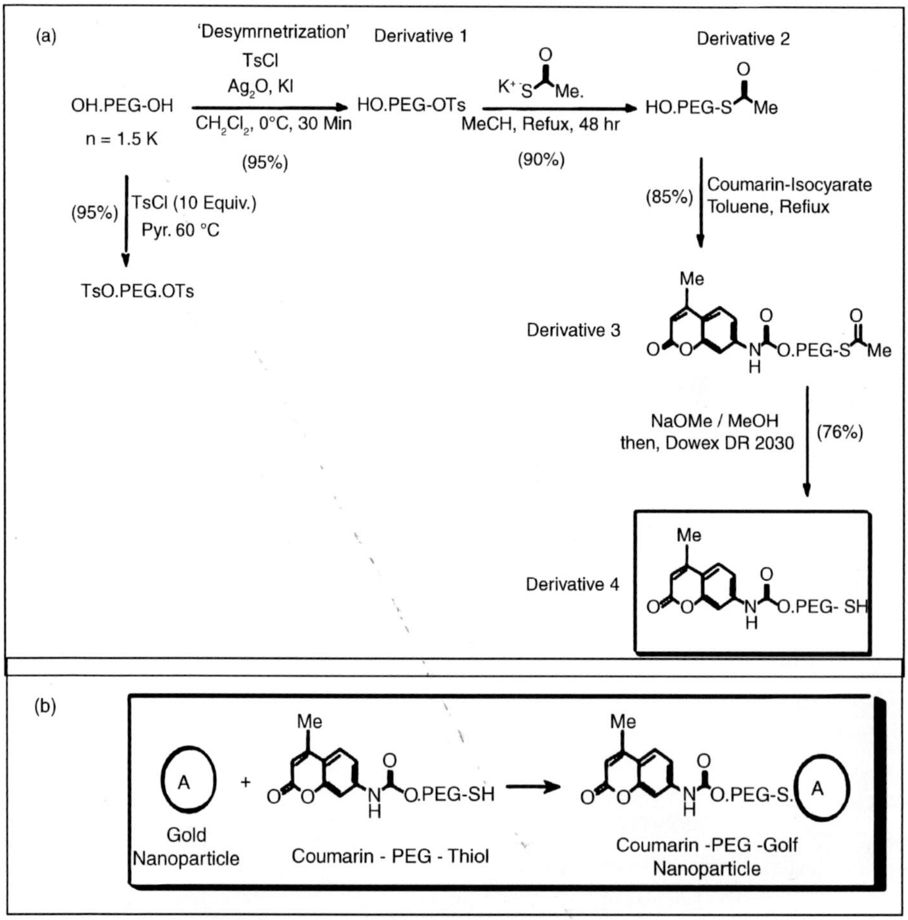

Figure : Schematic Illustration for the Chemical Synthesis of Hetero-bifunctional Poly(Ethylene Glycol) having a Thiol Group on one Terminus and Coumarin Carbamate on the other for Attachment to Gold Nanoparticles.

Synthesis of Derivative

To a chilled (0 °C) solution of the PEG (M_w = 1427 g/mol, 12.0 g; 8.40 mmol) in methylene chloride (distilled, 250 mL) was added freshly prepared Ag_2O (1.50 equiv, 2.92 g; 12.6 mmol) and recrystallized potassium iodide (0.40 equiv, 558 mg; 3.36 mmol). To this rapidly stirred solution in one portion, purified tosyl chloride (TsCl) (1.05 equiv, 1.68 g; 8.82 mmol) was added. The reaction was deemed to be complete in 1–2 hours by TLC (alumina). This was done by UV monitoring of the consumption of the TsCl reagent and the appearance of the tosylate product as a weakly UV active compound. The mixture was then carefully filtered over a pad of celite, and the clear filtrate obtained was then evaporated under reduced pressure to give a colourless oily product. Trituration with portions of ether to remove any trace TsCl reagent, followed by recrystallization (acetone/dry-ice bath) using methylene chloride– ether mixture, provided the monotosyl-PEG derivative as a white amorphous solid. Yield: 12.5 g (95%); ^{1}H NMR ($CDCl_3$, 300 MHz) δ7.80 (d, *J* = 8.1 Hz, Ar), 7.35 (d, *J* = 8.1 Hz, Ar), 4.16 (t, CH_2OTs), 3.86–3.39 (m, 29 × CH_2), 3.43 (t, 2H, CH_2), 2.83 (t, 2H, CH_2), 2.63 (br, s, OH), 2.45 (s, M*e tosyl*); ^{13}C NMR ($CDCl_3$, MHz) δ143.95, 132.41, 129.10, 127.12, 125.4, 71.85, 69.74, 69.49, 68.60, 67.83, 60.66, 20.83 ($MeSO_2$); IR (υ, cm^{-1}) 3620–3192 (br, m), 2879 (br, vs), 1646, 1598, 1468, 1358 (asymmetric SO_2 stretch), 1279, 1234, 1145 (symmetric SO_2 stretch), 1112 (br, vs), 945, 927, 840; ESIMS calculated for $MH^+ + H_2O$ (0.81) *m*/*z* 1598.95, found 1598.91.

A solution of the monotosyl-PEG (12.0 g; 7.59 mmol), 10 equiv of potassium thioacetate (8.67 g; 75.9 mmol) in dry MeOH (300 mL) was heated to reflux for 48 hours under an atmosphere of nitrogen. The solution was allowed to cool to room temperature, and the MeOH removed using a Buchi rotary evaporator. The solid reaction mixture was treated with methylene chloride and the solution filtered. Concentration under reduced pressure gave the crude product, which was purified by chromatography using alumina and by eluting with 1% acetic acid/1% MeOH in methylene chloride. This purification yielded the product as a pale-yellow oil; and crystallization using ether (dry ice/acetone bath) provided the PEG thioacetate derivative as a pale-yellow solid. Yield: 10.1 g (90%); ^{1}H NMR ($CDCl_3$, 300 MHz) δ4.18 (t, 2H, CH_2), 3.94–3.62 (m, 58H, 29 × CH_2), 3.41 (t, 2H, CH_2), 2.83 (m, 2H, CH_2), 2.81 (br s, 1H, OH), 2.04 (s, 3H, SAc); ^{13}C NMR ($CDCl_3$, MHz) δ227.78 (SC[O]), 72.12, 70.05, 69.83, 69.21, 61.04, 28.33 (SC[O]Me); IR (υ, cm^{-1}) 3603–3277 (br, m), 2888 (br, vs), 1697, 1466, 1359, 1347, 1277, 1242, 1112 (br, vs), 1061 (s, SO stretch), 948, 842; MALDI calculated for $M + Na^+ + H_2O$ (1.67) *m*/*z* 1538.90, found 1538.88.

To a solution of the monothioacetate-PEG (2.50 g; 1.68 mmol) in toluene (70 mL) was added freshly prepared coumarin isocyanate (1.05 equiv, 355 mg; 76 mmol). The solution was heated to reflux for 12 hours under an atmosphere of nitrogen. Then, the mixture was allowed to cool and the toluene was removed using a Buchi rotary evaporator. The crude product mixture was then treated with methylene chloride and the solution filtered over a pad of celite. Evaporation of the filtrate followed and then purification using alumina, by eluting with 1%–2% MeOH in methylene chloride), gave the product as a pale-yellow oil. Finally, crystallization using ether (dry ice/acetone bath) provided the coumarin-PEG-thioacetate derivative as a pale-yellow solid. Yield: 2.41 g (85%); ^{1}H NMR ($CDCl_3$, 300 MHz) δ7.87 (br s, 1H, NH), 7.53–7.41 (m, 3H, Ar), 6.17 (s, 1H, vinyl), 4.20 (t, 2H, CH_2), 3.87–3.40 (m, 60H, 30 × CH_2), 2.83 (m, 2H, CH_2), 2.40 (s, 3H, vinyl Me), 2.08 (s, 3H, SAc); ^{13}C NMR ($CDCl_3$, MHz) δ227.48,0020160.15, 153.59, 152.55, 151.64, 141.82, 124.33, 114.07, 113.92, 111.81, 104.97, 71.83, 70.03, 69.71, 69.57, 68.50, 63.42, 60.71, 29.14 (SC[O] Me), 17.76 (Me); IR (υ, cm^{-1}) 3558.04 (w), 2870 (s), 1732 (C = O, carbamate), 1694, 1620, 1575, 1533, 1463, 1350, 1291, 1231, 1108, 943; MALDI calcd, for M + Na^+ + H_2O (2.45) *m/z* 1754.13, found 1754.15.

A solution of the bifunctional coumarin-PEG-thioacetate compound (0.50 g; 0.30 mmol) in degassed methanol (50 mL) was treated with 5 equivalents of degassed 0.5 M NaOMe/MeOH. The mixture was allowed to stir overnight at room temperature. Then, the mixture was acidified using DOWEX-Dr. 2030 resin to pH 1–2. The solution was quickly filtered over an overhead stream of nitrogen, and the solvent was removed using a rotary evaporator to give the crude product. Purification by silica gel chromatography and by eluting with 2%–6% MeOH in methylene chloride gave the coumarin-PEG-thiol bifunctional derivative, which was obtained as a light-yellow solid upon further drying under diminished pressure. Yield: 375 mg (76%); ^{1}H NMR ($CDCl_3$, 300 MHz) δ8.0 (br s, 1H, NH), 7.52–7.42 (m, 3H, Ar), 6.17 (s, 1H, vinyl), 4.35 (t, 2H, CH_2), 3.94–3.63 (m, 58H, 29 × CH_2), 3.50 (t, 2H, CH_2), 2.83 (m, 2H, CH_2), 2.41 (s, 3H, vinyl Me), 1.25 (br s, 1H, SH); ^{13}C NMR ($CDCl_3$, MHz) δ227.48, 160.15, 153.62, 152.58, 151.64, 141.85, 124.34, 114.09, 113.96, 111.82, 104.99, 71.82, 70.05, 69.76, 69.55, 68.51, 63.43, 60.69, 29.53, 17.74; IR (υ, cm^{-1}) 3421 (br, vs), 2918 (SH, m), 2882, 1651, 1463, 1351, 1297, 1251, 1100, 949; MALDI calculated for M + Na^+ + H_2O (3.99) *m/z* 1739.84, found 1739.91.

Functionalized Gold Nanoparticles

Gold nanoparticles were synthesized by reduction of gold chloride ($HAuCl_4$) with freshly prepared sodium citrate and allowed to boil under reflux conditions. The pale-yellow solution turned to deep red as the gold nanoparticles were formed and stabilized by adsorbed citrate ions. Particle size analysis was performed with a Coulter® N-4 sub-micron particle size analyser. The nanoparticle suspension was air-dried on the specimen grid and observed with a transmission electron microscope (TEM).

Coumarin-PEG-thiol was added to the aqueous dispersion of gold nanoparticles and the reaction for conjugation of the fluorescence dye to the gold surface through the thiol functionality was carried out at room temperature. The coumarin dye was used primarily as a demonstration of the covalent attachment, and confirms two important issues. First, because of the covalent bonding, the fluorescence observed is not from detached dye molecules, but from the fluorescent entity covalently attached to the gold nanoparticles. Second, the fluorescence is not quenched by the presence of the gold nanoparticles, probably due to the presence of the PEG spacer.

Cytotoxicity Studies

The in vitro cytotoxicity of the coumarin-PEG-thiol functionalized gold nanoparticles was measured using a commercially available CellTiter 96® AQueous Non-Radioactive Cell Proliferation Assay kit. According to the manufacturer's instructions, the tetrazolium compound (3-[4, 5-dimethylthiazol-2-yl]-5-[3-carboxy-methoxyphenyl]-2-[4-sulfophenyl]-2H-tetrazolium, [MTS]) is bioreduced by viable cells into a formazan product. Approximately 10 000 MDA-MB-231 human breast carcinoma xenograft cells, suspended in Dulbecco's minimum essential medium (DMEM) modified with 10% (v/v) fetal bovine serum and other essential nutrients, were seeded per well into 96-well plates and allowed to adhere for 24 hours at 37 °C and 5% CO_2 atmosphere. The medium was then replaced with graded concentrations of poly-ethyleneimine (MW ~ 10000) as a positive control; unfunctionalized gold nanoparticles and nanoparticles functionalized with methoxy-PEG-thiol (OMe-PEG-SH) as negative controls; and gold nanoparticles functionalized with coumarin-PEG-thiol as the test preparation in serum-free medium (SFM). Incubation then continued for another 24 hours. Additional control wells received only SFM. After the incubation time, the cells were washed once with sterile phosphate-buffered saline (PBS, pH 7.4), followed by the addition of 180 μl of DMEM per well. The plates were incubated for 4 hours after the addition of 20 μl of MTS solution to each well and the absorbance of the formed formazan product was read at 490 nm using a microplate reader. The absorbance values were converted to the percentage viability against control and reported as mean and standard deviation.

Fusion Microscope System

The Keck 3-D Fusion microscope system was developed at Northeastern University based on a commercially available Nikon TE2000U inverted microscope. It has 5 different modes of operation on one common platform: quadrature microscopy (QM); differential interference contrast (DIC); confocal fluorescence microscopy (CFM); 2-photon confocal laser scanning microscopy with second-harmonic generation (2PM); and confocal reflectance microscopy (CRM).

A known number of cells were grown on circular glass cover slips in a 6-well plate containing DMEM. Once they had reached suitable confluency, the medium was replaced with SFM and the controls and test preparations were added. The final concentration of coumarin-PEG-thiol functionalized gold nanoparticles was 100 μg/mL per well and the samples were incubated for 1 hour. The cells were washed with sterile PBS twice and fixed with a para-formaldehyde solution (0.5% [w/v] in PBS). The cover slips were mounted (with cell side down) onto clean glass slides using Fluoromount G mounting medium. The fluorescent images were registered using an in-house built Keck 3-D Fusion microscope system using a 100 × oil immersion objective.

We used the 2-photon laser scanning microscopy to measure the fluorescent signatures of coumarin-PEG-gold nanoparticles in the cells. The sample slide was placed on the stage with sample side down. The immersion oil was required for the objective. The tunable titanium – sapphire laser was set to 780 nm for illumination. Detection was completed with a sensitive head-on photo multiplier tube (PMT) (HC1024-02, Hamamatsu).

SYNTHESIS PATHWAY

We have synthesized hetero-bifunctional PEG having a thiol group on one terminus and an alcohol carbamate on the other for covalent attachment of coumarin. The PEG derivative was synthesized in high efficiency and characterized during each step of the synthesis pathway. Surface functionalization through PEG spacers would allow for flexibility in the attached molecule for more efficient interactions with the biological target. Otsuka et al (2001) reported on the synthesis of hetero-bifunctional PEG synthesis having acetal-PEG-SH functionalities for attachment of lactose and other sugar molecules. They have also evaluated the interactions of the surface-attached sugar residues on gold nanoparticles with specific lectins for the development of biosensors.

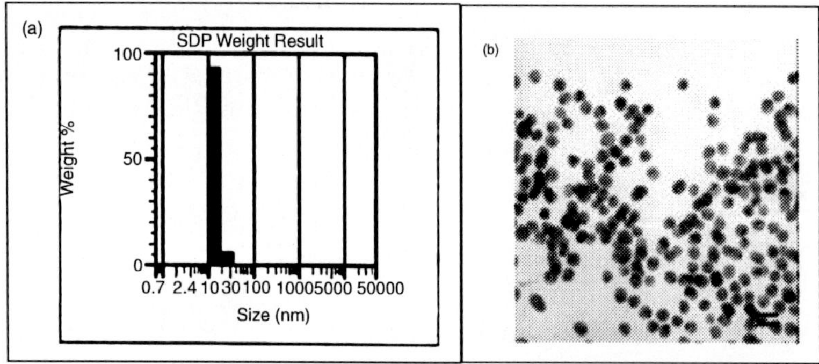

Figure : Coulter Particle-size Analysis (*a*) and Transmission Electron Microscopy Image of Gold Nanoparticles (*b*). Scale bar in (*b*) = 20 nm.

The Coulter particle-size analysis, we can reproducibly prepare 10–15 nm gold nanoparticles with a narrow size distribution. A TEM image shows that the nanoparticles have a spherical shape. The surface of gold nanoparticles was functionalized with coumarin-PEG-thiol by a covalent conjugation reaction. The unbound coumarin-PEG-thiol was removed by multiple centrifugation and washing steps, and the emission from bound dye was measured with a fluorescence spectrophotometer. Although gold nanoparticles are known to be very efficient fluorescence quenchers, the attachment of coumarin through PEG spacers did show significant emission intensity. The PEG-modified gold nanoparticles (with methoxy-PEG-thiol and coumarin-PEG thiol) did not aggregate even in high-ionic-strength solutions, confirming that the stable surface presence of PEG was effective in steric repulsion of the gold nanoparticles.

Cellular cytotoxicity studies, that gold nanoparticles, gold nanoparticles functionalized with methoxy-PEG-thiol (*i.e.*, gold-PEG-OMe nanoparticles), and coumarin-PEG-thiol functionalized gold nanoparticles were not cytotoxic at the concentrations studied. In contrast, polyethyleneimine (PEI), a known cytotoxic compound, did reduce cell viability to 40% of the control at the same concentrations.

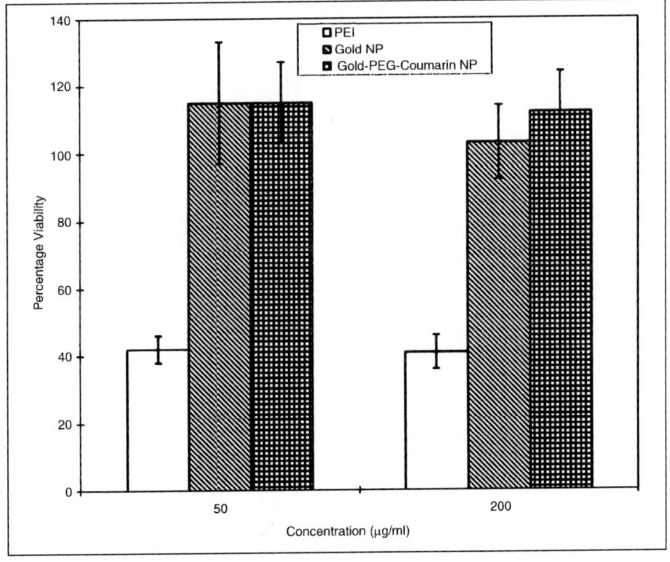

Figure : Cytotoxicity Analysis of the Controls and Coumarin-poly (Ethylene Glycol)-thiol Functionalized Gold Nanoparticles in MDA-MB-231 Human Breast Adenocarcinoma Cells.

Upon incubation with MDA-MB-231 cells in culture, the coumarin-PEG-thiol functionalized gold nanoparticles were internalized by non-specific endocytosis within the first few minutes. Within 30 minutes, a large fraction of the administered dose was found to be in the early endosomes. As time progressed, the nanoparticles traversed through the cytosol and reached the perinuclear region within 1 hour of incubation. The hetero-bifunctional PEG derivative with a molecular weight of 1500 daltons was sufficient to allow flexibility in the attached fluorescent ligand, but did not prevent cellular entry of the nanoparticles.

In addition, we did not observe any of the gold nanoparticles inside the nucleus even after 24 hours of incubation. Several recent studies, including some from our group, have shown that when nanoparticles are internalized in cells through endocytotic process, they tend to converge to the perinuclear area. Suh *et. al.* attributed this directed intracellular movement of nano-carriers as an active transport process driven by molecular motor proteins. Using the Keck 3-D Fusion microscope system, fast data acquisition, and custom particle tracking software, we can track functionalized gold nanoparticles inside the cells.

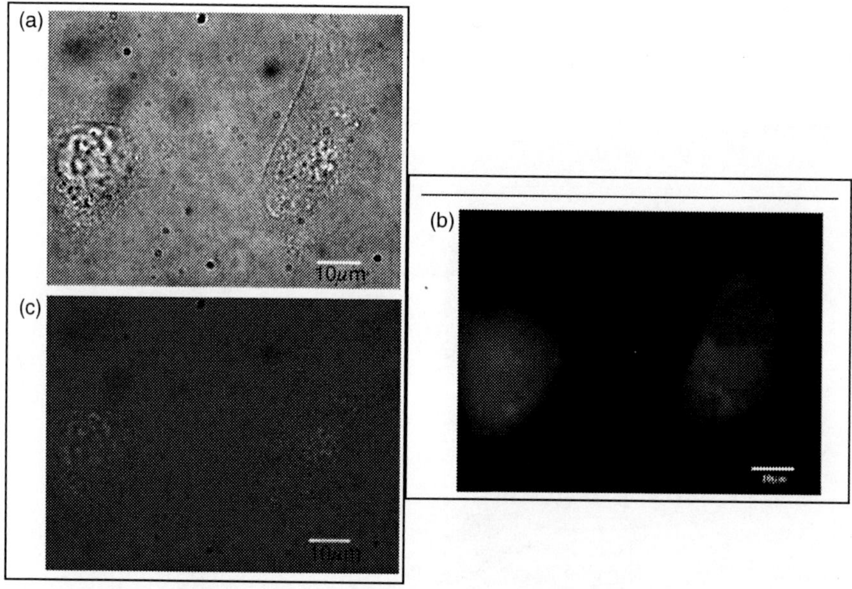

Figure : Cellular Uptake and Distribution of Coumarin-poly(ethylene glycol)-thiol Functionalized Gold Nanoparticles in MDA-MB-231 Human Breast Adenocarcinoma Cells. Differential Interference Contrast (*a*), Epifluorescence (*b*), Merged Images (*c*), and 3D Plot of Fluorescence Intensity Distribution as a Function of the X and Y Coordinates was Plotted

Gold nanoparticle surfaces were functionalized with coumarin, a fluorescent dye, through PEG spacers and used for intracellular tracking in MDA-MB-231 cells. Using the Keck 3-D Fusion microscope system, we can track functionalized nanoparticles with nanometre accuracy. The results show that these nanoparticles were rapidly internalized in the cells through non-specific endocytotic pathways and were found in the perinuclear region within 1 hour of incubation. Using the PEG spacer, the gold nano-platform can be conjugated with a variety of stimuli (*e.g.*, cell-penetrating peptides, endosomal buffering agents), probes (*e.g.*, radiolabel, fluorescent dye), and moieties to study cellular transport pathways and barriers to efficient gene transfection.

SELECTIVE CHEMISTRY OF SINGLE-WALLED NANOTUBES

Selective chemistry of single-walled nanotubes is a field in Carbon nanotube chemistry devoted specifically to the study of functionalization of single-walled carbon nanotubes.

STRUCTURE AND REACTIVITY

Reactivity of fullerene molecules with respect to addition chemistries is strongly dependent on the curvature of the carbon framework. Their outer surface (exohedral) reactivity increases with increase in curvature. In comparison with fullerene molecules single-walled nanotubes (SWNTs) are moderately curved. Consequently nanotubes are expected to be less reactive than most fullerene molecules due to their smaller curvature, but more reactive than a graphene sheet due to pyramidalization and misalignment of pi-orbitals. The strain of a carbon framework is also reflected in the pyramidalization angle (θ_p) of the carbon constituents. Trigonal carbon atoms (sp^2 hybridized) prefer a planar orientation with $\theta_p = 0°$ (*i.e.* graphene) and fullerene molecules have $\theta_p = 11.6°$. The (5,5) SWNT has $\theta_p \sim 6°$ for the sidewall. Values for other (n,n) nanotubes show a trend of increasing θ_p (sidewall) with decrease in n. Therefore generally the chemical reactivity of SWNT increases with decrease in diameter (or n, diameter increases with n). Apart from the curvature SWNT reactivity is also highly sensitive to chiral wrapping (n,m) which determine its electronic structure. Nanotubes with $n\text{-}m = 3i$ (i is an integer) are all metals and rest are all semiconducting (SC).

Sidewall Functionalization

Carbon nanotubes are metallic or semiconducting, based upon delocalized electrons occupying a 1-D density of states. However, any covalent bond on SWNT sidewall causes localization of these electrons. In the vicinity of localized electrons, the SWNT can no longer be described using a band model that assumes delocalized electrons moving in a periodic potential.

Two important addition reactions of SWNT sidewall are: (1) Fluorination, and (2) Aryl diazonium salt addition. These functional groups on SWNT improve solubility and processibility. Moreover, these reactions allow for combining unique properties of SWNTs with those of other compounds. Above all, the selective diazonium chemistry can be used to separate the semiconducting and metallic nanotubes.

Fluorination

The first extensive SWNT sidewall reaction was fluorination in 1998 by Mickleson *et al.* These fluorine moieties can be removed from the nanotube by treatment in hydrazine and the spectroscopic properties of the SWNT can be restored completely.

Diazonium Chemistry

One of the most important SWNT sidewall reaction is that with diazonium reagent which if done under controlled conditions can be used to do selective covalent chemistry. Water-soluble diazonium salts react with carbon nanotubes via charge transfer in which they extract electrons from SWNT and form a stable covalent aryl bond. This covalent aryl bond forms with extremely high affinity for electrons with energies near the Fermi level, E_f of the nanotube. Metallic SWNT have a greater electron density near E_f resulting in their higher reactivity over semiconducting nanotubes. The reactant forms a charge-transfer complex at the nanotube surface, where electron donation from the latter stabilizes the transition state and accelerates the forward rate. Once the bond symmetry of the nanotube is disrupted by the formation of this defect, adjacent carbons increase in reactivity and initial selectivity for metallic SWNT is amplified. Under carefully controlled conditions this behaviour can be exploited to obtain highly selective functionalization of metallic nanotubes to the near exclusion of the semiconductors.

Selective Reaction Conditions

Primary condition is addition of reactant molecules at a very small rate to SWNT solution for a sufficient long time. This ensures reaction with only metallic SWNTs and with no semiconducting SWNTs as all the reactant molecules are taken up by the metallic SWNTs. Long time injection ensures that all metallic tubes are reacted. For example one highly selective condition is: addition of 500 μL of 4-hydroxybenzene diazonium tetrafluoroborate solution in water (0.245 mM) at an injection rate of 20.83 μL/h into 5 mL of SWNT solution (1 wt % sodium dodecyl sulfate (SDS)) over 24 hrs. However, if the entire diazonium solution is added all instantaneously then semiconducting SWNTs will also react due to presence of excess reactant.

SPECTROSCOPY AND FUNCTIONALIZATION

SWNTs have unique optical and spectroscopic properties largely due to one-dimensional confinement of electronic and phonon states, resulting in so-called van Hove singularities in the nanotube density of states (DOS).

Probing Selective Chemistry via Optical Absorption

Optical absorption monitors the valence (v) to conduction (c) electronic transitions denoted E_{nn} where n is the band index. The E_{11} transitions for the metallic nanotubes occur from ~440 to 645 nm. The E_{11} and E_{22} transitions for the semiconducting nanotubes are found from 830 to 1600 nm and 600 to 800 nm, respectively. These separated absorption features allow for the monitoring of valence electrons in each distinct nanotube. Reaction at the surface result in localization of valence electrons makes them no longer free to participate in photoabsorption which results in decay of the spectrum features. Selective diazonium chemistry abruptly decreases the peak intensities that represent the first Van Hove transition of metallic species (E_{11}, metal), while the peak intensities representing the second (E_{22}, semiconducting) and first (E_{11}, semiconducting) Van Hove transition of the semiconducting species show little or no change. A relative decrease in metallic SWNT absorption features over semiconducting features represents a highly preferential functionalization of the metallic nanotubes.

Raman Spectroscopy

Raman spectroscopy is a powerful technique with wide ranging applications in carbon nanotube studies. Some important Raman features are radial breathing mode (RBM), tangential mode (G-band), and disorder-related mode. RBM features correspond to the coherent vibration of the C atoms in the radial direction of the nanotube. These features are unique in carbon nanotubes and occur with the frequencies ω_{RBM} between 120 and 350 cm^{-1} for SWNT in the diameter range (0.7 nm^{-2} nm). They can be used to probe the SWNT diameter, electronic structure through their frequency and intensity (I_{RBM}) respectively and hence perform an (n, m) assignment to their peaks. The addition of the moiety to the sidewall of the nanotube disrupts the oscillator strength that gives rise to RBM feature and hence causes decay of these features. These features are distinct for species of a particular nanotube (n, m) and hence enables to probe which SWNTs are functionalized and to what extent.

Two main components of the tangential mode include G^+ at 1590 cm^{-1} and G^- at 1570 cm^{-1}. G^+ feature is associated with carbon atom vibrations along the nanotube axis. The G^- feature is associated with vibrations of carbon atoms along the circumferential direction. The G-band frequency can be used (1) to distinguish between metallic and semiconducting SWNTs, and (2) to probe charge transfer arising from doping a SWNT. Frequency of G^+ is sensitive to charge transfer. It upshifts for acceptors and downshifts for donors. Lineshape of G^- is highly sensitive to the whether SWNT is metallic (Breit-Wigner-Fano lineshape) or semiconducting (Lorentzian lineshape).

The disorder-related mode (D peak) is a phonon mode at 1300 cm^{-1} and involves the resonantly enhanced scattering of an electron via phonon emission by a defect that breaks the basic symmetry of the graphene plane. This mode corresponds to the conversion of a sp^2-hybridized carbon to a sp^3-hybridized on the surface. Intensity of D peak measures covalent bond made with the nanotube surface. This feature does not increase as a result of surfactant or hydronium ion adsorption on the nanotube surface.

Selective Reaction and Raman Features

Selective functionalization increases the intensity of the D peak due to formation of aryl-nanotube bond and decreases the tangential mode due loss of electronic resonance. These two effects are generally summed together as increase in their peaks ratio (D/G). RBM peaks of metallic nanotubes decay and the peaks corresponding to those of semiconducting nanotubes remain almost unchanged.

Reaction Mechanism

Diazonium reagent and SWNT reaction has a two step mechanism. First, the diazonium reagent adsorbs noncovalently to an empty site on the nanotube surface, forming a charge-transfer complex. This is a fast, selective noncovalent adsorption and diazonium group in this complex partially dopes the nanotube, diminishing the tangential mode in the Raman spectrum. Desorption of A from nanotube is negligible ($k_{-1} \sim 0$). In second step complex B then decomposes to form a covalent bond with the nanotube surface. This is a slower step that need not be selective and is represented by the restoration of the G peak and increase in the D band.

REVERSIBILITY OF DIAZONIUM CHEMISTRY

Nanotubes reacted with the diazonium reagent can be converted back into pristine nanotubes when thermally treated at 300 °C in an atmosphere of inert gas. This cleaves the aryl hydroxyl moieties from the nanotube sidewall and restores the spectroscopic feature (Raman and absorption spectra) of pristine nanotube.

Chemical Separation of Metallic and Semiconducting SWNTs

Metallic and semiconducting carbon nanotubes generally coexist in as-grown materials. To get only semiconducting or only metallic nanotubes selective functionalization of metallic SWNTs via 4-hydroxybenzene diazonium can be used. Separation can be done in solution by deprotonation of the p-hydroxybenzene group on the reacted nanotubes (metallic) in alkaline solution followed by electrophoretic separation of these charged species from the neutral species (semiconducting nanotubes). This followed by annealing would give separated pristine semiconducting and metallic SWNT.

MOLECULAR SWITCHES, NANOELEVATORS, CATENANES AND ROTAXANES CREATED USING NANOTECHNOLOGY

Molecular self-assembly is widely used in biological systems to form higher-order structures. An example is the formation of the tobacco mosaic virus (TMV). The TMV viral particle consists of a single strand of RNA encased in a protein sheath. The protein sheath is formed from 2130 identical protein monomers and the isolated components may be reconstructed *in vitro* to reform the intact virus. Chemists have taken the paradigm of molecular self-assembly to construct well-defined molecular nanostructures. These nanostructures are formed spontaneously by molecules which assemble under the influence of inter-molecular forces.

SUPRAMOLECULAR STRUCTURES

The interactions between π-electron deficient aromatic units and π-electron rich aromatic rings in the template-directed synthesis of mechanically interlocked compounds, specifically so-called catenanes and rotaxanes. Catenanes comprise a molecular ring that is mechanically interlinked with a different ring. Rotaxanes comprise a molecular ring threaded onto a molecular axle.

Supramolecular Structures Exemplified by a Nanoelevator

The nanoelevator is a rig-like construct with three legs embracing an interlocked deck-like component which can be made to move between two levels. It is about 3.5 nanometres in diameter and 2.5 nanometres in height. The elevator has been made to go up and down 10 times by the consecutive addition of acid and base, respectively. Such nanoscale robotic devices could find use in slow-release drug delivery systems and in the control of chemical reactions within nanofluidic systems conducted in laboratories on a chip.

THE FORMATION OF MOLECULAR SWITCHES

The same technology has also been used to form molecular switches. A molecular switch is a nanoscale machine which switches reversibly between two or more positions. The goal of this research is to fabricate molecular electronic components by "engineering up" from a molecule to a functioning electronic device. The so-called "bottom up" approach. Molecular electronics require switchable devices as well as the interfaces to exchange information with the outside world. Furthermore, these assemblies must be controllable, reversible and readable at the molecular level. One approach involves stringing together dozens of molecular switches and nanowires into logic circuits and memory circuits in which the molecular switches are able to communicate with one another. The molecular switches which Stoddart *et al.* are working towards involve redox-active catenanes and rotaxanes. In both structures, the ring can assume two different positions that represent the digital states "1" and "0", and it can be switched between those states using applied voltages.

Use of Silicon Nanowires and Carbon Nanotubes

To connect molecular switches, Stoddart *et al.* are exploring the use of silicon nanowires and carbon nanotubes arranged in a grid. This architecture is derived from that used in a unique silicon-based computer known as Teramac, which was built at Hewlett-Packard a few years ago. At each junction of this grid, the nanowires are connected by a monolayer of the molecular switches.

INCORPORATING MOLECULAR SWITCHES INTO LOGIC AND MEMORY CIRCUITS

Recently, Stoddart *et al.* demonstrated that rotaxane-based molecular switches could be strung together to form logic gates, although the state of the switches could only be changed one time. In August, the group took the next step and reported on catenane-based molecular switches that can be reconfigured, *i.e.* switched on and off, many times. Although the difference between the "on" and "off" states is much too small (in terms of resistance) to be useful for logic circuits, the switches could be useful for memory. Importantly is that it is the first time that a solid-state molecular switch has been shown to work repeatedly under ambient conditions. The challenge now is to incorporate such structures into logic and memory circuits.

MOLECULAR SWITCH

A molecular switch is a molecule that can be reversibly shifted between two or more stable states. The molecules may be shifted between the states in response to environmental stimuli, such as changes in pH, light, temperature, an electrical current, microenvironment, or in the presence of a ligand. In some cases, a combination of stimuli is required. The oldest forms of synthetic molecular switches are pH indicators, which display distinct colours as a function of pH. Currently synthetic molecular switches are of interest in the field of nanotechnology for application in molecular computers. Molecular switches are also important to in biology because many biological functions are based on it, for instance allosteric regulation and vision. They are also one of the simplest examples of molecular machines.

PHOTOCHROMIC MOLECULAR SWITCHES

A widely studied class are photochromic compounds which are able to switch between electronic configurations when irradiated by light of a specific wavelength. Each state has a specific absorption maximum which can then be read out by UV-VIS spectroscopy. Members of this class include azobenzenes, diarylethenes, dithienylethenes, fulgides, stilbenes, spiropyrans and phenoxynaphthacene quinones.

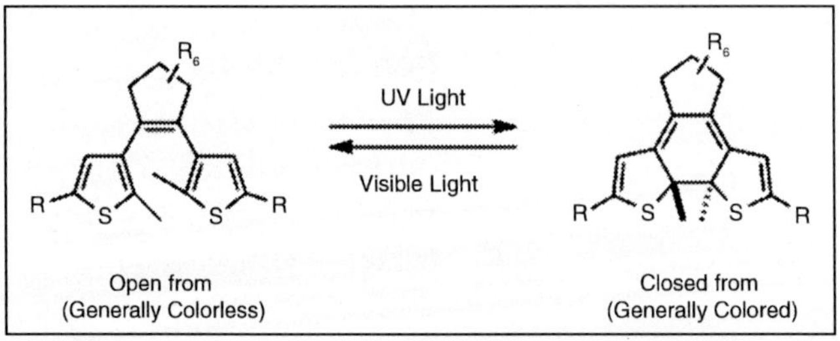

UV Light

Visible Light

Open from
(Generally Colorless)

Closed from
(Generally Colored)

Chiroptical molecular switches are a specific subgroup with photochemical switching taking place between an enantiomeric pairs. In these compounds the readout is by circular dichroism rather than by ordinary spectroscopy. Hindered alkenes such as the one depicted below change their helicity as response to irradiation with right or left-handed circularly polarized light.

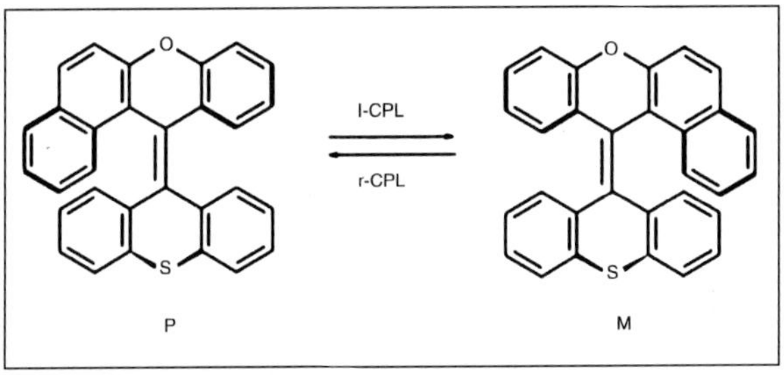

Chiroptical molecular switches that show directional motion are considered synthetic molecular motors:

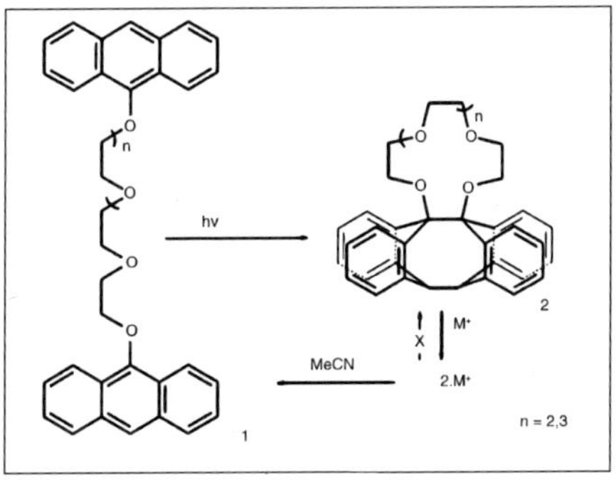

HOST-GUEST MOLECULAR SWITCHES

In host-guest chemistry the bistable states of molecular switches differ in their affinity for guests. Many early examples of such systems are based on crown ether chemistry.

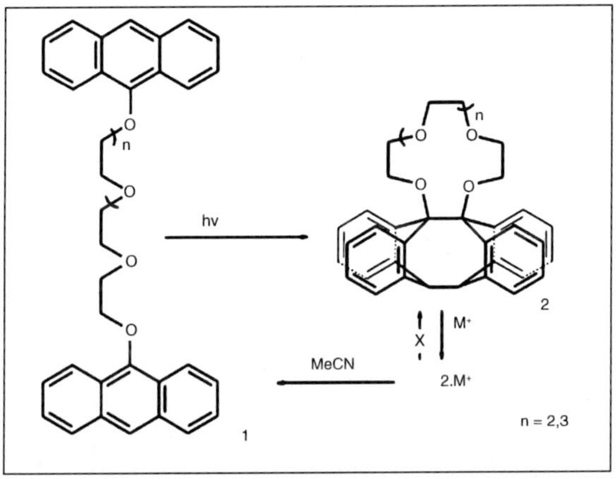

The first switchable host is described in 1978 by Desvergne & Bouas-Laurent who create a crown ether via photochemical anthracene dimerization. Although not strictly speaking switchable the compound is able to take up cations after a photochemical trigger and exposure to acetonitrile gives back the open form.

In 1980 Yamashita *et al.* construct a crown ether already incorporating the anthracene units (an anthracenophane) and also study ion uptake vs photochemistry.

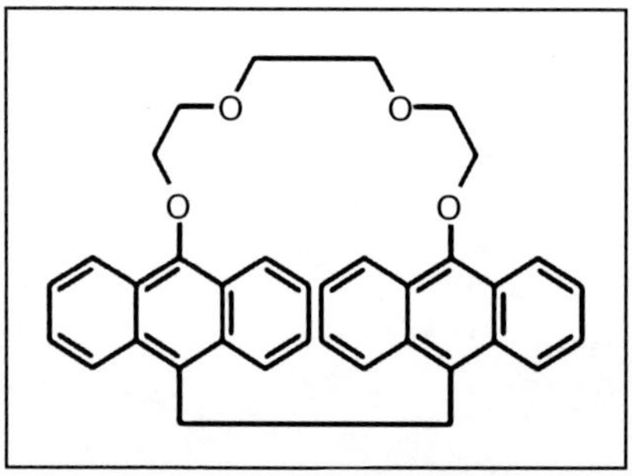

Also in 1980 Shinkai throws out the anthracene unit as photoantenna in favour of an azobenzene moiety and for the first time envisions the existence of molecules with an on-off switch. In this molecule light triggers a trans-cis isomerization of the azo group which results in ring expansion. Thus in the trans form the crown binds preferentially to ammonium, lithium and sodium ions while in the cis form the preference is for potassium and rubidium (both larger ions in same alkali metal group). In the dark the reverse isomerization takes place.

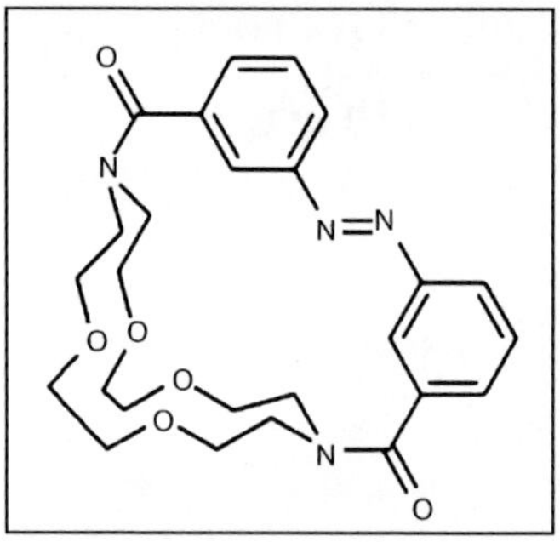

Shinkai employs this devices in actual ion transport mimicking the biochemical action of monensin and nigericin: in a biphasic system ions are taken up triggered by light in one phase and deposited in the other phase in absence of light.

Mechanically-interlocked Molecular Switches

Some of the most advanced molecular switches are based on mechanically-interlocked molecular architectures where the bistable states differ in the position of the macrocycle. In 1991 Stoddart devices a molecular shuttle based on a rotaxane on which a molecular *bead* is able to shuttle between two *docking stations* situated on a molecular *thread*. Stoddart predicts that when the stations are dissimilar with each of the stations addressed by a different external stimulus the shuttle becomes a molecular machine. In 1993 Stoddart is scooped by supramolecular chemistry pioneer Fritz Vögtle who actually delivers a switchable molecule based not on a rotaxane but on a related catenane

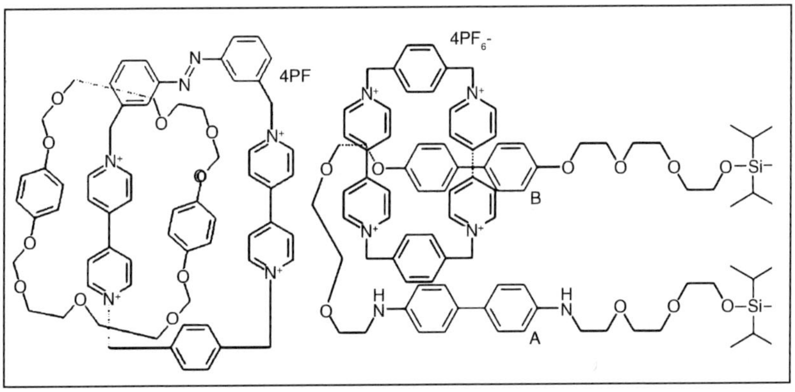

Figure: Photo switchable catenane Vögtle 1993 and Molecular switch Kaifer and Stoddart 1994

This compound is based on two ring systems: one ring holds the photoswichable azobenzene ring and two paraquat docking stations and the other ring is a polyether with to arene rings with binding affinity for the paraquat units. In this system NMR spectroscopy shows that in the azo trans-form the polyether ring is free to rotate around its partner ring but then when a light trigger activates the cis azo form this rotation mode is stopped

Kaifer and Stoddart in 1994 modify their molecular shuttle such a way that an electron-poor tetracationic cyclophane bead now has a choice between two docking stations: one biphenol and one benzidine unit. In solution at room temperature NMR spectroscopy reveals that the bead shuttles at a rate comparable to the NMR timescale, reducing the temperature to 229K resolves the signals with 84% of the population favouring the benzidine station. However on addition of trifluoroacetic acid, the benzidine nitrogen atoms are protonated and the bead is fixed permanently on the biphenol station. The same effect is obtained by electrochemical oxidation (forming the benzidine radical ion) and significantly both processes are reversible.

In 2007 molecular shuttles are utilized in an experimental DRAM circuit. The device consists of 400 bottom silicon nanowire electrodes (16 nanometre (nm) wide at 33 nm intervals) crossed by another 400 titanium top-nanowires with similar dimensions sandwiching a monolayer of a bistable rotaxane depicted below:

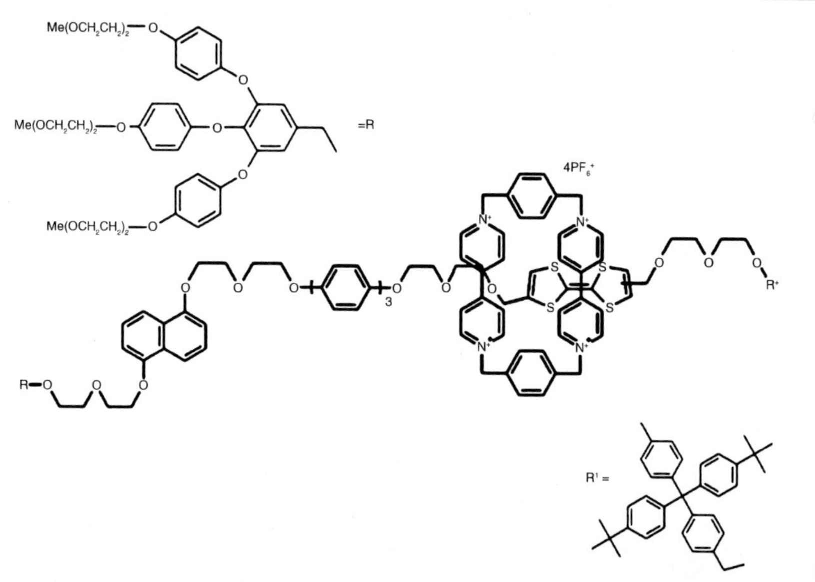

Each bit in the device consists of a silicon and a titanium crossbar with around 100 rotaxane molecules filling in the space between them at perpendicular angles. The hydrophilic diethylene glycol stopper on the left (gray) is specifically designed to anchor to the silicon wire (made hydrophilic by phosphorus doping) while the hydrophobic tetraarylmethane stopper on the right does the same to the likewise hydrophobic titanium wire. In the ground state of the switch, the paraquat ring is located around a tetrathiafulvalene unit (in red) but it moves to the dioxynaphthyl unit (in green) when the fulvalene unit is oxidized by application of a current. When the fulvalene is reduced back a metastable high conductance '1' state is formed which relaxes back to the ground state with a chemical half-life of around one hour. The problem of defects is circumvented by adopting a defect-tolerant architecture also found in the Teramac project. In this way a circuit is obtained consisting of 160,000 bits on an area the size of a white blood cell translating into 10^{11} bits per square centimetre.

REFERENCES

- Awasthi, K.P. : *Advanced Magnetic Nanostructures,* Cyber Tech Publications, Delhi, 2008.
- Bainbridge, W.S. : *Societal Implications of Nanoscience and Nanotechnology,* UK: Intellect Books, 2001.
- Ipsita Roy: *Bio-Nano-Geo Sciences : The Future Challenge,* Ane Books, Delhi, 2009.

5

Nanowire

A nanowire is a nanostructure, with the diameter of the order of a nanometre (10^{-9} meters). Alternatively, nanowires can be defined as structures that have a thickness or diameter constrained to tens of nanometres or less and an unconstrained length. At these scales, quantum mechanical effects are important — which coined the term "quantum wires". Many different types of nanowires exist, including metallic (*e.g.*, Ni, Pt, Au), semiconducting (*e.g.*, Si, InP, GaN, etc.), and insulating (*e.g.*, SiO_2, TiO_2). Molecular nanowires are composed of repeating molecular units either organic (*e.g.* DNA) or inorganic (*e.g.* $Mo_6S_{9-x}I_x$). The nanowires could be used, in the near future, to link tiny components into extremely small circuits. Using nanotechnology, such components could be created out of chemical compounds.

OVERVIEW

Typical nanowires exhibit aspect ratios (length-to-width ratio) of 1000 or more. As such they are often referred to as one-dimensional (1-D) materials. Nanowires have many interesting properties that are not seen in bulk or 3-D materials. This is because electrons in nanowires are quantum confined laterally and thus occupy energy levels that are different from the traditional continuum of energy levels or bands found in bulk materials.

Peculiar features of this quantum confinement exhibited by certain nanowires manifest themselves in discrete values of the electrical conductance. Such discrete values arise from a quantum mechanical restraint on the number of electrons that can travel through the wire at the nanometre scale. These discrete values are often referred to as the quantum of conductance and are integer multiples of

$$\frac{2e^2}{h} \simeq 77.41\,\mu S$$

They are inverse of the well-known resistance unit h/e^2, which is roughly equal to 25812.8 ohms, and referred to as the von Klitzing constant R_K (after Klaus von Klitzing, the discoverer of exact quantization). Since 1990, a fixed conventional value $R_{K\text{-}90}$ is accepted.

Examples of nanowires include inorganic molecular nanowires ($Mo_6S_{9-x}I_x$, $Li_2Mo_6Se_6$), which can have a diameter of 0.9 nm and be hundreds of micrometers long. Other important examples are based on semiconductors such as InP, Si, GaN, etc., dielectrics (*e.g.* SiO_2,TiO_2), or metals (*e.g.* Ni, Pt).

There are many applications where nanowires may become important in electronic, opto-electronic and nanoelectromechanical devices, as additives in advanced composites, for metallic interconnects in nanoscale quantum devices, as field-emitters and as leads for biomolecular nanosensors.

SYNTHESIS OF NANOWIRES

There are two basic approaches to synthesizing nanowires: top-down and bottom-up. A top-down approach reduces a large piece of material to small pieces, by various means such as lithography or electrophoresis. A bottom-up approach synthesizes the nanowire by combining constituent adatoms. Most synthesis techniques use a bottom-up approach.

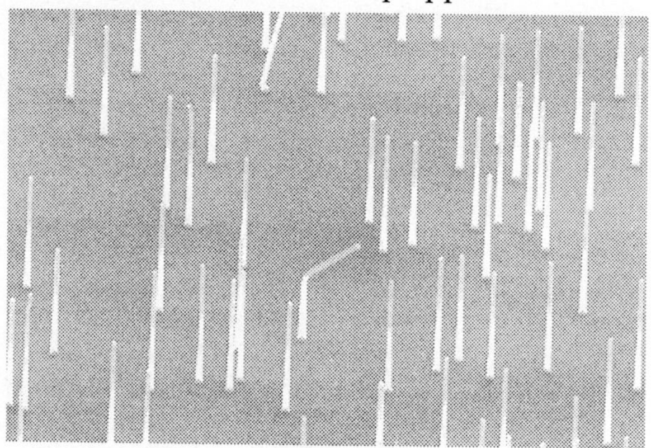

Figure: An SEM Image of Epitaxial Nanowire Heterostructures Grown from Catalytic Gold Nanoparticles.

Nanowire production uses several common laboratory techniques, including suspension, electrochemical deposition, vapour deposition, and VLS growth. Ion track technology enables growing homogeneous and segmented nanowires down to 8 nm diameter.

SUSPENSION

A suspended nanowire is a wire produced in a high-vacuum chamber held at the longitudinal extremities. Suspended nanowires can be produced by:

- The chemical etching of a larger wire
- The bombardment of a larger wire, typically with highly energetic ions

- Indenting the tip of a STM in the surface of a metal near its melting point, and then retracting it.

VLS GROWTH

A common technique for creating a nanowire is Vapour-Liquid-Solid (VLS) synthesis. This process can produce crystalline nanowires of some semiconductor materials. It uses as source material either laser ablated particles or a feed gas such as silane. VLS synthesis requires a catalyst. For nanowires, the best catalysts are liquid metal (such as gold) nanoclusters, which can either be self-assembled from a thin film by dewetting, or purchased in colloidal form and deposited on a substrate. The source enters these nanoclusters and begins to saturate them. On reaching supersaturation, the source solidifies and grows outward from the nanocluster. Simply turning off the source can adjust the final length of the nanowire. Switching sources while still in the growth phase can create compound nanowires with super-lattices of alternating materials. A single-step vapour phase reaction at elevated temperature synthesises inorganic nanowires such as $Mo_6S_{9-x}I_x$. From another point of view, such nanowires are cluster polymers.

SOLUTION-PHASE SYNTHESIS

Solution-phase synthesis refers to techniques that grow nanowires in solution. They can produce nanowires of many types of materials. Solution-phase synthesis has the advantage that it can produce very large quantities, compared to other methods. In one technique, the polyol synthesis, ethylene glycol is both solvent and reducing agent. This technique is particularly versatile at producing nanowires of lead, platinum, and silver.

PHYSICS OF NANOWIRES

Conductivity of Nanowires

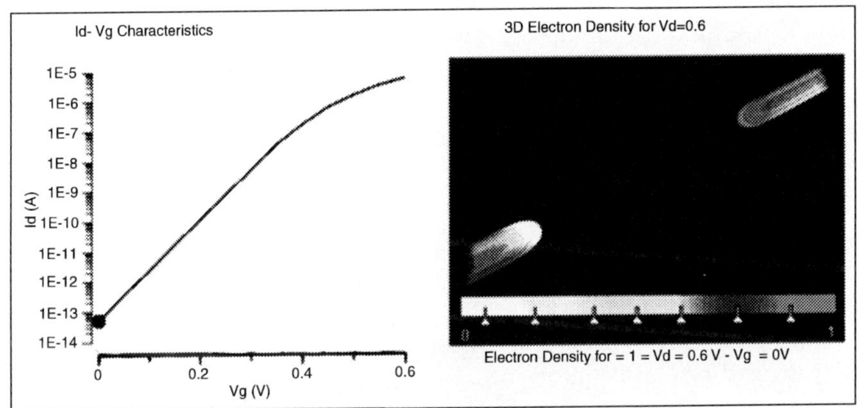

Figure: Atomistic Simulation Result for Formation of Inversion Channel (Electron Density) and Attainment of Threshold Voltage (IV) in a Nanowire MOSFET.

Several physical reasons predict that the conductivity of a nanowire will be much less than that of the corresponding bulk material. First, there is scattering from the wire boundaries, whose effect will be very significant whenever the wire width is below the free electron mean free path of the bulk material. In copper, for example, the mean free path is 40 nm. Copper nanowires less than 40 nm wide will shorten the mean free path to the wire width.

Nanowires also show other peculiar electrical properties due to their size. Unlike carbon nanotubes, whose motion of electrons can fall under the regime of ballistic transport (meaning the electrons can travel freely from one electrode to the other), nanowire conductivity is strongly influenced by edge effects. The edge effects come from atoms that lay at the nanowire surface and are not fully bonded to neighbouring atoms like the atoms within the bulk of the nanowire. The unbonded atoms are often a source of defects within the nanowire, and may cause the nanowire to conduct electricity more poorly than the bulk material. As a nanowire shrinks in size, the surface atoms become more numerous compared to the atoms within the nanowire, and edge effects become more important.

Furthermore, the conductivity can undergo a quantization in energy: *i.e.* the energy of the electrons going through a nanowire can assume only discrete values, which are multiples of the Von Klitzing constant $G = 2e^2/h$ (where e is the charge of the electron and h is the Planck constant).

The conductivity is hence described as the sum of the transport by separate *channels* of different quantized energy levels. The thinner the wire is, the smaller the number of channels available to the transport of electrons.

This quantization has been demonstrated by measuring the conductivity of a nanowire suspended between two electrodes while pulling it: as its diameter reduces, its conductivity decreases in a stepwise fashion and the plateaus correspond to multiples of G.

The quantization of conductivity is more pronounced in semiconductors like Si or GaAs than in metals, due to their lower electron density and lower effective mass. It can be observed in 25 nm wide silicon fins, and results in increased threshold voltage. In practical terms, this means that a MOSFET with such nanoscale silicon fins, when used in digital applications, will need a higher gate (control) voltage to switch the transistor on.

Welding Nanowires

To incorporate nanowire technology into industrial applications, researchers in 2008 developed a method of welding nanowires together: a sacrificial metal nanowire is placed adjacent to the ends of the pieces to be joined (using the manipulators of a scanning electron microscope); then an electric current is applied, which fuses the wire ends. The technique fuses wires as small as 10 nm.

For nanowires with diameters less than 10 nm, existing welding techniques, which requiring precise control of the heating mechanism and may introducing the possibility of damage, will be not practical. Recently scientists discovered that single-crystalline ultrathin gold nanowires with diameters ~3-10 nm can be "cold-welded" together within seconds by mechanical contact alone, and under remarkably low applied pressures (unlike macro- and micro-scale cold welding process). High-resolution transmission electron microscopy and in situ measurements reveal that the welds are nearly perfect, with the same crystal orientation, strength and electrical conductivity as the rest of the nanowire. The high quality of the welds is attributed to the nanoscale sample dimensions, oriented-attachment mechanisms and mechanically assisted fast surface diffusion. Nanowire welds were also demonstrated between gold and silver, and silver nanowires (with diameters ~5-15 nm) at near room temperature, indicating that this technique may be generally applicable for ultrathin metallic nanowires. Combined with other nano- and microfabrication technologies, cold welding is anticipated to have potential applications in the future bottom-up assembly of metallic one-dimensional nanostructures.

USES OF NANOWIRES

Nanowires still belong to the experimental world of laboratories. However, they may complement or replace carbon nanotubes in some applications. Some early experiments have shown how they can be used to build the next generation of computing devices. To create active electronic elements, the first key step was to chemically dope a semiconductor nanowire. This has already been done to individual nanowires to create p-type and n-type semiconductors.

The next step was to find a way to create a p-n junction, one of the simplest electronic devices. This was achieved in two ways. The first way was to physically cross a p-type wire over an n-type wire. The second method involved chemically doping a single wire with different dopants along the length. This method created a p-n junction with only one wire. After p-n junctions were built with nanowires, the next logical step was to build logic gates. By connecting several p-n junctions together, researchers have been able to create the basis of all logic circuits: the AND, OR, and NOT gates have all been built from semiconductor nanowire crossings.

It is possible that semiconductor nanowire crossings will be important to the future of digital computing. Though there are other uses for nanowires beyond these, the only ones that actually take advantage of physics in the nanometre regime are electronic. Nanowires are being studied for use as photon ballistic waveguides as interconnects in quantum dot/quantum effect well photon logic arrays. Photons travel inside the tube, electrons travel on the outside shell. When two nanowires acting as photon waveguides cross each other the juncture acts as a quantum dot. Conducting nanowires offer the possibility of connecting molecular-scale entities in a molecular computer. Dispersions of conducting nanowires in different polymers are being investigated for use as transparent electrodes for flexible flat-screen displays.

Because of their high Young's moduli, their use in mechanically enhancing composites is being investigated. Because nanowires appear in bundles, they may be used as tribological additives to improve friction characteristics and reliability of electronic transducers and actuators. Because of their high aspect ratio, nanowires are also uniquely suited to dielectrophoretic manipulation.

FABRICATION OF NANOWIRES AT SURFACES

The goal of this project is the design of artificial materials that consist of ultrafine wires or linear arrays of dots, ten to hundred times finer than those produced with commercial micro-structure fabrication techniques. In fact, we have gone all the way down to atom chains which may be viewed as the ultimate nanowires (scroll to the bottom for those). These patterns are formed by self-assembly, where atoms arrange themselves naturally at stepped silicon surfaces.

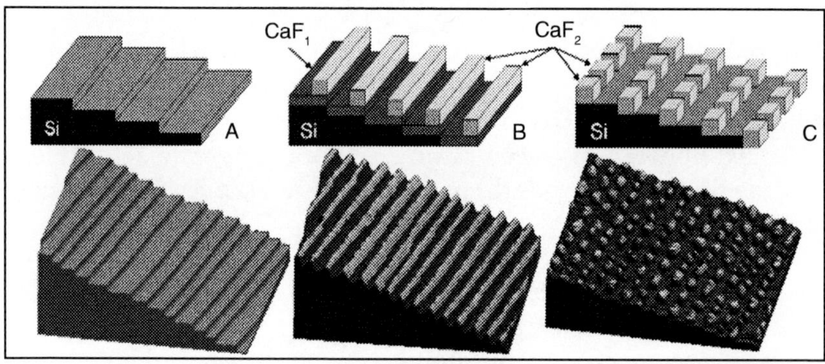

Figure: The Preparation of Calcium Fluoride Masks in Schematic Form (top), Together with Actual Data (bottom).

An important aspect in fabricating nanowires is the ability to prepare wires of an any material on any substrate with any thickness. In particular, using silicon wafers as substrate is highly-desirable. To achieve this goal we suggest the following "universal" process. First, a silicon substrate with a regular array of steps is prepared (A). Then, stripes (B) or dots (C) of a passivating material are attached to the step edges. This part is analogous to creating a photoresist mask in traditional lithography. As mask material we use calcium fluoride, which is lattice-matched to silicon and chemically inert. Eventually, the desired material is deposited on the remaining silicon, for example by substrate-selective chemical vapour deposition (CVD) or electroplating. Alternatively, calcium fluoride could become useful as an etch mask for producing trenches in the silicon that can be filled with new materials to achieve a planar structure.

To start along this pathway, we determine the conditions for obtaining highly-regular step structures on silicon. The images below demonstrate the range of step arrays that can be formed on silicon surfaces by self-assembly. Typically, the step spacing is comparable to the size of a virus. These images are taken with a scanning tunnelling microscope (STM). They show the derivative of the tip height. That gives the impression of a surface illuminated from the left, with the steps casting dark shadows to the right.

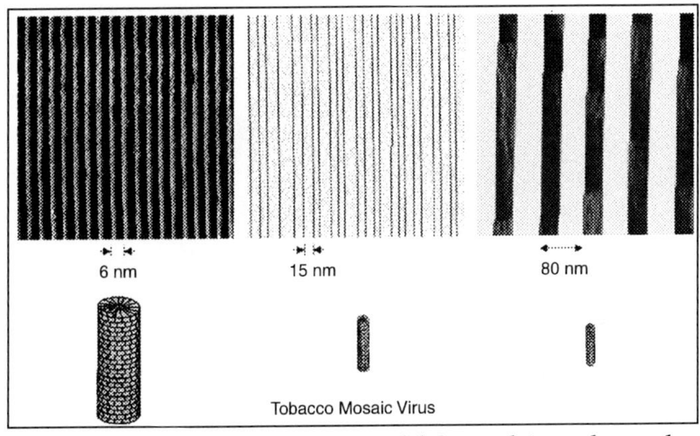

Particularly perfect step arrays could be achieved on the Si(111)7x7 surface. The 7×7 structure causes steps running along the [011]-direction to become extremely straight, because each kink requires generating 14 new rows of silicon atoms (7 rows, two layers deep).

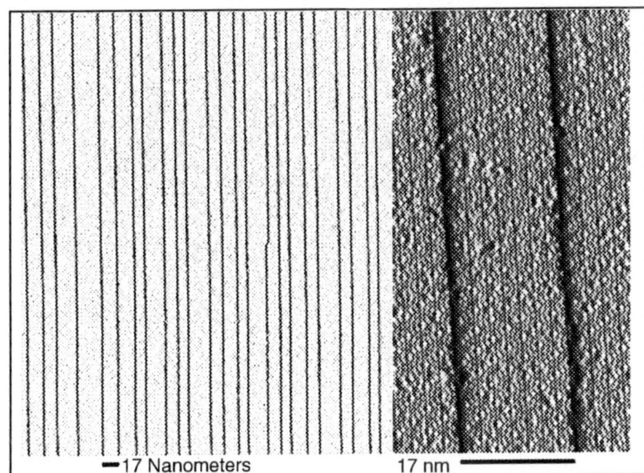

The step edges are atomically-straight with a kink spacing as low as a single kink in 20,000 atoms, as seen in the image below. These are taken with a scanning tunnelling microscope (STM). The x-derivative of the topography is displayed, which makes steps appear as dark lines. The image on the right zooms in on the terrace between two steps (heavy dark lines). The atomic pattern of the 7×7 structure is resolved, which has fine grooves built in that run parallel to the step edges. These grooves are 2.3 nm apart and determine the location of possible step edges. Our "universal process" can be carried further by producing a calcium fluoride mask, as shown in the image below. A stepped silicon surface is coated with a layer of CaF_1 and CaF_2 stripes are formed on top of that layer. These stripes are continuous and do not touch each other, because adjacent stripes cannot bond to each other. The stripe width of 7 nm achieved here is well below the resolution of 180 nm achieved in commercial lithography for chip fabrication.

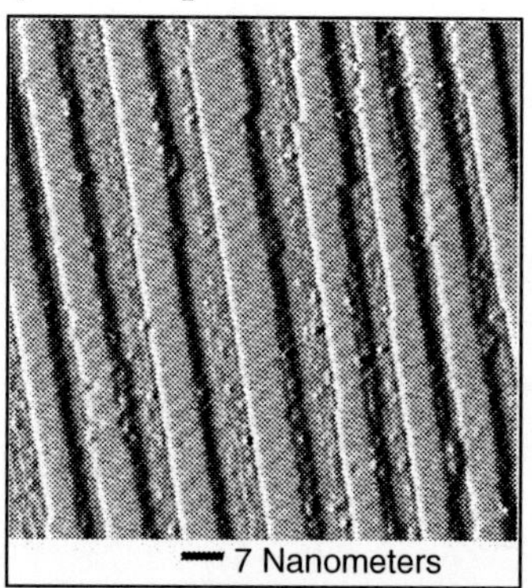

The third step of the "universal process" involves selective deposition or etching between the masked areas. The picture below shows that molecules can be deposited selectively in the CaF_1 grooves between CaF_2 stripes.

Using organometallic molecules, such as ferrocene, it is possible to fabricate iron wires 3 nanometres wide.

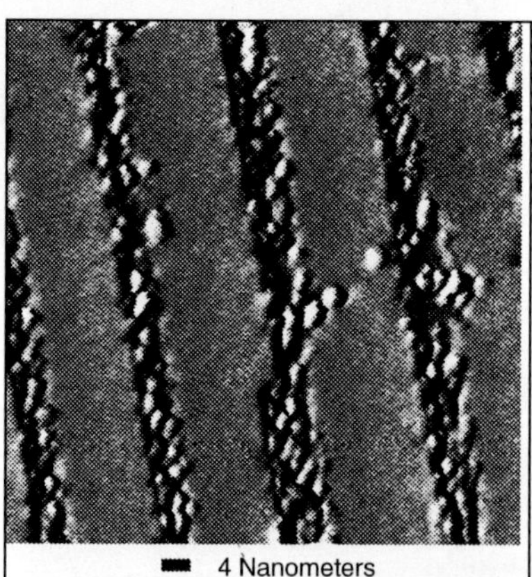

Images (derivative of the topography, "illuminated from the side")
Si(111)7 × 7 closeup, showing the corner holes and grooves of the 7 × 7 cell.
Step array on silicon with only a single kink in 20,000 edge atoms.
Steps and kinks on silicon.
Large array of CaF_2 stripes.
3 nm copper wires on molybdenum, made visible by image state contrast.

ATOM CHAINS, THE ULTIMATE NANOWIRES

Self-assembly can reach atomic precision for very small structures (up to 10 nm in size). It is possible to go all the way to the ultimate limit for nanowires, *i.e.* chains of single atoms with a single set of orbitals connecting them. Such atomic wires are obtained by depositing a fraction of a monolayer of metal atoms onto a stepped silicon surface. An example is the Si(557)-Au surface shown below. It contains a step every five silicon atom rows and a row of gold atoms in the middle of the terrace. The STM image below shows two rows of fine white dots, which are magnified in the inset. They correspond to silicon atoms with dangling bonds.

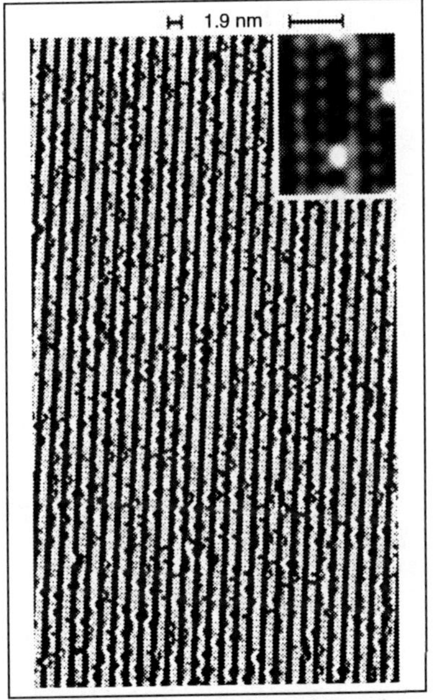

Being able to fabricate atomic wires opens up the possibility of studying one-dimensional electrons moving along the atom chains. The properties of electrons become more and more exotic when progressing from the three-dimensional world into lower dimensions. In a two-dimensional electron gas one observes surprising phenomena already, such as fractional charge and statistics with the fractional quantum Hall effect. Predictions for one-dimensional electrons are even farther out. The concept of a single electron fails because electrons constantly penetrate each other when moving along the same one-dimensional line. They lose their identity and separate into two quasiparticles, a spinon that carries spin without charge, and a holon that carries the positive charge of a hole without its spin. That is the theoretical prediction, and the hunt is on whether such strange behaviour can be observed experimentally. The complete set of quantum numbers for electrons in a solid is revealed by angle-resolved photoemission with tunable synchrotron radiation. The two principal quantum numbers energy and momentum are combined in a band structure plot, as in the picture below. The dark bands represent regions with high photoemission intensity. With multitedection techniques for energy and angle it is now possible to see the band structure directly on a television screen.

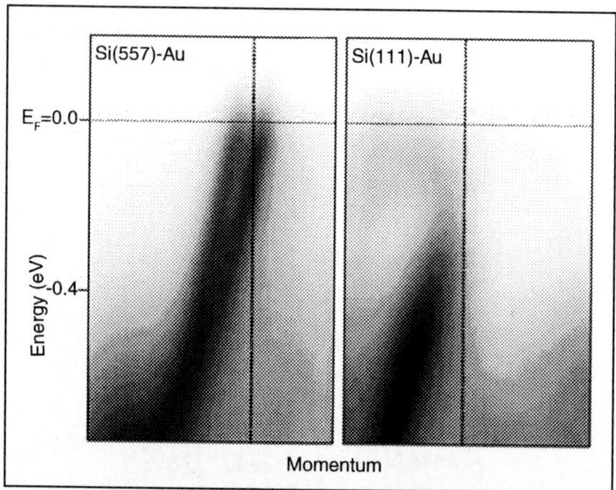

Some of the atomic wires are found to be metallic. The metallic behaviour is deduced from the observation that the bands extend all the way up to the Fermi level E_F for Si(557)-Au, as in a metal. By way of contrast, the flat Si(111)-Au surface in the panel on the right shows a band that does not reach E_F. Even though the metal atoms are strongly coupled to the substrate, metallic electrons do not interact with the silicon substrate because their energy lies in the band gap of silicon.

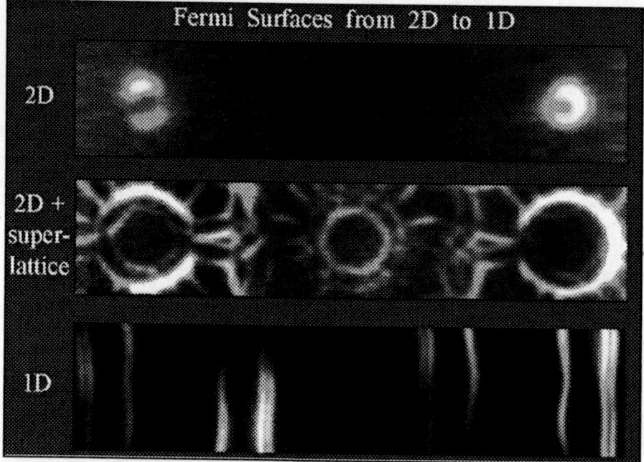

Electrons at the Fermi level are particularly interesting because they assemble themselves into exotic phases in low dimensions. These electrons are best characterized by the Fermi surface, which describes their momentum distribution. Such Fermi surfaces are shown below for two- and one-dimensional electrons at silicon surfaces with extra metal atoms (typically a monolayer for a 2D structure and 1/5 of a monolayer for a 1D structure). Two-dimensional Fermi surfaces typically consist of circles, which may form intricate patterns by replication at a superlattice. One-dimensional Fermi surfaces are straight lines, which become modulated by small wiggles if a weak two-dimensional coupling is present between the atom chains.

Currently, we are exploring silicon surfaces with a variety of step spacings. That makes it possible to vary the dimensionality between 2D and 1D. For example, the coupling ratio parallel/perpendicular to the chains can be varied from 10:1 to >70:1. In addition, structures with fractional filling can be produced, such as for the one-dimensional Fermi surface. Fractional filling is achieved by embedding the chains into a 2D superlattice where silicon atoms act as dopants, as in modulation doping of semiconductors. That raises interesting questions about an analog to the fractional quantum Hall effect in a one-dimensional electron gas. Chains can be formed not only with gold atoms, but also with a large variety of other metal atoms. These include magnetic atoms, such as Gd and other rare earths, which open up possibilities of fabricating spin chains. In fact, even non-magnetic gold chains form spin-split Fermi surfaces. That is due to the spin-orbit interaction in the reduced symmetry of a surface. Another chain structure spontaneously decomposes into metallic and semiconducting segments, which can be viewed as the one-dimensional analog of stripes, which have been a hot topic in the context of high temperature superconductivity. All these atomic chain structures form an ideal playground to search for interesting new phases of matter in one dimension.

SPIN CHAINS FOR SINGLE SPIN ELECTRONICS

The ultimate limit for making electronic circuits smaller and smaller is reached when one bit of data is shrunk to a single electron, carrying a single spin and sitting on a single atom.

Figure: Intrinsic magnetism at silicon surfaces, Nature Communications.

We are still far away from this limit, but theoretical work about the arrangement of atomic wires at silicon step edges provides a first glimpse. Self-assembly of silicon and gold atoms produces a very unusual structure at the step edge. The silicon atoms form a stripe of graphitic silicon at the step edge, and the gold atoms form two rows in the middle of the terrace. Broken bonds at the step edge contain a single electron each, and with it comes a single spin. These unpaired spins occur exactly at every third edge atom, while the two atoms in between pair up their spins and are non-magnetic. This tripling of the period is indeed observed experimentally. Recently, it has become possible to detect single spins at surfaces with scanning tunnelling microscopy. This sets the stage for atomic-scale spin memory and logic.

HOW NANOWIRES WORK

In 1965, engineer Gordon Moore predicted that the number of transistors on an integrated circuit—a precursor to the microprocessor—would double approximately every two years. Today, we call this prediction Moore's Law, though it's not really a scientific law at all. Moore's Law is more of a self-fulfilling prophecy about the computer industry. Microprocessor manufacturers strive to meet the prediction, because if they don't, their competitors will.

NANOTECHNOLOGY IMAGE GALLERY

To fit more transistors on a chip, engineers have to design smaller transistors. The first chip had about 2,200 transistors on it. Today, hundreds of millions of transistors can fit on a single microprocessor chip. Even so, companies are determined to create increasingly tiny transistors, cramming more into smaller chips. There are already computer chips that have nanoscale transistors (the nanoscale is between 1 and 100 nanometres-a nanometre is one billionth of a meter). Future transistors will have to be even smaller.

Enter the nanowire, a structure that has an amazing length-to-width ratio. Nanowires can be incredibly thin-it's possible to create a nanowire with the diameter of just one nanometre, though engineers and scientists tend to work with nanowires that are between 30 and 60 nanometres wide. Scientists hope that we will soon be able to use nanowires to create the smallest transistors yet, though there are some pretty tough obstacles in the way.

NANOWIRE PROPERTIES

Depending on what it's made from, a nanowire can have the properties of an insulator, a semiconductor or a metal. Insulators won't carry an electric charge, while metals carry electric charges very well. Semiconductors fall between the two, carrying a charge under the right conditions. By arranging semiconductor wires in the proper configuration, engineers can create transistors, which either acts as a switch or an amplifier. Some interesting — and counterintuitive — properties nanowires possess are due to the small scale. When you work with objects that are at the nanoscale or smaller, you begin to enter the realm of quantum mechanics. Quantum mechanics can be confusing even to experts in the field, and very often it defies classical physics.

For example, normally an electron can't pass through an insulator. If the insulator is thin enough, though, the electron can pass from one side of the insulator to the other. It's called electron tunnelling, but the name doesn't really give you an idea of how weird this process can be. The electron passes from one side of the insulator to the other without actually penetrating the insulator itself or occupying the space inside the insulator. You might say it teleports from one side to the other. You can prevent electron tunnelling by using thicker layers of insulator since electrons can only travel across very small distances. Another interesting property is that some nanowires are ballistic conductors. In normal conductors, electrons collide with the atoms in the conductor material. This slows down the electrons as they travel and creates heat as a byproduct. In ballistic conductors, the electrons can travel through the conductor without collisions. Nanowires could conduct electricity efficiently without the byproduct of intense heat.

At the nanoscale, elements can display very different properties than what we've come to expect. For example, in bulk, gold has a melting point of more than 1,000 degrees Celsius. By reducing bulk gold to the size of nanoparticles, you decrease its melting point, because when you reduce any particle to the nanoscale, there's a significant increase in the surface-to-volume ratio. Also, at the nanoscale, gold behaves like a semiconductor, but in bulk form it's a conductor. Other elements behave strangely at the nanoscale as well. In bulk, aluminum isn't magnetic, but very small clusters of aluminum atoms are magnetic. The elemental properties we're familiar with in our everyday experience — and the ways we expect them to behave — may not apply when we reduce those elements down to the size of a nanometre.

BUILDING NANOWIRES FROM THE TOP DOWN

Nanoscience specialists talk about two different approaches to building things in the nanoscale: the top-down approach and the bottom-up approach. A top-down approach essentially means that you take a bulk amount of the material you plan on using for nanowires and carve away until you are down to the right size. A bottom-up approach is an assembly process where smaller particles join to make a larger structure. Although we can build nanowires using either approach, no one has found a way to make mass production feasible. Right now, scientists and engineers would have to spend a lot of time to make a fraction of the number of nanowires they would need for a microprocessor chip. An even greater challenge is finding a way to arrange the nanowires properly once they are built. The small scales make it very difficult to build transistors automatically — right now, engineers usually manipulate wires into place with tools while observing everything through a powerful microscope.

An example of a top-down approach is the way scientists make fibre-optic nanowires. Fibre-optic wires carry information in the form of light. To make a fibre-optic nanowire, engineers first start with a regular fibre-optic cable. There are a few different approaches to reduce a fibre-optic cable to the nanoscale. Scientists could heat up a rod made out of sapphire, wrap the cable around the rod, and pull the cable, stretching it thin to create a nanowire. Another method uses a tiny furnace made from a small cylinder of sapphire. Scientists draw the fibre-optic cable through the furnace and stretch it into a thin nanowire. A third procedure called flame brushing uses a flame under the fibre-optic cable while scientists stretch it.

GROWING NANOWIRES

Chemical vapour deposition (CVD) is an example of a bottom-up approach. In general, CVD refers to a group of processes where solids form out of a gaseous phase. Scientists deposit catalysts (such as gold nanoparticles) on a base, called a substrate. The catalysts act as an attraction site for nanowire formation. Scientists put the substrate in a chamber with a gas containing the appropriate element, such as silicon, and the atoms in the gas do all the work. First, atoms in the gas attach to atoms in the catalysts, then additional gas atoms attach to those atoms, and so on, creating a chain or wire. In other words, the nanowires assemble themselves.

A new way to build nanowires is to print them directly to the appropriate substrate. A team of researchers in Zurich pioneered this method. First, they carved a silicon wafer so that the raised portions on the wafer coincided with the way they wanted the nanowires arranged. They used the wafer like a stamp, pressing it against a synthetic rubber called PDMS. They then drew a liquid filled with gold nanoparticles, called a colloidal suspension, across the PDMS. The gold particles settled into the channels created by the silicon wafer stamp. Now the PDMS became a mold capable of transferring a "print" of gold nanowires onto another surface. PDMS molds can be used repeatedly and may play a role in the mass production of nanowire circuitry in the future.

Several laboratories have created transistors using nanowires, but their creation requires a lot of time and manpower. Nanowire transistors perform as well or better than current transistors. If scientists can find a way to design a way to produce and connect nanowire transistors together efficiently, it will pave the way to smaller, faster microprocessors, which will allow the computer industry to keep up with Moore's Law. Computer chips will continue to get smaller and more powerful. Research in nanowire production continues across the world. Many scientists believe it's just a matter of time before someone comes up with a viable way to mass produce nanowires and nanowire transistors. Hopefully, if and when we reach that point, we'll also have a way to arrange nanowires the way we want so that we can use them to their full potential.

NANOWIRE APPLICATIONS

Perhaps the most obvious use for nanowires is in electronics. Some nanowires are very good conductors or semiconductors, and their miniscule size means that manufacturers could fit millions more transistors on a single microprocessor. As a result, computer speed would increase dramatically.

Nanowires may play an important role in the field of quantum computers. A team of researchers in the Netherlands created nanowires out of indium arsenide and attached them to aluminum electrodes. At temperatures near absolute zero, aluminum becomes a superconductor, meaning it can conduct electricity without any resistance. The nanowires also became superconductors due to the proximity effect. The researchers could control the superconductivity of the nanowires by running various voltages through the substrate under the wires. Nanowires may also play an important role in nano-size devices like nanorobots. Doctors could use the nanorobots to treat diseases like cancer. Some nanorobot designs have onboard power systems, which would require structures like nanowires to generate and conduct power.

Using piezoelectric material, nanoscientists could create nanowires that generate electricity from kinetic energy. The piezoelectric effect is a phenomenon certain materials exhibit — when you apply physical force to a piezoelectric material, it emits an electric charge. If you apply an electric charge to this same material, it vibrates. Piezoelectric nanowires might provide power to nano-size systems in the future, though at present there are no practical applications. There are hundreds of other potential nanowire applications in electronics. Researchers in Japan are working on atomic switches that might some day replace semiconductor switches in electronic devices. Scientists with the National Renewable Energy Laboratory hope that coaxial nanowires will improve the energy efficiency of solar cells. Because we are still learning about the properties of nanowires and other nanoscale structures, there could be thousands of applications we haven't even considered yet.

NIOBATES NANOWIRES: SYNTHESIS, CHARACTERIZATION AND APPLICATIONS

Perovskite oxides such as alkaline niobates crystal possess many interesting properties including piezoelectricity, pyroelectricity, electro-optic and nonlinear optical response. The most common alkaline niobates material is lithium niobate ($LiNbO_3$). Since the discovery of $LiNbO_3$ ferroelectricity, its properties are widely exploited by electronic devices particularly in telecom applications. Those devices are made from bulk or thin films material and serves as sensors, actuators, detectors or filters. Potassium niobate ($KNbO_3$) is most known for its large nonlinear coefficients ideal for wavelength conversion like second-harmonic generation (SHG), sum frequency mixing, as material in optical parametric oscillator, or lead-free piezoceramics. Sodium niobate ($NaNbO_3$), less studied than $LiNbO_3$ and $KNbO_3$, also belongs to the alkaline niobates. Generally associated with potassium, $NaNbO_3$ is a very promising lead-free piezoelectric ceramics.

Besides bulk and thin films structures, zero- (0D) and one-dimensional (1D) alkaline niobates nanostructures were synthesized recently to combine the dimensional confinement with the other known properties of perovskite materials. Different synthesis routes have been explored to obtain 0D nanoparticles or nanoflakes from alkaline niobates such as mechano-chemical milling, nonaqueous route, sol-gel method or hydrothermal route. Almost simultaneously, anisotropic alkaline niobates 1D structure were synthesized with various methods such as template assisted pyrolysis resulting in regular arrays of tubes, solution-phase synthesis resulting in rod-like structures, or hydrothermal route giving free-standing nanowires with high aspect ratio.

Up to now, the nanomaterials properties have been well characterized using standard materials sciences methods like X-ray diffraction (XRD), scanning electron (SEM) or transmission electron (TEM) microscopy. However, nonlinear optical or electro-optic properties have been rarely studied. Moreover, few applications have used these types of nanowires while combining the various physical properties of perovskite alkaline materials and the anisotropic shape at the nanoscale level. Nanometric SHG light probe manipulated by optical tweezers and capable of guiding light has been already demonstrated, as well as localized SHG light source in optofluidics environment. In this chapter, we focus on the synthesis, optical characterization and applications of Li-, Na-, $KNbO_3$ nanowires. Some crystal and second-order optical properties of the investigated alkaline niobates are summarized. We will describe hydrothermal and molten salt synthesis. We will measure some optical and physical properties of these nanowires. Then we propose the use of plasmonics gold nanoshells to enhance the SHG signal and for best biocompatibility with biological applications. Finally we show interesting applications using optical tweezers and microfluidics chips by combining dielectrophoresis and SHG.

SYNTHESIS METHODS

Several research groups developed different chemical synthetic approaches to fabricate crystalline alkaline niobates nanowires such as exploiting sol-gel route, hydrothermal route and Molten Salt Synthesis (MSS).

Among the three studied alkaline niobate materials, $LiNbO_3$ nanowires synthesis is the most challenging one. So far only two groups reported successful synthesis of free-standing $LiNbO_3$. In contrast multiple synthesis routes can be found for other alkaline materials as potassium niobate $KNbO_3$ and sodium niobate $NaNbO_3$ nanowires. We have investigated two chemical synthesis methods for the nanowires fabrication: hydrothermal synthesis and molten salt synthesis (MSS).

Hydrothermal synthesis is a technique used to crystallize substances under moderate temperatures (200-250°C) and high pressures. Thanks to the use of an autoclave, a thick-walled steel cylinder with an hermetic seal, the wanted crystalline materials can be obtained in one step. The large amount of material that can be fabricated, the easiness (one step synthesis) and the speed of the synthesis make this approach very convenient. To optimize the synthesis parameters one can adjust the temperature, the time, the pressure (by an external pressure or the degree of the autoclave filling), the caustic soda concentration, the solid-liquid ratio and additives to control the properties of the end product. Thus, hydrothermal synthesis is promising because many operation parameters can be modulated to control the particle size and morphology. Due to its simplicity the hydrothermal technique has been widely studied and employed in inorganic synthesis for many years.

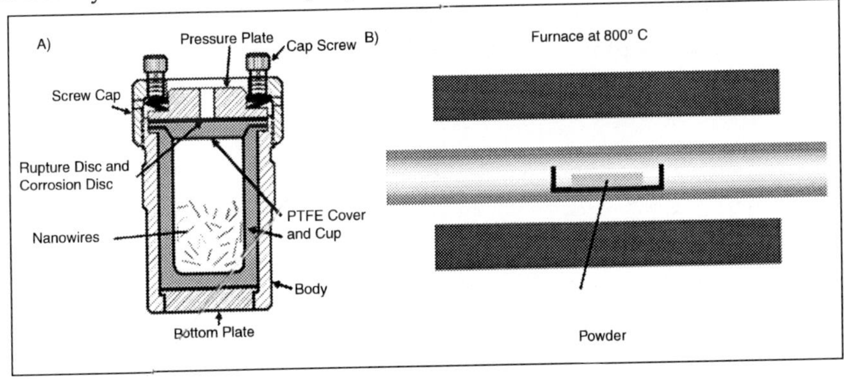

Figure : (*a*) Schematic Diagram of a Typical Laboratory Autoclave from Parr (Acid digestion bomb, 125ml, Model 4748). (*b*) Schematic of the Experimental Setup for Molten Salt Synthesis.

Molten salt synthesis (MSS) is used to prepare complex oxides from their constituent oxide. It is a multi steps synthesis and it is way longer than hydrothermal synthesis approach but more adaptable to other alkaline materials than $KNbO_3$ as shown in the literature. Oxides corresponding to a perovskite compound are mixed with one or two kinds of salt sand then heated at a temperature above the melting point of the salt to form a flux of the salt composition. At this temperature, the oxides are rearranged and then diffused rapidly in a liquid state of the salt. With further heating, particles of the perovskite phase are formed through the nucleation and growth processes. A huge advantage of chemical synthesis in respect to chemical vapour deposition (CVD) or lithographic fabrication of nanowires is the ability to synthesize free-standing nanowires. Thus no further step is needed to isolate or detached the nanowires, because no substrate is involved in the synthesis. Typical results of $KNbO_3$ nanowires hydrothermally synthesized, $NaNbO_3$ nanowires hydrothermal synthesized following recipe and $LiNbO_3$ nanowires molten salt synthesized. Due to the low solubility of lithium hydroxide, molten salt synthesis is preferred to fabricate $LiNbO_3$ nanowires even if it is a multistep synthesis.

Figure : SEM Images of (*a*) $KNbO_3$ Nanowires with an Aspect Ratio up to 25. (*b*) $NaNbO_3$ Nanowires with an Aspect Ratio up to 50. (*c*) $LiNbO_3$ Nanowires with an Aspect Ratio up to 10.

MATERIAL PROPERTIES AND CHARACTERIZATION

To investigate the structure of produced nanowires XRD characterization was employed, confirming the $KNbO_3$ orthorhombic structure in a percentage of 100%, $NaNbO_3$ orthorhombic phase in a percentage of 15% and $LiNbO_3$ trigonal phase in a percentage of 75%.

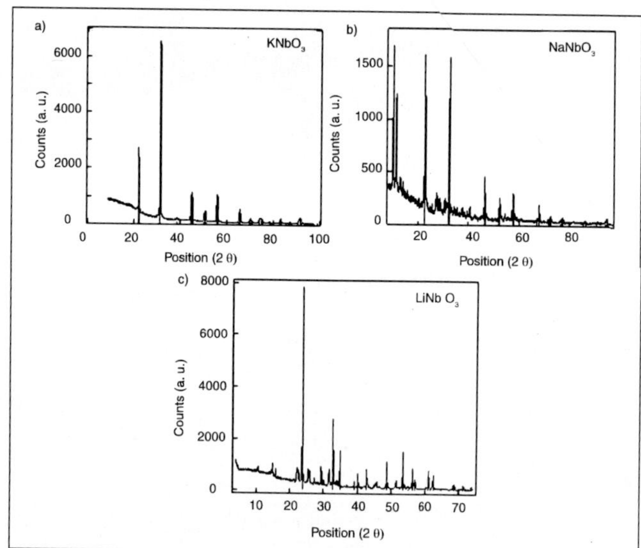

Figure : X-ray Diffraction (XRD) Measurements Together with Theoretical Peaks Expected for Desired Phases of (*a*), $KNbO_3$ (*b*) $NaNbO_3$, (*c*) $LiNbO_3$.

In the hydrothermal synthesis procedure, to increase the percentage of preferred nanowire crystal structure, a calcination step (annealing at high temperatures such as 550°C) can be performed. This calcination step increases the amount of ordered material phase in respect to other phases present in the sample, without changing the shape of the sample. For $NaNbO_3$ synthesis, this additional step resulted in the drastic increase in the amount of pure material switching from 15% to 64% of presence of $NaNbO_3$ ordered phase in the sample. Besides XRD, that will give results over a high number of nanowires, it is possible to perform Raman scattering measurements on single nanowires that allow to distinguish materials and reveal unexpected phenomena. As a preliminary result, we performed Raman measurements on $KNbO_3$ nanowires and for comparison we show $LiNbO_3$ nanoflakes Raman measurements too. Typical $LiNbO_3$ traces show peaks between 300Cm^{-1} and 480Cm^{-1}. $KNbO_3$ has one strong peak right below 300Cm^{-1} and nothing else between 300 and the silicon substrate peak. The spectra are then different enough even between two alkaline niobates to easily distinguish the materials.

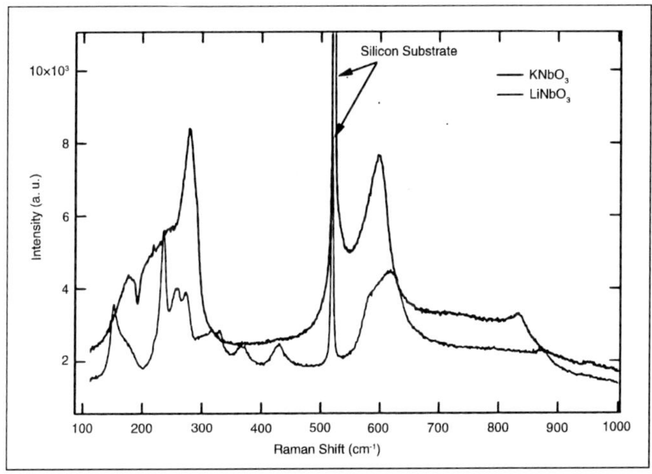

Figure : Raman Spectrum of $KNbO_3$ Nanowires at 80K showing the Typical Shape known for this Material and of $LiNbO_3$ Nanoflakes.

NONLINEAR OPTICAL CHARACTERIZATION

The perovskite nanowires with their non centrosymmetric crystal structures exhibit second-order optical effects which open up a wide range of applications even at the nanoscale level. Indeed, this nonlinear effect scales with the square of the electric field, but it is a volume effect which is measureable down to a single nanoparticle level even in far field microscopy, contrary to weak surface SHG effect of centrosymmetric materials. In the nonlinear regime, the optical response is expressed by the polarization P as a power series in the electric field E as

$$P^{\rightarrow} = \varepsilon_0\chi_1E^{\rightarrow} + \varepsilon_0\chi_2E^{\rightarrow} 2 + \varepsilon_0\chi_3E^{\rightarrow} 3+...$$

where ε_0 is the permittivity of free space and χ_i is the i th-order nonlinear optical susceptibility tensor. Each χ_i represents a different optical effect that can be summarized as follows for a physical understanding of Equation. χ_1, the linear susceptibility, is related to absorption and reflection of light. χ_2 encompasses sum and difference frequency generation such as SHG. χ_3 describes multiphoton absorption, third harmonic generation or coherent anti-Stokes Raman scattering.

Our applications are related to SHG. When a nanocrystal of non centrosymmetric structure is optically excited at a fundamental frequency, it emits the optical signal at the exact doubled frequency. Only materials with crystalline structures lacking a centre of symmetry are capable of efficient SHG. Not only SHG will be scattered but the fundamental frequency too, thus efficient filters are needed to cut the fundamental.

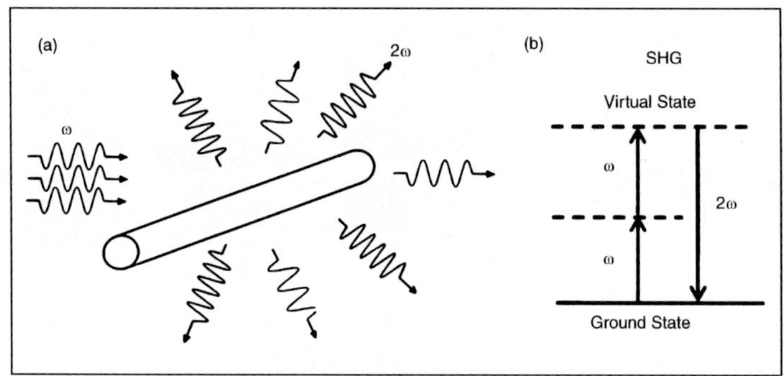

Figure : (a) Schematic Diagram of the SHG Mechanism.
(b) Energy Diagram of the Physical SHG Mechanism.

The setup for the optical characterization of the SHG signal. A near infrared laser light (100 fs Ti:Sa oscillator) is focused onto a sample by lens L1 and the objective (OBJ) collects the signal imaged through a 4f configuration with lens L2. Filters are used to cut the fundamental frequency and detect only a narrow band around the SHG frequency onto an electron multiplying charges coupled device (EMCCD).

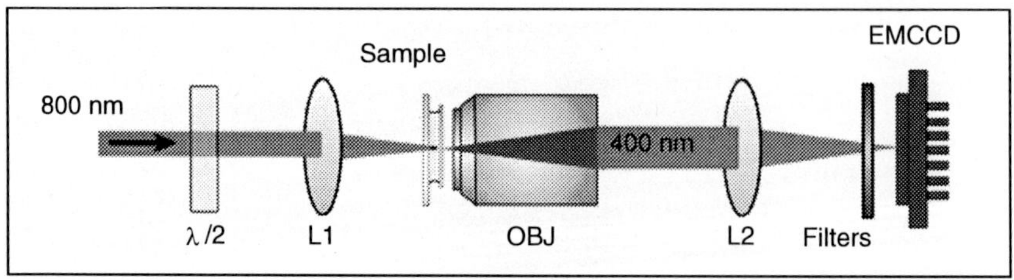

Figure : Setup for Measuring SHG from Nanowires Dried on a Microscope Glass Slide. L_1, L_2, Lenses; λ/2 Half Wave Plate; OBJ, Objective.

Typical SHG measurements at different incident polarization angle are display. When the polarization is parallel to the nanowire the signal is the strongest, which is expected for a nanowires with an optical axis along the nanowire. Full polar SHG characterization is then possible and a fit of the experimental data can confirm the crystal orientation of a nanowire, which is not a priori known for chemically bottom-up synthesized nanowires.

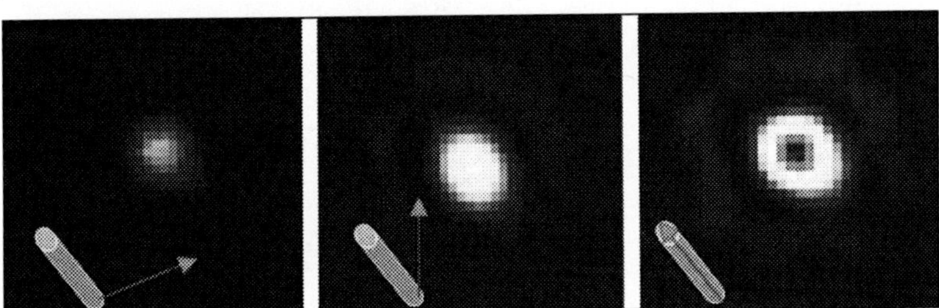

Figure : EMCCD Images of a Single Niobate Nanowire under different Incident Light Polarization. The Insets Sketched the Wire Position and the Incident Light Polarization.

In addition to conventional measurements of SHG from niobate nanowires we have carried out SHG characterization by using Optical Tweezers (OT). An OT uses a focused laser beam to provide an attractive or repulsive force (typically on the order of pN), depending on the refractive index mismatch, to physically hold and move microscopic dielectric objects. Optical tweezers are routinely used tools in both physical and life sciences for manipulating objects from micron to the atomic scale and as force transducers in the pN range.

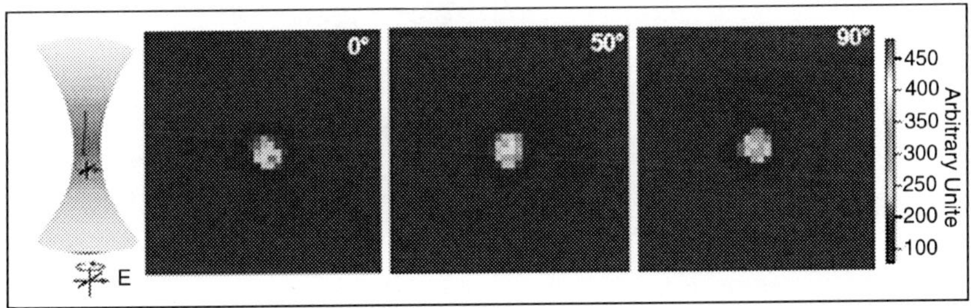

Figure : EMCCD Images of Trapped Single Niobate Nanowire under different incident Light Polarization.

The ability to control and monitor the position of a mesoscopic object with nanometric precision is important for the rapid progress of nanoscience and it is perfectly suited for the study of biological phenomena. Besides its application in biology, OT are also an appealing tool for semiconductor nanowire integration owing to their ability to act in situ in closed aqueous chambers, their potential applicability to a broad range of dielectric materials, their spatial positioning accuracy (<1 nm), and the degree to which their intensity, wavelength and polarization can be controlled using tuneable lasers.

All trapping experiments and subsequent SHG characterization were carried out on a home-built single beam optical tweezers system.

As in the conventional setup, we can probe SHG from individual niobate nanowire under different incident light polarization. The SHG signal under different incident polarization is much less sensitive in this orientation due to the tweezing geometry. However, it may still slightly differ from wire to wire due to the variability of crystalline phases of the bottom-up synthesized nanowires.

PLASMONICS NANOSHELLS FOR ENHANCED SHG

We showed that non centrosymmetric nanowires exhibit SHG signal. However, nonlinear optical processes such as harmonic generation are generally inefficient at very small scale (the intensity goes down with the square of the particle volume). By developing nonlinear optical plasmonics core shell cavities, it was possible to strongly enhanced the SHG response of $BaTiO_3$ nanoparticles. Similarly, $KNbO_3$ nanowires are covered by a thin layer of gold to reach a near infrared plasmonics resonance and enhance the SHG signal for optimized used as imaging probes or localized light source.

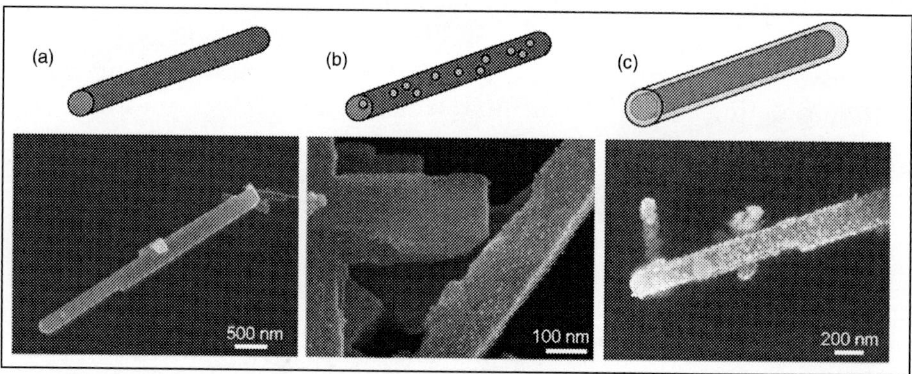

Figure : Synthesis Steps of $KNbO_3$/Au Core-shell Nanowires with their Corresponding SEM Images. (*a*) Bare $KNbO_3$ Nanowires (*b*) Surface Seeding with 2-3 nm Colloidal Gold Particles (*c*) Growth of a Gold Complete shell Around the Wires.

The gold coating process for $KNbO_3$/Au core-shell nanowires involved three main steps. First of all, primary amines are coated on the surface of the $KNbO_3$ nanowires. The amino complex (Aminopropyltriethoxysilane) presents NH_2 complex at the surface of the wires. Then, 2-3 nm gold particles are adsorbed onto the surface of $KNbO_3$ wire thanks to the silane-amine functionalization (Au^{3+} cations are attracted by the primary amine NH_2 which become secondary amine). Finally, the Au shell is grown all around the particle from the seeded particles by a reaction of reduction of gold by hydroxylamine.

The gold reduction process to coat the wire surface still needs some improvement for a better and more uniform coverage of the surface. However, inhomogeneities may even enhance the plasmonics effects as commonly used in photovoltaic devices based on plasmonics nanoparticles. Further measurements will be performed to compare SHG signal from coated and uncoated wires.

APPLICATIONS IN OPTOFLUIDICS ENVIRONMENT

We show applications of nanowires in an optofluidics environment. First we describe the devices fabrication, then a simulation of the device and finally experimental results with the nanowires.

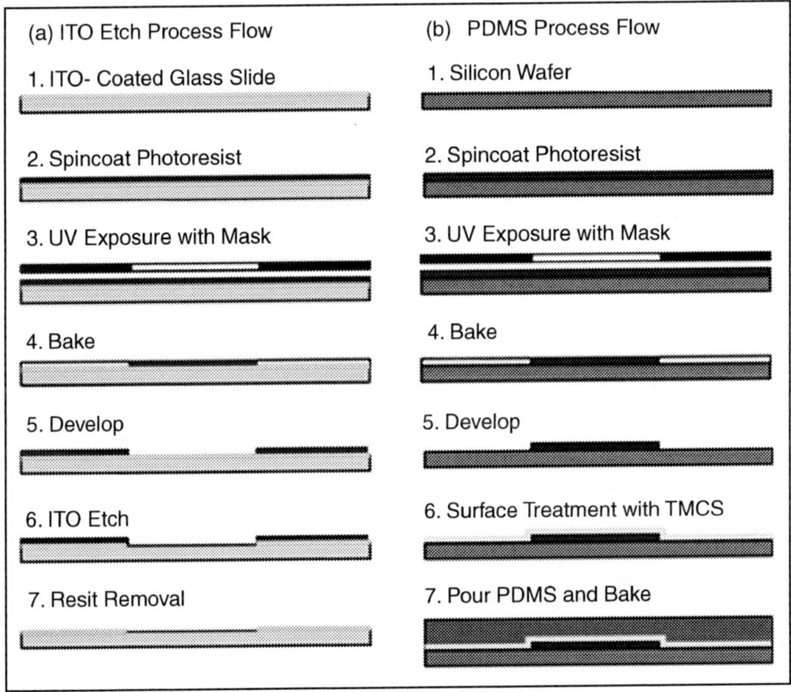

Figure : (*a*) ITO Process Flow), (*b*) PDMS Process Flow.

The optofluidics device fabrication implies different process flows depending on the applications. For dielectrophoresis applications, a 1 mm thick glass plate with a conductive electrode pattern in indium tin oxide (ITO) is used as described in. The sample is fabricated using contact photo lithography. First, we obtain ITO coated glass plates rated at 30-60 Ω/square. Next, positive photoresist is spincoated onto the plates. A pattern is exposed using a UV mask aligner and developed. Finally, the ITO is etched to create the electrode pattern. In case of microfluidic channels that are made of polydimethylsiloxane (PDMS), the fabrication uses the replica moulding technique. A master mold is produced through UV lithography on a silicon wafer. After trimethylchlorosilane (TMCS) treatment of the master mold for 5 min, the PDMS is poured onto the mold with a 10 : 1 base-to-curing agent ratio. After curing in an oven at 80°C for 1 h, the silicone is released from the mold.

For the first type of device, a floating dielectrophoresis electrode design was used to prevent shorting the nanowire after being trapped. This concept is best demonstrated by the COMSOL simulation. A voltage is placed between the top and bottom electrodes.

Due to the geometry of the floating electrodes, high electric fields are generated at the tips in the middle floating electrodes. The direction and magnitude of the electric field is indicated by the arrows. The high electric fields serve as dielectrophoretic traps and the direction of the electric field will align the nanowires due to electro-orientation. For the specific design, triangular ITO electrodes were designed with a 2 μm gap and the triangular ITO electrodes were separated by the contact electrodes with a 5 μm gap. The application of an external electric field induces an additional electro-orientation force caused by dielectrophoresis (DEP). DEP forces have been utilized to orient and manipulate several different types of dielectric nanowires.

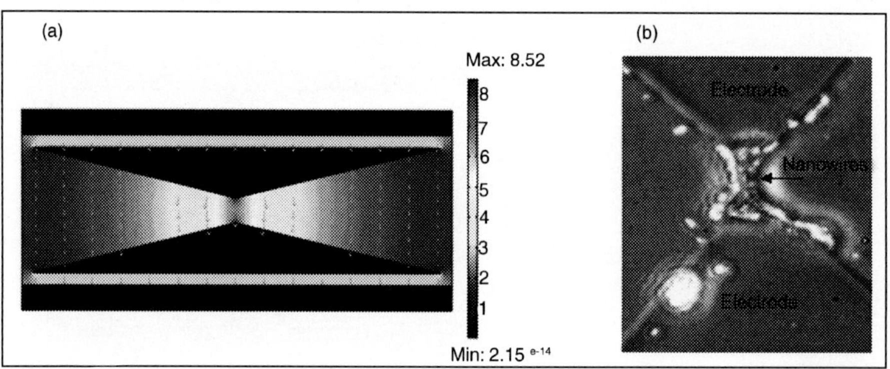

Figure : (*a*) COMSOL Simulation of the Electrode Design. Normalized Surface Electric Field (V/m), Arrow Electric Field (V/m).(*b*) White Light Image of the Trapping of Lithium Niobate Nanowires at the Tips of the Floating Electrodes.

A standard ITO etch with a 5:6:1 ratio of H_2O:HCl (40%):HNO_3 (60%) at room temperature was utilized. By changing the etching time, we were able to control the size of the gap between the triangular electrodes. With a 5 minutes etch, the electrodes were shorted and there was no gap. At 10 minutes, a 1 μm gap existed. By 15 minutes, a 2 μm gap was formed. Results utilized a 10 to 15 minutes electrode etch. A PDMS microfluidic channel with the dimensions of 100 μm wide, 10 μm deep and 1.5 cm long was placed to overlap with the electrode gap. The microfluidic channel can hold 15 nL of liquid. To obtain at least a single nanowire in the entire channel, a concentration of 6.7 x 10^4 particles/mL is required. The trapping of several nanowires with the floating electrodes design.

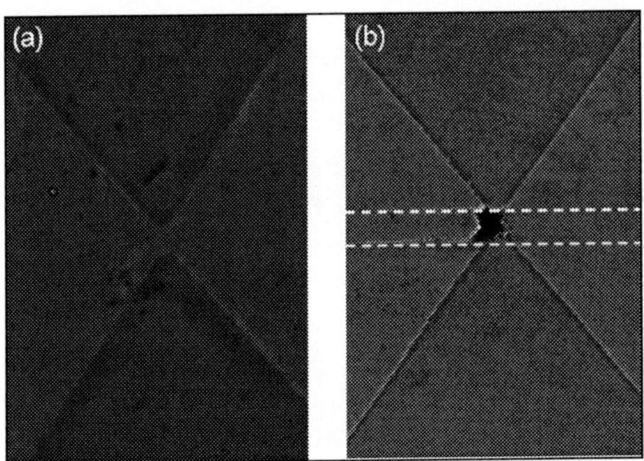

Figure : (*a*) Empty Electrodes (gap = 2 μm width)(*b*) Electrodes with PDMS Channel (indicated with dashed lines) to Concentrate the Nanowires at the Tip.

On top of the previous electrode device, we placed a PDMS chip with a single channel to orient the nanowires between the two electrodes. The purpose of such design is to be able to detect electro-optic effects. Indeed, the bulk photovoltaic effect present in $LiNbO_3$ might be of interest to generate locally an electric field under laser illumination. It was possible to concentrate nanowires at the electrodes tip.

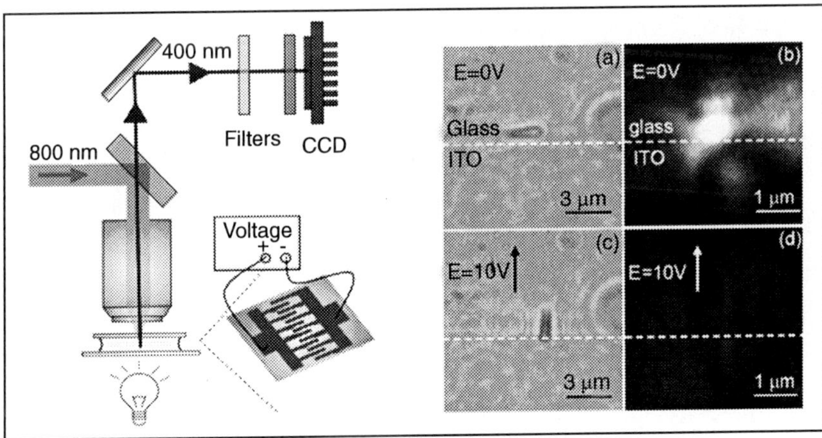

Figure : *Left*: Schematic of the Setup for Tweezing Optically and Electrically Nanowires and Detecting their SHG Signal on a CCD after Passing through Filters to Eliminate the Incident Light. *Right*: DEP Response of a $LiNbO_3$ Nanowire Suspended in 170 μS/cm Conductivity de-ionized Water (a and b) without and (c and d) with an Electric Field of 10 V_{pp} at 150 kHz. (*a*) and (*c*) White Light Images; (*b*) and (*d*) SHG Response.

The last device we developed was to manipulate lithium niobate nanowires in a fluidic environment and monitor the second-harmonic generation (SHG) response using an optical trapping setup. The conductive electrode pattern is interdigitated with specific spacing and width that were previously determined. The optical tweezer is generated with a specific polarization that is determined with a half wave plate and located outside of the fluidic region, close to either substrate. Therefore, the nanowire is not allowed to orient along the direction of propagation and it locates itself orthogonal to the beam. This polarization imparts a force to orient the nanowire. Furthermore, an external electric field is applied to the nanowires using the pattern of electrodes. This external electric field induces an additional electro-orientation force caused by the dielectrophoretic (DEP) response of the nanowires. Note that the torque on the nanowire due to the external electric field is almost ten times greater in magnitude than the torque on the nanowire due to the optical polarization of the laser for the particle orientation denoted above.

White light images (a and c) of a nanowire suspended in deionised water. At time $t = 0$, no electric field ($E_{electrodes}$) is applied and the nanowire is only tweezed by the laser (E_{laser}) with a polarization parallel to the electrodes. Then a 10 V_{pp} electric field with a frequency of 150 kHz is applied and after 1 second, the wire is oriented along this electric field due to the stronger field than the one from the laser. The SHG response measured with and without the applied electric field.

The nanowires are aligned with the field due to positive DEP forces. Increasing the conductivity of the suspension as well as changing the frequency of the applied electric field can change the type of DEP force on the nanowire which will result in the nanowire to not be aligned with the field. We notice the crossover conductivity between negative and positive DEP when the conductivity of the solution is around 170 μS/cm. The combination provided by the applied external electric field and the polarization of the laser beam serves useful for several measurements in the nanometre scale. For instance, an estimation of the conductivity near the nanowire may be obtained. This may be useful if there is a conductivity gradient in the sample. Moreover, the polarization dependency of the nanowire allows for the detection of its position. The SHG signal can then be useful for detecting smaller wires close to the diffraction limit of a bright field microscope.

APPLICATIONS OF OPTICALLY TWEEZED NANOWIRES

Recently, Nakayama *et al.* proved that several different types of nanowires can be stably trapped in optical traps; and that optical traps can thus be used to manipulate assembled three-dimensional nanowire heterostructures. Individual optically trapped nanowire can be placed and held in direct contact with living cells; in addition, one class of nanowires ($KNbO_3$ wires) exhibits efficient SHG and act as frequency converters, raising the possibility for a new type of scanning light microscopy. Due to their small cross-section, nanowires represent ideal probes for mechanical and optical stimulation of cells and even organelles without overloading the sample with photons.

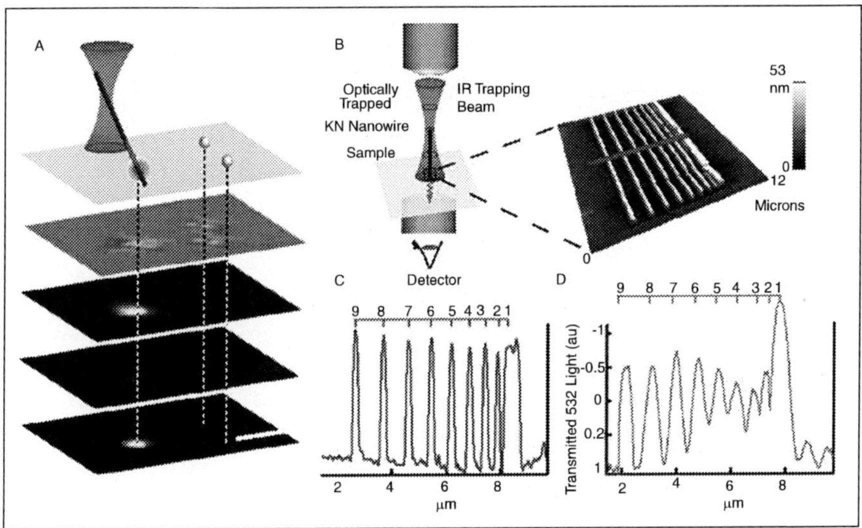

Figure : Schematic of the Experimental Geometry where Laser Tweezers are used to Accurately Position the Nanowire and Excite Fluorescence (A) Excitation of a Fluorescent bead by Waveguided SHG Signal from an Optically trapped $KNbO_3$ Nanowire. (B) Schematic of inverted Optical Scanning Configuration band AFM Topographic Image of Thermally Evaporated Pattern of Gold Stripes on a Glass Coverslip. (C) AFM Line Scan from Region Indicated in B). (D) Optical Transmission Profile Captured by Scanning a Single $KNbO_3$ Nanowire over the Metallic Surface Structure. The Nanowire Dimensions used to Create the Transmission Line Scan was measured by AFM: width=122 nm, length =1.4 μm and height=53 nm.

The nanowires can be also used as complementary biomarkers for long term cell tracking experiments. Two important nanowire features such as subwavelength waveguiding and frequency conversion capability with the ability to use optical tweezers to manipulate nanowires in realistic physiological environments can be used for various biological applications as local light and force sources. Due to the subwavelength optical waveguiding nature of niobate nanowires, it is possible in combination with laser tweezers to create highly localized excitation source with the size which is determined by the nanowire tip diameter and to achieve the high spatial accuracy of nanowire position. This method is not only limited to the membrane proteins but it can be extended also to the imaging of intracellular compartments such as mitochondria. The success of intracellular imaging depends mostly on the nanowires functionalization with appropriate surface chemistries which will decrease the force required to introduce wires in the cell.

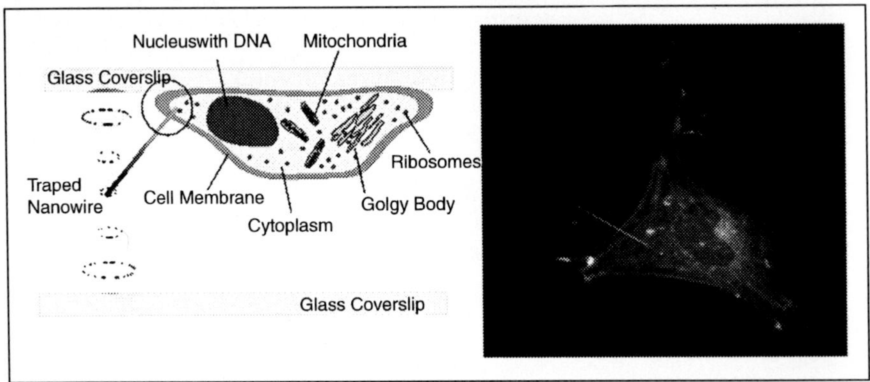

Figure : Schematic of the Experimental Geometry where Laser Tweezers are used to Accurately Position the Nanowire on the Cell Membrane while the Nanowire through efficient SHG Creates Local Excitation Source.

How most common types of alkaline niobates nanowires, $KNbO_3$, $NaNbO_3$ and $LiNbO_3$ can be synthesized via hydrothermal or molten salt synthesis. We performed materials characterization to determine the composition of the end product of the synthesis as well as electron imaging to determine the aspect ratio of the three types of nanowires. As a further characterization step, we provided nonlinear optical measurements on single niobates nanowires with different setup geometries: transmission setup or optical tweezers.

A SHG signal was measured for all types of nanowires as it is well known in similar non centrosymmetric bulk materials and it was possible to vary the SHG intensity by changing the polarization. Observed nonlinear properties of alkaline niobate nanowires suggest that all three types of nanowires could be used as frequency converters, mechano-optical probes or logic components at the nanoscale.

Then, plasmonics nanoshells were synthesized around $KNbO_3$ nanowires. The thin gold layer serves, first, as enhancing the SHG signal through the plasmon resonance in the near infrared wavelength range of the excitation laser light. And secondly, it makes niobates nanowires compatible with biological entities and it can be an excellent starting point for further DNA or protein functionalization.

Several applications are described and performed to best use the nanowires properties at the overlap between the micro- and the nanoscale. Nanowires are demonstrated as trapped SHG probes for localized illumination. Using the polarization dependency of the SHG signal allow us to determine the orientation of nanowires in optofluidics environment too.

To conclude, applications using the alkaline niobates nanowires are only at the beginning and they are not yet using all the physical properties known for these perovskite materials. Therefore, there is room to improve the synthesis, for instance aspect ratio, to better know the physical properties and to be able to measure small signal as electro-optic effects on single nanowire.

RATIO GOLD NANOWIRES WITH HIGHLY POROUS MORPHOLOGY

Gold nanoparticles, discrete ensembles of gold atoms with one or more dimensions on the submicron scale, are materials that have attracted considerable interest in recent decades owing to their unique-size-dependant electronic and optical properties as well as their high unit surface areas ; the combination of which makes them particularly suited to use in a variety of catalytic and chemical sensor applications.

Typically, these applications utilize small solid nanoparticles ranging between 2–50nm in diameter, as this size range is where such nanoparticles exhibit their useful properties. However, the use of such small nanoparticles presents a number of difficulties.

For instance, gold nanoparticles with these dimensions possess rather high surface energies. Therefore, in order to maintain a high unit surface area, the implementation of methods to stabilize them against aggregation is required. These methods must be carefully selected so that useful (in particular surface) properties are retained. Such small nanoparticles are also relatively difficult to separate from solution, making their reuse in applications costly.

One recent development that promises to help overcome these issues has been the preparation of porous gold nanowires. Owing to their high aspect (length to diameter) ratios, such particles can be easily isolated from solution, while their porosity yields exposed surface features (known as ligaments) with dimensions suitable for use in applications, even when the particles as a whole are relatively large.

Such porosity also ensures that these particles have much higher surface areas relative to solid nanowires with equivalent dimensions. Furthermore, unlike solid nanoparticles, aggregation is not a significant problem, as the particle ligaments and surfaces are largely internal and, therefore, remain accessible following aggregation.

Very few methods have been published to date by which porous gold nanowires can be synthesized. Current methods involve the electrodeposition of gold into a nanoporous template; either as a pure gold, followed by galvanic exchange reaction, or as a single-phase, two-component gold alloy, which is subsequently dealloyed through chemical etching of the nongold metal. Of these methods, electrodeposition of gold alloys into nanoporous templates followed by selective chemical etching is perhaps the most advantageous. It is a relatively simple synthetic technique that is amenable to the formation of large quantities of nanoparticles, and it provides a high level of control over the particle dimensions (through choice of template pore diameter and the quantity of deposited metal) as well as morphology and composition (through modulation of various deposition parameters and the use of multiple plating steps to incorporate different components, both porous and solid, along the length of the particles).

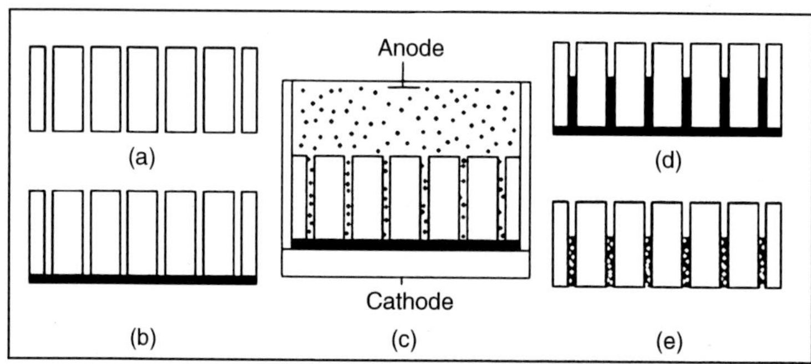

Figure : The Process for Synthesizing Porous Gold Nanowires by Template Electrodeposition with Chemical Etching: A Nanoporous Template *(a)* is Coated on one Side with a Conductive Layer *(b)* Gold and the Sacrificial Metal are then Simultaneously Electroplated Exclusively into the Template Pores onto the Conductive Layer *(c)* The Resulting Alloy Filled Template *(d)* is then Treated to de-alloy The Nanowires (and thereby form pores) *(e)*, Followed by Removal of the Template to Isolate the Nanowires.

To date, however, the vast majority of the literature has only involved large porous nanowires (≥200nm in diameter), which are not ideal for use in the aforementioned applications as they suffer from either comparatively small surface areas (in the case of wires with large pores) or relatively hard to access internal surfaces due to the large diameter of these nanowires (in the case of particles with small pores).

Nanowires with smaller diameters would help remedy these problems; however, the only report to date of such narrow (<60nm diameter) porous gold nanowires involve particles with what has been described as a "hinged" morphology, the formation of which has been ascribed to the limits placed on the nanostructure growth mechanism (spinodal decomposition of the deposited gold upon etching of the silver) by the narrow membrane pores. Porous gold nanowires with this morphology are characterized by large ligaments (relative to the nanowire diameter) and correspondingly small pores, making these particles nonideal for use in catalytic and sensing applications using gold.

Here, we report the synthesis of narrow (diameter ≤60nm), high-aspect-ratio porous gold nanowires (through electrodeposition into nanoporous templates) which possess a unique porous morphology that is characterized by small ligament dimensions. These nanowires were unexpectedly fabricated using a modified version of an established procedure, involving the electrodeposition of silver and gold within the narrow pores of a commercially available filtration membrane in discrete steps, rather than through the simultaneous electrodeposition of both metals from a single plating solution. This approach is advantageous not only in that it yields highly porous gold nanowires with an open pore structure, but also in that it grants a larger degree of freedom in the choice of electroplating bath chemistries: a feature that is not present in current literature methods, which require the metal species to have similar reduction potentials.

NANOROD SYNTHESIS

A commercially available radiation track-etched, hydrophilic polycarbonate membrane is first prepared for use as a nanoporous template by sputter-coating one side with a 10nm thick layer of chromium (adhesion layer) followed by a 200nm thick layer of silver (conducting layer) using a Cressington 208HR sputter coater.

The silver-coated membrane is then placed coated side down onto the cathode; a steel contact plate (adhesion of the membrane to the plate is enhanced by placing 1-2 drops of water onto the contact plate), and the electrochemical cell is assembled by clamping a rubber O-ring (34mm diameter) and glass tube onto the membrane. This setup ensures that metal deposition occurs exclusively into the pores of the membrane. The cell is then filled with Millipore Nanopure water (18MΩ) and the membrane left to soak for 30min in order to completely wet the membrane pores so subsequent electrodeposition will occur in all pores simultaneously.

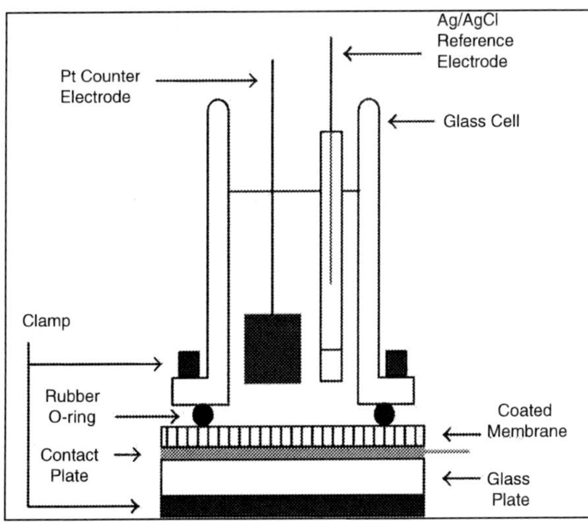

Figure : Diagram of the Electroplating Cell used to Deposit Metal into the Pores of a Nanoporous Membrane.

Following this, the water is replaced with a silver electroplating solution (Silver Cyless II (silver oxide), Technic). An Ag/AgCl reference electrode and a platinum counter electrode (anodic surface area ~2cm2) are then suspended in the solution close to the membrane. 100mC of silver (0.112mg) is then deposited potentiostatically (–1.0V versus Ag/AgCl) using a BAS 100b electrochemical analyser. Following silver deposition, the cell is thoroughly rinsed three times with nanopure water, and the cell is then filled with a gold plating solution (434 neutral soft gold (gold cyanide), Technic). This electrodeposition and rinsing process is then carried out with the gold solution in the same manner.

NANOWIRE COLLECTION

The filled membrane is first subjected to a brief (10sec) treatment with 4M nitric acid at room temperature in order to selectively dissolve the silver. The membrane is then rinsed with distilled water, blotted dry, and placed into a 10mL polyethylene centrifuge tube. The tube is then filled with 6mL of dichloromethane and the membrane is dissolved with brief sonication and shaking.

The nanowires are then sedimented into a pellet by centrifuging the suspension at 3200RPM (1090g) for 30min. The supernatant is removed and replaced with 3mL of pure dichloromethane and the pellet is broken up with sonication. This process is repeated several times to remove the residual polymer from the membrane. This process is then further repeated twice more using ethanol instead of dichloromethane, with the final product being stored in 3mL of ethanol.

NANOROD CHARACTERISATION

Nanowire samples are imaged by transmission electron microscopy (TEM) using a Phillips CM200 TEM equipped with a Gatan 832 SC1000 CCD camera and energy-dispersive X-ray analysis (EDAX) apparatus at an accelerating voltage of 200kV. The spot size setting for general imaging is 3, but it is set as small as spot size 7 for collecting elemental composition (EDAX) data at specific locations of the nanowires. Nanowire samples are prepared for TEM by briefly sonicating nanowire suspensions in ethanol, and then placing one drop (~5μL) of the resulting suspension onto a Cu grid (200mesh) that is coated with a forever membrane. The ethanol is then allowed to evaporate before collecting images.

Results

TEM images (along with EDAX analysis) of the nanowires that result from the deposition of 25mC of gold (0.051mg) into the pores of polycarbonate membranes under the conditions used in this work. It is observed that highly porous nanowires that are comprised entirely of gold have formed. These nanowires have a mean diameter of 52.4nm (standard deviation: 6.3nm); wider than the rated 10nm pore diameter of these nanoporous templates due to the well-known "cigar shaped" pores of commercially available PCTE membranes. They are also noted to be substantially longer than their diameter (a relatively high aspect ratio).

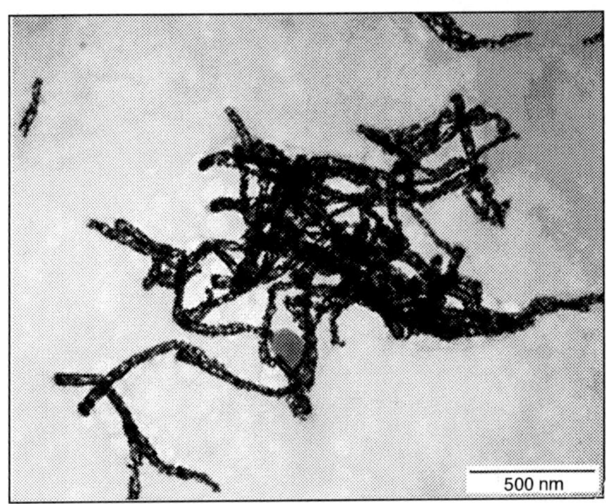

Figure : TEM Images of Porous Gold Nanowires (25mC).

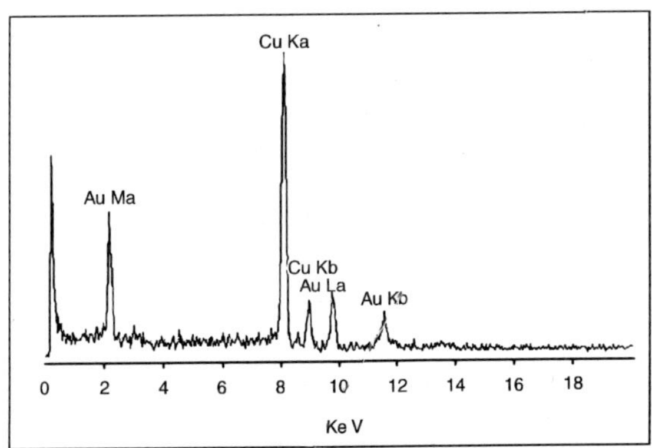

Figure : EDAX Analysis of Porous Gold Nanowires.

Examination of the fine structure of these nanowires by high-resolution TEM possess a structural morphology that is markedly different from the so called "hinged" morphology that is characteristic of similarly sized porous gold nanowires as previously cited in the literature.

While the hinged morphology is comprised of stacks of large ligaments (relative to the nanowire diameter) that are aligned perpendicular to the length of the nanowires (giving them a lumpy appearance), the nanowires produced in this work possess a porous structure that is defined by relatively small ligaments (most often aligned either parallel or perpendicular to the length of the nanowires), with large open pores extending through the width of the nanowires.

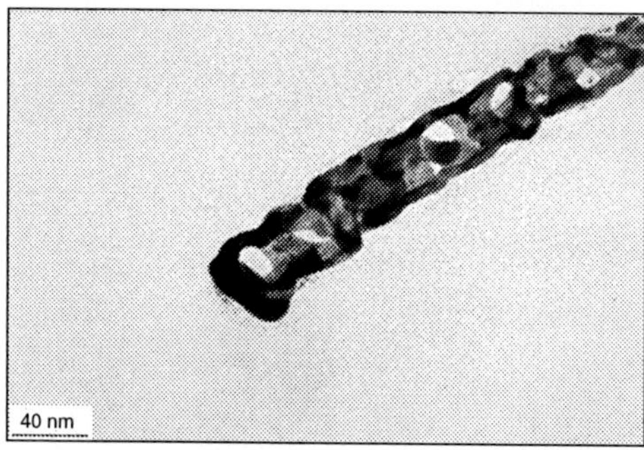

Figure : High-resolution TEM Image of Porous Gold Nanowires.

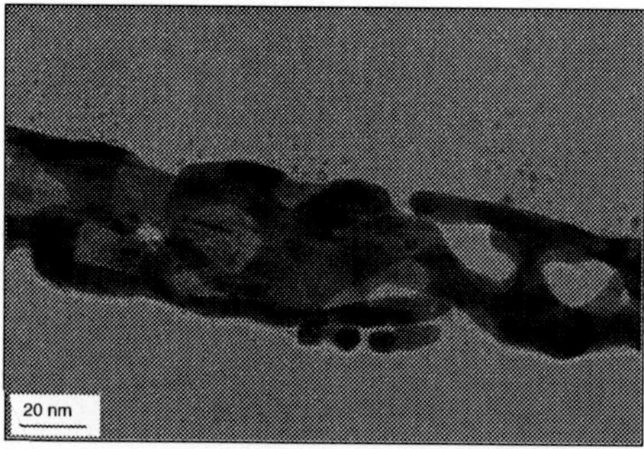

Figure : High-resolution TEM Image of Porous Gold Nanowires that Demonstrates their Unique Morphology.

The formation of such high-porosity gold nanowires, while an improvement over previous work, is noted as being most unusual, as the procedure used in their synthesis involves the deposition of gold under electroplating conditions typically used to form solid gold nanowires rather than porous gold. In order to explain this result, several possibilities were considered. Firstly, it was considered that the current density during electrodeposition of the gold may have been high enough that hydrogen gas was evolved at the cathode during the electrodeposition process, with the resulting bubbles remaining trapped in the narrow membrane pores, thereby creating voids in the deposited gold. To determine if this may be the case, current versus time data for the gold deposition was thus examined.

As can be seen, there is an initial peak in current which rapidly drops, followed by a more gradual decrease towards a steady state. The initial current spike is ascribed to rapid charging of the cathode electric double layer upon application of the nonequilibrium potential difference between the anode and cathode, as the increase in potential difference leads to an increase in the charge separation in the diffuse double layer; the movement of these charges being experienced as a (capacitive) current. This charging process is completed rapidly, yielding the drop in current to a level commensurate with that due to the electrodeposition process. The subsequent gradual current decrease towards a steady value (2.98mA/cm^2) is indicative of the formation of a diffusion layer of growth species within the pores; a result of the rate of reduction of metal species at this overpotential being greater than the rate of replenishment of the metal species near the cathode by mass transport into the pores from the bulk solution. This then suggests that the deposition process is mass transport controlled, which is not unexpected given the membrane pore dimensions and negative charge on the gold species. Importantly, it can be seen from this data that the actual current density at the cathode (excluding the capacitive "current") is well within the limits for gold deposition with high current efficiency for this plating solution (1.1–10.8mA/cm^2) for effectively the whole deposition period. This means that essentially all of the charge passed was used to reduce the gold species rather than evolve hydrogen gas.

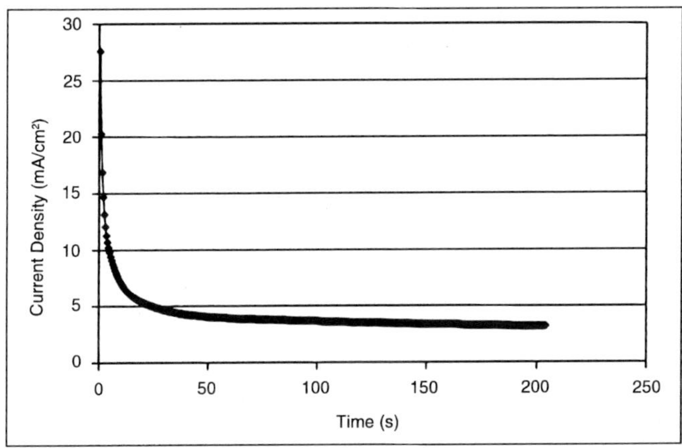

Figure : Current versus Time Data for the Electrodeposition of Gold into Nanoporous Templates.

A second possible explanation is that these structures are in fact incompletely formed gold nanotubes, which can come about due to the reduction of gold species that are preferentially adsorbed on the walls of the membrane pores. This explanation was contemplated as the prevailing conditions during the gold deposition; namely, low overpotentials (the reduction potential of gold cyanide is –0.6V versus Ag/AgCl) and a low gold concentration near the cathode are conducive to the formation of such gold structures. In order to test this hypothesis, syntheses were carried out with larger amounts of gold being deposited into the membrane pores (50, 75, and 100mC). The reason for doing this is that the growth mechanism of these structures involves the formation of long complete nanotubes followed by inward growth towards solid nanowires (starting at the cathode surface and advancing towards the pore entrance). Therefore, greater amounts of deposited gold should first yield complete nanotubes, followed by thicker nanotubes that ultimately grow into solid nanowires. However, TEM imaging

depositing more gold instead yields nanowires that retain a porous gold segment (similar to those produced after the deposition of 25mC of gold) attached to a solid segment comprised of gold, which gets progressively longer with increased quantities of deposited gold.

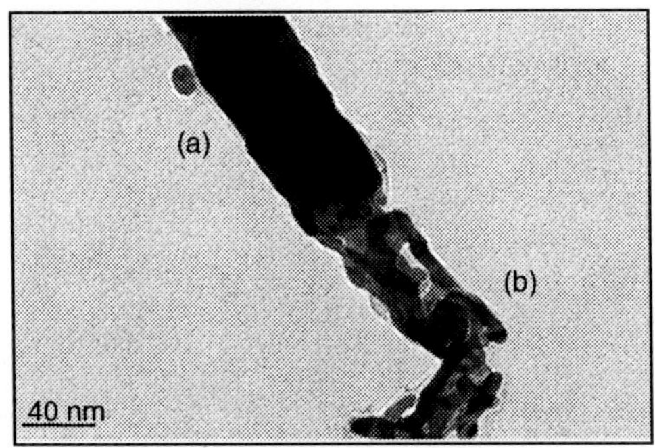

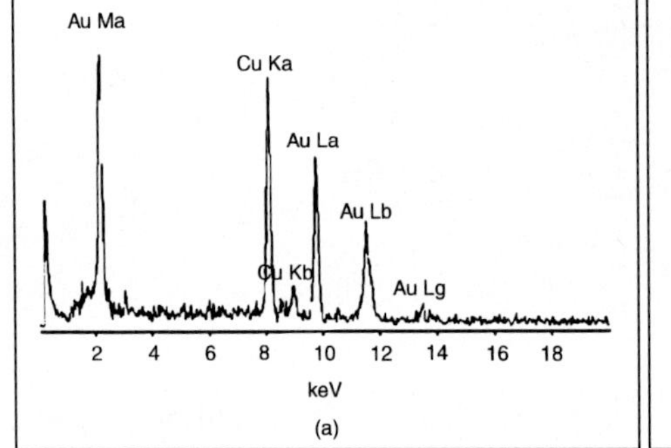

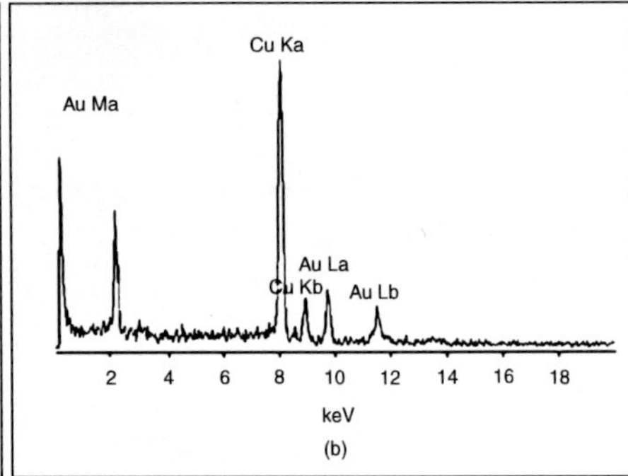

Figure : High-resolution TEM Image and EDAX Analysis of a bi-segmented Nanowire Comprised of (a) a Solid Gold Segment and (b) a Porous Gold Segment.

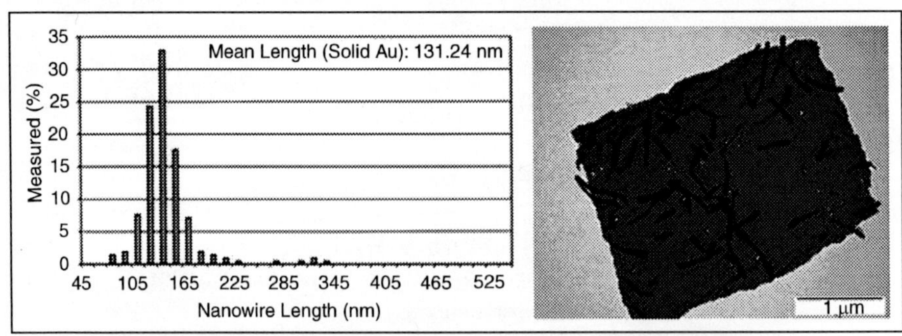

Figure : (*a*) Bisegmented Gold Nanowires Resulting from 0.05C of Gold being Deposited. (*b*) Bisegmented Gold Nanowires Resulting from 0.075C of Gold being Deposited. (*c*) Bisegmented Gold Nanowires Resulting from 0.1C of Gold being Deposited.

Based on this result, it was then hypothesized that these porous gold structures may in fact be due to codeposition of the gold with residual silver species from the previous electrodeposition, which remain within the pores even following the use of established pore rinsing methods. This explanation would account for the formation of both the porous segments (coarsening of the gold upon chemical removal of the silver from the resulting gold-silver alloy) and the additional solid gold segments upon further gold deposition (residual silver species are eventually depleted yielding only gold deposition). To test this explanation, TEM images were collected of nanowires (50mC of charge passed) prior to treatment with nitric acid. As predicted, nanowires with two distinct segments are present; a solid gold segment, and a lower electron density segment that is confirmed by EDAX to be comprised of a mixture of both gold and silver.

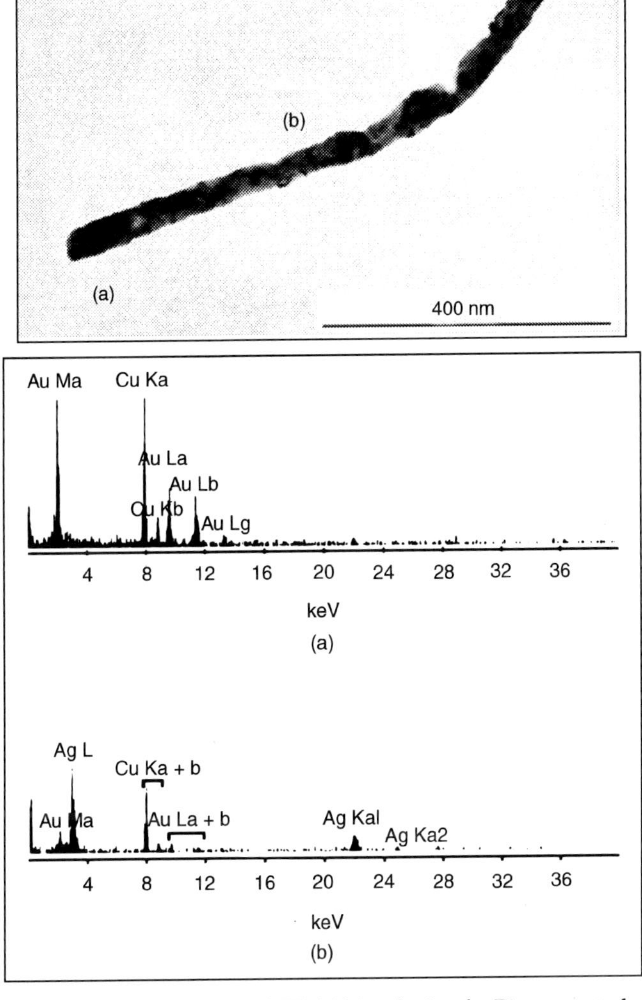

Figure : TEM Image and EDAX Analysis of a Bisegmented Nanowire Prior to Dealloying.

Overall, given that gold-silver alloyed nanowires are formed by this procedure, and that porous gold nanostructures are formed following treatment with nitric acid, it is reasonable to conclude that these structures are a result of the coarsening of the gold during chemical removal of the silver.

REFERENCES

- Chacko, Philip; Christabelle Noronha and Sujata Agrawa: *Small Wonder : The Making of the Nano*, Westland, Delhi, 2010.
- Devoret, M.H. : *Single Charge Tunneling in Coulomb Blockade Phenomena in Nanostructures*, Plenum Press, New York/London, 1992.
- Hillebrands, B. : *Processes and Characterization of Advanced Nanostructured Materials*, Macmillan Publishers India, Delhi, 2011.
- Mathur, K.P. : *Nanotechnology Applications*, Rajat, Delhi, 2007.
- Phani, N. : *Principles of Nanotechnology : Materials, Tools and Process at Nanoscale*, Scitech Pub, Delhi, 2010.
- Raghavan, Y.S. : *Nanostructures and Nanomaterials: Synthesis, Properties and Applications*, Arise Pub, Delhi, 2010.

6

Nanobiomaterials

BIOMATERIALS

A biomaterial is any matter, surface, or construct that interacts with biological systems. The development of biomaterials, as a science, is about fifty years old. The study of biomaterials is called biomaterials science. It has experienced steady and strong growth over its history, with many companies investing large amounts of money into the development of new products. Biomaterials science encompasses elements of medicine, biology, chemistry, tissue engineering and materials science.

SELF-ASSEMBLY

Self-assembly is the most common term in use in the modern scientific community to describe the spontaneous aggregation of particles (atoms, molecules, colloids, micelles, etc.) without the influence of any external forces. Large groups of such particles are known to assemble themselves into thermodynamically stable, structurally well-defined arrays, quite reminiscent of one of the 7 crystal systems found in metallurgy and mineralogy (*e.g.* face-centred cubic, body-centred cubic, etc.). The fundamental difference in equilibrium structure is in the spatial scale of the unit cell (or lattice parameter) in each particular case.

Molecular self-assembly is found widely in biological systems and provides the basis of a wide variety of complex biological structures. This includes an emerging class of mechanically superior biomaterials based on microstructural features and designs found in nature. Thus, self-assembly is also emerging as a new strategy in chemical synthesis and nanotechnology. Molecular crystals, liquid crystals, colloids, micelles, emulsions, phase-separated polymers, thin films and self-assembled monolayers all represent examples of the types of highly ordered structures which are obtained using these techniques. The distinguishing feature of these methods is self-organization.

STRUCTURAL HIERARCHY

Nearly all materials could be seen as hierarchically structured, especially since the changes in spatial scale bring about different mechanisms of deformation and damage. However, in biological materials this hierarchical organization is inherent to the microstructure. One of the first examples of this, in the history of structural biology, is the early X-Ray scattering work on the hierarchical structure of hair and wool by Astbury and Woods. In bone, for example, collagen is the building block of the organic matrix—a triple helix with diameter of 1.5 nm. These tropocollagen molecules are intercalated with the mineral phase (hydroxyapatite, a calcium phosphate) forming fibrils that curl into helicoids of alternating directions. These "osteons" are the basic building blocks of bones, with the volume fraction distribution between organic and mineral phase being about 60/40. In another level of complexity, the hydroxyapatite crystals are platelets that have a diameter of approximately 70–100 nm and thickness of 1 nm. They originally nucleate at the gaps between collagen fibrils.

Similarly, the hierarchy of abalone shell begins at the nanolevel, with an organic layer having a thickness of 20–30 nm. This layer proceeds with single crystals of aragonite (a polymorph of $CaCO_3$) consisting of "bricks" with dimensions of 0.5 and finishing with layers approximately 0.3 mm (mesostructure).

Crabs are arthropods whose carapace is made of a mineralized hard component (which exhibits brittle fracture) and a softer organic component composed primarily of chitin. The brittle component is arranged in a helical pattern. Each of these mineral 'rods' (1 μm diameter) contains chitin-protein fibrils with approximately 60 nm diameter. These fibrils are made of 3 nm diameter canals which link the interior and exterior of the shell.

Applications

Biomaterials are used in:

- Joint replacements
- Bone plates
- Bone cement
- Artificial ligaments and tendons
- Dental implants for tooth fixation
- Blood vessel prostheses
- Heart valves
- Skin repair devices (artificial tissue)
- Cochlear replacements
- Contact lenses
- Breast implants.

Biomaterials must be compatible with the body, and there are often issues of biocompatibility which must be resolved before a product can be placed on the market and used in a clinical setting. Because of this, biomaterials are usually subjected to the same requirements as those undergone by new drug therapies. All manufacturing companies are also required to ensure traceability of all of their products so that if a defective product is discovered, others in the same batch may be traced.

Heart Valves

In the United States, 45% of the 250,000 valve replacement procedures performed annually involve a mechanical valve implant. The most widely used valve is a bileaflet disc heart valve, or St. Jude valve. The mechanics involve two semicircular discs moving back and forth, with both allowing the flow of blood as well as the ability to form a seal against backflow. The valve is coated with pyrolytic carbon, and secured to the surrounding tissue with a mesh of woven fabric called Dacron™ (du Pont's trade name for polyethylene terephthalate). The mesh allows for the body's tissue to grow while incorporating the valve.

Skin Repair

Most of the time "artificial" tissue is grown from the patients own cells. However, when the damage is so extreme that it is impossible to use the patient's own cells, artificial tissue cells are grown. The difficulty is in finding a scaffold that the cells can grow and organize on. The characteristics of the scaffold must be that it is biocompatible, cells can adhere to the scaffold, mechanically strong and biodegradable. One successful scaffold is a copolymer of lactic acid and glycolic acid.

COMPATIBILITY

Biocompatibility is related to the behaviour of biomaterials in various environments under various chemical and physical conditions. The term may refer to specific properties of a material without specifying where or how the material is to be used. For example, a material may elicit little or no immune response in a given organism, and may or may not able to integrate with a particular cell type or tissue). The ambiguity of the term reflects the ongoing development of insights into how biomaterials interact with the human body and eventually how those interactions determine the clinical success of a medical device (such as pacemaker or hip replacement). Modern medical devices and prostheses are often made of more than one material—so it might not always be sufficient to talk about the biocompatibility of a specific material.

Also, a material should not be toxic unless specifically engineered to be so—like "smart" drug delivery systems that target cancer cells and destroy them. Understanding of the anatomy and physiology of the action site is essential for a biomaterial to be effective. An additional factor is the dependence on specific anatomical sites of implantation. It is thus important, during design, to ensure that the implement will fit complementarily and have a beneficial effect with the specific anatomical area of action.

BIOPOLYMERS

Biopolymers are polymers produced by living organisms. Cellulose and starch, proteins and peptides, and DNA and RNA are all examples of biopolymers, in which the monomeric units, respectively, are sugars, amino acids, and nucleotides. Cellulose is both the most common biopolymer and the most common organic compound on Earth. About 33% of all plant matter is cellulose.

Some biopolymers are biodegradable. That is, they arc broken down into CO_2 and watcr by microorganisms. In addition, some of these biodegradable biopolymers are compostable. That is, they can be put into an industrial composting process and will break down by 90% within 6 months. Biopolymers that do this can be marked with a 'compostable' symbol, under European Standard EN 13432 (2000). Packaging marked with this symbol can be put into industrial composting processes and will break down within 6 months (or less). An example of a compostable polymer is PLA film under 20 μm thick: films which are thicker than that do not qualify as compostable, even though they are biodegradable. A home composting logo may soon be established: this will enable consumers to dispose of packaging directly onto their own compost heap.

BIOMINERALIZATION

Biomineralization is the process by which living organisms produce minerals, often to harden or stiffen existing tissues. Such tissues are called mineralized tissues.

Figure: Calcitic Skeletal Parts of Belemnites

It is an extremely widespread phenomenon; all six taxonomic kingdoms contain members that are able to form minerals, and over 60 different minerals have been identified in organisms. Examples include silicates in algae and diatoms, carbonates in invertebrates, and calcium phosphates and carbonates in vertebrates. These minerals often form structural features such as sea shells and the bone in mammals and birds. Organisms have been producing mineralised skeletons for the past 550 million years. Other examples include copper, iron and gold deposits involving bacteria. Biologically-formed minerals often have special uses such as magnetic sensors in magnetotactic bacteria (Fe_3O_4), gravity sensing devices ($CaCO_3$, $CaSO_4$, $BaSO_4$) and iron storage and mobilization ($Fe_2O_3 \bullet H_2O$ in the protein ferritin).

In terms of taxonomic distribution, the most common biominerals are the phosphate and carbonate salts of calcium that are used in conjunction with organic polymers such as collagen and chitin to give structural support to bones and shells. The structures of these biocomposite materials are highly controlled from the nanometre to the macroscopic level, resulting in complex architectures that provide multifunctional properties. Because this range of control over mineral growth is desirable for materials engineering applications, there is significant interest in understanding and elucidating the mechanisms of biologically controlled biomineralization.

BIOLOGICAL ROLES

Biominerals perform a variety of roles in organisms, the most important being support, defence and feeding.

Biology

If present on a super-cellular scale, biominerals are usually deposited by a dedicated organ, which is often defined very early in the embryological development. This organ will contain an organic matrix that facilitates and directs the deposition of crystals. The matrix may be collagen, as in deuterostomes, or based on chitin or other polysaccharides, as in molluscs.

Shell Formation in Molluscs

The mollusc shell is a biogenic composite material that has been the subject of much interest in materials science because of its unusual properties and its model character for biomineralization. Molluscan shells consist of 95–99% calcium carbonate by weight, while an organic component makes up the remaining 1–5%. The resulting composite has a fracture toughness ~3000 times greater than that of the crystals themselves. In the biomineralization of the mollusc shell, specialized proteins are responsible for directing crystal nucleation, phase, morphology, and growths dynamics and ultimately give the shell its remarkable mechanical strength. The application of biomimetic principles elucidated from mollusc shell assembly and structure may help in fabricating new composite materials with enhanced optical, electronic, or structural properties.

Chemistry

Because extracellular iron is strongly involved in inducing calcification, its control is essential in developing shells; the gene *ferritin* plays an important role in controlling the distribution of iron.

Evolution

The first evidence of biomineralization dates to some 750 million years ago, and sponge-grade organisms may have formed calcite skeletons 630 million years ago. But in most lineages, biomineralization first occurred in the Cambrian or Ordovician periods. Organisms used whichever form of calcium carbonate was more stable in the water column at the point in time when they became biomineralized, and stuck with that form for the remainder of their biological history. The stability is dependent on the Ca/Mg ratio of seawater, which is thought to be controlled primarily by the rate of sea floor spreading, although atmospheric CO_2 levels may also play a role.

Biomineralization evolved multiple times, independently, and most animal lineages first expressed biomineralized components in the Cambrian period. Interestingly, many of the same processes are used in unrelated lineages, which suggests that biomineralization machinery was assembled from pre-existing "off-the-shelf" components already used for other purposes in the organism. Although the biomachinery facilitating biomineralization is complex – involving signalling transmitters, inhibitors, and transcription factors – many elements of this 'toolkit' are shared between phyla as diverse as corals, molluscs, and vertebrates. The shared components tend to perform quite fundamental tasks, such as designating that cells will be used to create the minerals, whereas genes controlling more finely tuned aspects that occur later in the biomineralization process – such as the precise alignment and structure of the crystals produced – tend to be uniquely evolved in different lineages.

This suggests that Precambrian organisms were employing the same elements, albeit for a different purpose — perhaps to *avoid* the inadvertent precipitation of calcium carbonate from the supersaturated Proterozoic oceans. Forms of mucus that are involved in inducing mineralization in most metazoan lineages appear to have performed such an anticalcifatory function in the ancestral state. Further, certain proteins that would originally have been involved in maintaining calcium concentrations within cells are homologous to all metazoans, and appear to have been co-opted into biomineralization after the divergence of the metazoan lineages. The *galaxins* are one probable example of a gene being co-opted from a different ancestral purpose into controlling biomineralization, in this case being 'switched' to this purpose in the Triassic scleractinian corals; the role performed appears to be functionally identical to the unrelated pearlin gene in molluscs. Carbonic anhydrase serves a role in mineralization in sponges, as well as metazoans, implying an ancestral role. Far from being a rare trait that evolved a few times and remained stagnant, biomineralization pathways in fact evolved many times and are still evolving rapidly today; even within a single genus it is possible to detect great variation within a single gene family.

The homology of biomineralization pathways is underlined by a remarkable experiment whereby the nacreous layer of a molluscan shell was implanted into a human tooth, and rather than experiencing an immune response, the molluscan nacre was incorporated into the host bone matrix. This points to the exaptation of an original biomineralization pathway.

The most ancient example of biomineralization, dating back 2 billion years, is the deposition of magnetite, which is observed in some bacteria, as well as the teeth of chitons and the brains of vertebrates; it is possible that this pathway, which performed a magentosensory role in the common ancestor of all bilaterians, was duplicated and modified in the Cambrian to form the basis for calcium-based biomineralization pathways. Iron is stored in close proximity to magnetite-coated chiton teeth, so that the teeth can be renewed as they wear. Not only is there a marked similarity between the magnetite deposition process and enamel deposition in vertebrates but some vertebrates even have comparable iron storage facilities near their teeth.

Potential Applications

Most traditional approaches to synthesis of nanoscale materials are energy inefficient, requiring stringent conditions (*e.g.*, high temperature, pressure or pH) and often produce toxic byproducts. Furthermore, the quantities produced are small, and the resultant material is usually irreproducible because of the difficulties in controlling agglomeration. In contrast, materials produced by organisms have properties that usually surpass those of analogous synthetically manufactured materials with similar phase composition. Biological materials are assembled in aqueous environments under mild conditions by using macromolecules. Organic macromolecules collect and transport raw materials and assemble these substrates and into short- and long-range ordered composites with consistency and uniformity. The aim of biomimetics is to mimic the natural way of producing minerals such as apatites. Many man-made crystals require elevated temperatures and strong chemical solutions, whereas the organisms have long been able to lay down elaborate mineral structures at ambient temperatures. Often, the mineral phases are not pure but are made as composites that entail an organic part, often protein, which takes part in and controls the biomineralisation. These composites are often not only as hard as the pure mineral but also tougher, as the micro-environment controls biomineralisation.

Astrobiology

It has been suggested that biominerals could be important indicators of extraterrestrial life and thus could play an important role in the search for past or present life on Mars. Furthermore, organic components (biosignatures) that are often associated with biominerals are believed to play crucial roles in both pre-biotic and biotic reactions.

NANOFIBROUS BIOMATERIALS

An important class of nanostructured biomaterials on which intensive research has been carried out is nano-fibrous materials, especially biodegradable polymer nanofibres. This class of biomaterials mimics the nanofibrillar structure of extracellular matrix (ECM). Collagen is the major ECM component of many tissues and has been investigated as a substrate or scaffold for cell attachment, proliferation and differentiation. Importantly, the nanoscaled collagen fibrillar structure (50-500 nm in diameter) has been found to enhance cell/matrix interactions. To serve as a temporary ECM for regenerative cells, it may be beneficial for a biomaterial to emulate such nanofibrous features of collagen (and other natural ECM components such as fibronectin, proteoglycans, and so forth). Biomaterials that mimic collagen fibre bundles in nanometre size have been developed using a few different techniques: electrospinning, phase separation, and self-assembly. These technologies generate nanofibrous biomaterials with varying degrees of success in tissue regeneration applications.

ELECTROSPINNING

Electrospinning is a well-established process that has been used to produce ultrafine fibres including microfibres (>1 μm) or nanofibres (<1000 nm). In electrospinning, a high voltage is applied to a polymer solution or melt, which overcomes the surface tension to form a charged jet. The charged polymer solution or melt is ejected, dried and solidified onto a grounded substrate. The ejected polymer solutions repel each other during the travel to the grounded collector, which forms thin fibres after solvent evaporation. By controlling the spinning conditions, the resulting fibres can range from about 0.02 μm to about 20 μm. When a stationary collector is used, the resulting fibres deposit randomly due to the random motion of the electrospinning polymer solution. However, preferentially oriented fibres can be obtained by using an electrically grounded rotating drum as the collector.

Many electrospun nanofibrous biomaterials have been investigated as tissue regeneration scaffolds. Materials used in electrospinning can be natural macromolecules such as collagen, chitosan, silk fibroin; synthetic biodegradable polymers such as PGA, PLGA, PLLA, PCL; and combinations of these natural and synthetic polymers. In addition, various substances (proteins, growth factors, and hydroxyapatite) can be incorporated into nanofibrous materials during electrospinning. Therefore, electrospun nanofibrous biomaterials have been explored to engineer various tissues. The main advantage of electrospinning process is the relative quick and simple way to fabricate a variety of materials into nanofibrous structure. However, significant challenges still exist in using this technique to fabricate complex 3D scaffold shapes or to generate designed internal pore structures, limiting its potential for many tissue engineering applications.

Phase Separation

Phase separation techniques have been used to prepare porous polymer membranes for purification and separation purposes. The phase separation process can be induced either thermally or by using a non-solvent. Phase separation induced by a non-solvent results in heterogeneous pore formation, which are generally not suitable for uniform tissue engineering scaffold fabrication. In thermally induced phase separation (TIPS), a homogeneous polymer solution becomes thermodynamically unstable under certain temperature conditions and tends to separate into a multi-phase system (*e.g.* a polymer-rich phase and a polymer-lean phase) to lower the free energy. After removal of the solvent, the polymer-rich phase solidifies to form the structure while polymer-lean phase becomes pores. A variety of biodegradable polymers have been fabricated into three-dimensional porous scaffolds using phase separation techniques. Depending on the polymer system and phase separation conditions, various pore structures and morphologies have been reported and these scaffolds have been investigated for tissue regeneration applications.

A novel phase separation technique has been developed in our laboratory to generate nanofibrous structures by manipulating the phase separation process. For example, a PLLA solution is thermally induced to separate through a spinodal liquid-liquid phase separation and subsequent self assembly into nanofibres. After the removal of solvent a porous structure with 3D continuous nanofibrous network is formed. The fibrous scaffold can have a porosity of 98% and contains nanofibres ranging from 50-500 nm in diameter, which is similar to natural collagen fibres in size. PLLA nanofibrous scaffolds have surface area about two orders of magnitude higher than non-fibrous solid-walled scaffolds. Due to the substantial surface area difference, degradation is much more rapid in such nanofibrous scaffolds, in which the overall mass loss is 51% while mass loss in solid-walled non-fibrous foams is only 6% after 15 months. During degradation, both molar mass and surface area of nanofibrous scaffolds decrease more rapidly than those of solid-walled scaffolds, indicating that nanofibrous characteristics accelerate the rate of hydrolytic degradation of the polymer scaffolds.

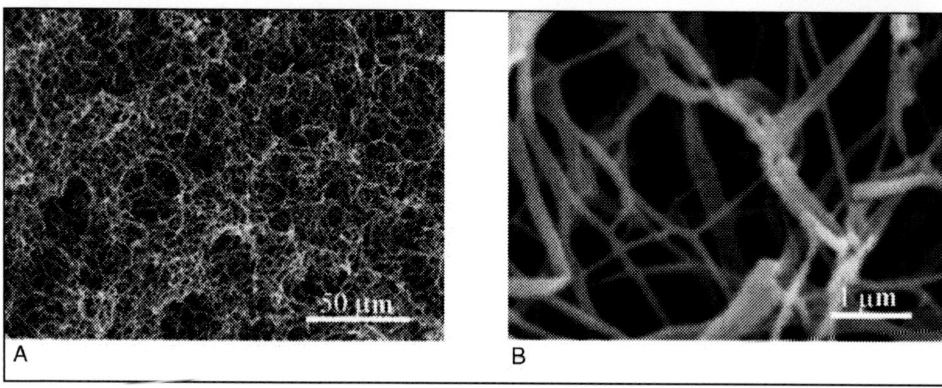

Figure : Scanning Electron Micrographs (SEM) of a PLLA Nanofibrous Scaffold Prepared from 2.5% PLLA/THF Solution at a Phase Separation Temperature of 8 °C: (A) 500x; (B) 20,000x.

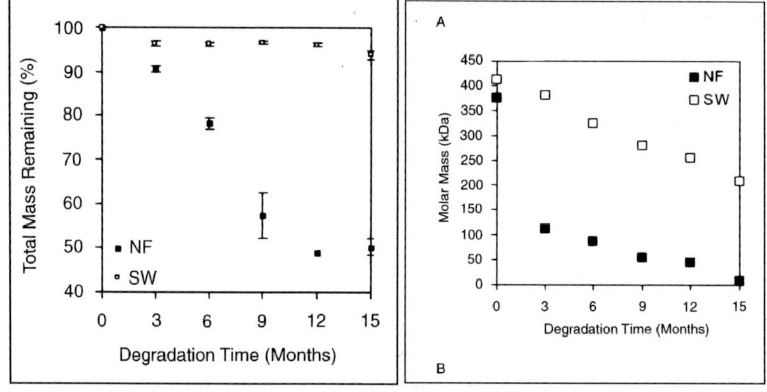

Figure : Normalized Change in Average Mass (A) and Molar Mass (B) of PLLA Nanofibrous (NF) and Solid-walled (SW) Scaffolds after in Vitro Degradation in PBS.

One limitation of the early nanofibrous materials generated using the phase-separation technique is the lack of interconnected macropores, which are critical for cell seeding and recruiting, mass transfer, vascularization and tissue organization. To overcome this problem, phase separation techniques are used in combination with other scaffold fabrication techniques such as porogen leaching and solid freeform fabrication (SFF). The combined technique provides broader control over porous architectures from macro, micro to nano scales. The hierarchical pore structure has been shown to substantially promote tissue regeneration.

A combined technique of phase separation and sugar sphere template leaching has been developed recently. Water-soluble sugar spheres are prepared and bound to form a template. Phase separation occurs when a polymer solution is cast on to a sugar template and the temperature is lowered.

Removal of sugar template leads to the formation of macroporous and nanofibrous scaffolds.

The technique advantageously controls macropore shape and size by sugar spheres, interpore opening size by assembly conditions (time and temperature of heat treatment for sugar spheres), and pore wall morphologies by phase separation parameters. Alternatively, sugar fibre template or solid freeform fabricated wax mold are used to prepare nanofibrous scaffolds with interconnected macro-tubular structures or channels, respectively. Using a computer-assisted-design (CAD) technique, external scaffold shape can be generated from computed-tomography scans or histological sections. For example, a nanofibrous polymer scaffold with the shape of a human ear is precisely created using the phase-separation technique and a mold reconstructed from its histological sections.

Despite the different external scaffold shapes and macroporous structures (spherical, channeled), all scaffolds have nanofibrous architectures on the pore walls. The nanofibrous scaffolds also have high porosity (>90%) and high surface area (around 100 m^2/g).

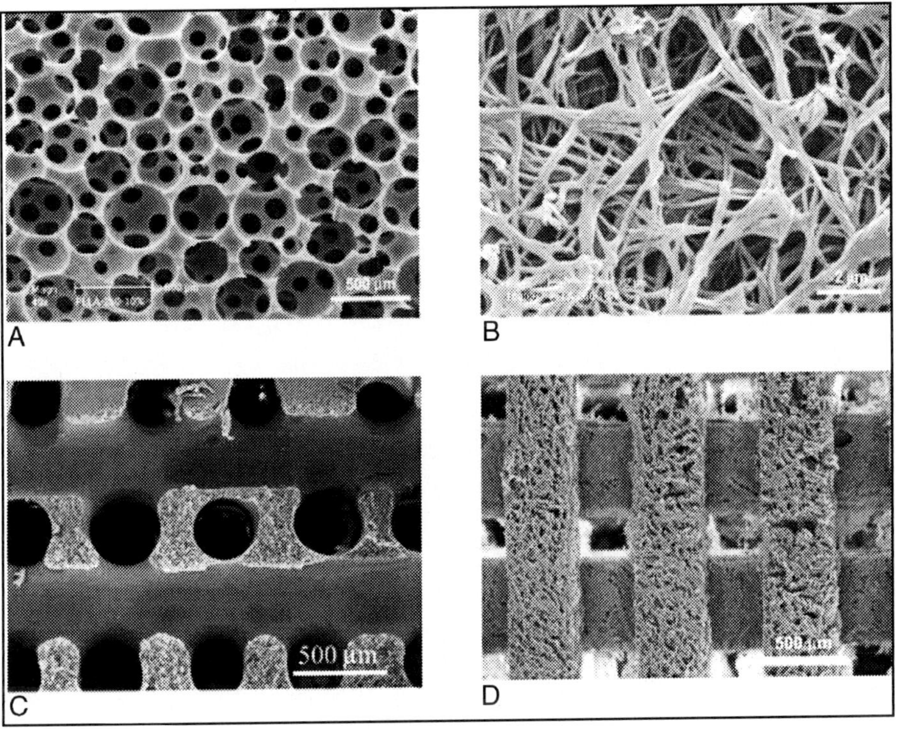

Figure: SEM Micrographs of 3D Macroporous and Nanofibrous PLLA Scaffolds. (A, B) Prepared from Sugar Sphere Template Leaching and Phase Separation; (C) Prepared from Sugar Fibre Template Leaching and Phase Separation; (D) Prepared from Solid Freeform Fabrication and Phase Separation.

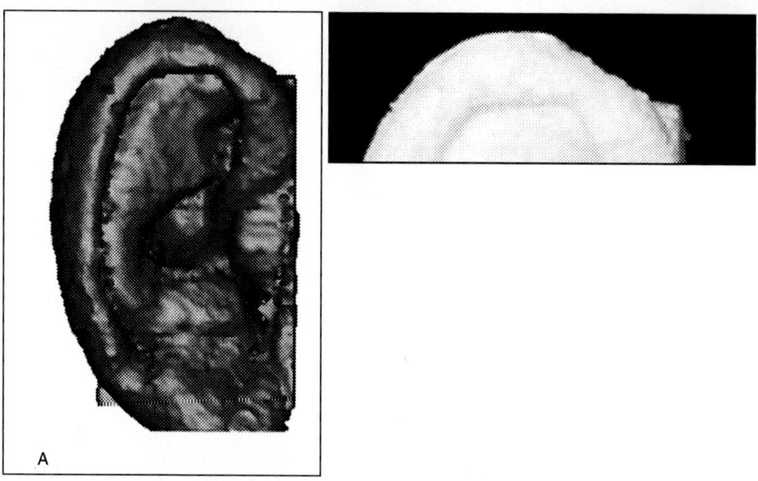

Figure: Nanofibrous Scaffolds Created from 3D Medical Images and a Phase-separation Technique. (A) Human Ear Template Reconstructed from Histological Sections; (B) Resulting Nanofibrous Scaffold of the Human Ear (scale bar: 10 mm); (C) the Nanofibrous Pore Wall Morphology (scale bar: 5 μm).

Self-assembly

Nanofibrous matrices can be created by a molecular self-assembly approach which involves the spontaneous organization of individual molecules into a well-defined and stable hierarchical structure with preprogrammed non-covalent interactions. As a natural process for several essential biological components including nucleic acid or protein synthesis, self-assembly technology usually incorporates some specific biological components of the extracellular matrix (ECM) and mimics ECM assembly process.

Self-assembling molecules require some specific configurations to be assembled into nanofibres. For example, nanofibre forming peptide-amphiphiles (PAs) are designed to have several critical structural features: a long alkyl tail that conveys hydrophobic characteristics to drive self-assembly; four consecutive cysteine residues that create disulfide bonds to polymerize the structure; a linker region of three glycine residues to provide the flexibility to the hydrophilic head group; a phosphorylated serine residue that interacts strongly with calcium ions intended to enhance mineralization; and an Arg-Gly-Asp (RGD) peptide to aid in cell adhesion. These PAs when subject to an acid are induced to self-assemble to form nanofibres with 5-8 nm in diameter and 1 μm in length. This involves treating PA solutions with dithiothreitol at a pH of 8 and then reducing the pH to 4. As the solution is acidified the PAs rapidly become insoluble and form nanofibres of a gel. Besides pH-driven, self-assembly of PAs can also be achieved by drying and introductions with divalent ions such as calcium. Some of the self-assembled hydrogels are being explored for cell entrapment utilizing polyvalent metal ions in the media, and one of them has been shown to allow for osteoblastic cells (MC3T3-E1) to survive for 3 weeks.

Zhang and coworkers have synthesized ionic self-complementary oligopeptides that consist of regular repeats of alternating ionic hydrophilic and hydrophobic amino acids. In water, these oligopeptides form a β-sheet and both a polar surface (with charged ionic chains) and a non-polar surface (with alanines) are present. Upon exposure to monovalent alkaline cations or under physiological conditions, the oligopeptides spontaneously assemble into hydrogels of various shapes. These hydrogel matrices are interwoven nanofibres approximately 10-20 nm in diameter with pores about 50-200 nm in diameter. The nanofibrous matrices support mammalian cell attachment, proliferation and differentiation. In addition to PAs and oligopeptides, synthetic diblock/triblock copolymers and dendrimers can be self-assembled into nanofibrous structures. However, as regenerative biomaterials, self-assembled nanofibrous scaffolds are currently limited to biological molecules such as peptides in the form of hydrogels. One of the limitations is their inability to controllably form mechanically stable 3D geometry. The peptide nanofibres can be fragmented and may be susceptible to endocytosis. In addition, their degradation has not been systematically addressed. The peptide backbone structure can be degraded by enzymes, making it difficult to control their *in vivo* degradation behaviour at will.

SURFACE MODIFICATION OF NANOFIBROUS SCAFFOLDS

Although a variety of biomaterials have been fabricated into nanofibrous scaffolds using various techniques, many of them have low biological compatibility with cells, which has limited their regeneration applications. Surface modification is an effective way to improve cellular interactions with the nanofibrous scaffolds. Different from bulk modification which usually changes the chemical and mechanical properties of the biomaterials, surface modification has the advantage of not altering the scaffold architectures significantly. This is especially desirable for the modification of ultrafine structures such as nanofibres.

Plasma exposure has been used to introduce desired functional groups and molecular chains onto the nanofibrous surface of an electrospun matrix. Non-woven poly(epsilon-caprolactone) (PCL) nanofibres were prepared by electrospinning and type-I collagen was then immobilized on the nanofibres after surface modification by plasma treatment. Collagen immobilization enhanced the attachment, spreading and proliferation of human dermal fibroblasts. Grafting gelatin on to plasma treated PCL nanofibres improved endothelial cell spreading and proliferation as well as cell orientation. Because of the limited depth of plasma penetration, this technique can be used for 2D films or very thin 3D structures but is not adequate for large 3D scaffolds with complex architectures. To mimic both the nanofibrous structure and chemical composition of collagen fibres, 3D nanofibrous scaffolds have been prepared using a phase separation technique and surface modified with gelatin, a biomacromolecule derived from ECM collagen. An entrapment method was developed to effectively incorporate gelatin onto nanofibrous wall surfaces of both interior and exterior pores. Compared to a simple coating of gelatin onto the scaffold, entrapped gelatin molecular chains entangle with the molecular chains of scaffold materials and are physically locked on the surface permanently. The entrapment modification method can be used for various geometries, morphologies, and thicknesses of 3D polymer scaffolds without interfering bulk properties and architectures of a scaffold. Surface modification can also be carried out during scaffold fabrication using a porogen-induced modification method. For example, gelatin spheres can act as both the porogen for scaffold fabrication and the agent of surface-modification, with which scaffold formation and surface modification are completed in a simple one-step process. The gelatin-modified porous nanofibrous scaffold has been demonstrated to significantly improve osteoblast cell adhesion and proliferation throughout the scaffold.

Since most biomacromolecules are charged cationic or anionic polyelectrolytes, the immobilization of these biomacromolecules onto 3D nanofibrous scaffolds can also be achieved by a layer-by-layer self-assembly process. For example, a macroporous and nanofibrous PLLA scaffold was activated in an aqueous poly(diallyldimethylammonium chloride) (PDAC) solution to obtain positively charged pore wall surface. The scaffold was subsequently immersed in a solution of negatively charged biomacromolecules (*e.g.* gelatin). Surface layers with positive or negative charges are developed by alternative immersion in the two different solutions. Self-assembly approach is carried out in aqueous solutions under mild conditions and offers a controlled way to regulate the surface charge type and the thickness of the surface modification layer. Growth factors and DNAs (as a gene-delivery strategy) may also be self-assembled on to nanofibrous scaffolds to regulate cellular response and gene expression.

BIOLOGICAL EFFECTS OF NANOFIBRES

Nanofibres have special properties arising from their dimensions and high surface area to volume ratio. Nanofibrous scaffolds serve as promising tissue engineering scaffolds because nanofibrous surfaces elicit special biological responses of cells during cultivation and after implantation. The nanofibres regulate protein adsorption, cell attachment, proliferation, and differentiation as well as extracellular matrix deposition.

The nanofibrous architectures built in a macroporous scaffold enhance protein adsorption in vitro. Compared to solid-walled scaffolds without nanofibrous structures, nanofibrous scaffolds were found to adsorb 4 times more human serum proteins. More interestingly, the profile of serum proteins adsorbed onto the nanofibrous scaffolds was different from that adsorbed onto the solid-walled scaffolds. With SDS-PAGE analysis, the protein with an approximate size of 150 kDa (arrow) adsorbed to both scaffolds similarly while the proteins with approximate sizes of 120, 45, or 40 kDa (arrowheads) exclusively adsorbed onto the nanofibrous scaffold. Based on the measurements of band intensities of proteins, the amounts of proteins adsorbed to nanofibrous scaffolds were 0.57–13.35 times those of corresponding proteins (bands) in control serum, whereas the amounts of proteins adsorbed to solid-walled scaffolds were 0.24–0.79 times those of corresponding proteins in the control serum. Furthermore, Western blot analysis showed that nanofibrous scaffolds adsorbed large amounts of fibronectin and vitronectin from serum, while these cell-adhesion proteins were barely detectable on the solid-walled scaffolds. The findings suggest that nanofibres alone may have certain features enhancing protein-affinity and binding strength.

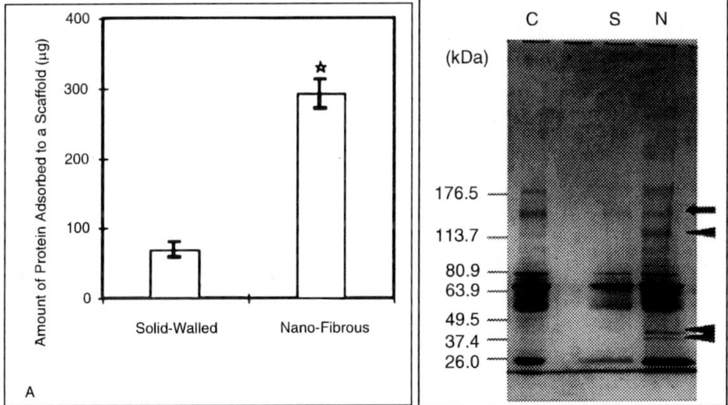

Figure: (A) Amounts of Adsorbed Serum Proteins onto the Scaffolds. Significantly different from Solid-walled Scaffolds, $p < 0.05$, $n = 4$. (B) Protein Adsorption Profile on Nanofibrous and Solid-walled Scaffolds. Polyacrylamide Gels Stained with Coomassie. Lane C, Bovine Serum Proteins; lane S, Adsorbed Bovine Serum Proteins to the Solid-walled Scaffold; lane N, Adsorbed Bovine Serum Proteins to the Nanofibrous Scaffold.

Cells behave differently on nanofibrous scaffolds from those on solid-walled scaffolds, both pretreated with fetal bovine serum. About 70% more osteoblastic progenitor cells (MC3T3-E1) were attached on a nanofibrous scaffold than on a solid-walled scaffold (24 h after seeding). Cell filopodia were reported to direct to certain particulate structures. The nanofibrous architecture may allow filopodia to anchor more tightly, and this mode of anchorage could also contribute to the stronger cell adhesion to the nanofibrous scaffolds. Osteoprogenitor cells also proliferated more rapidly on the nanofibrous scaffolds compared with solid-walled scaffolds. Since both types of substrates (3D nanofibrous vs. solid-walled scaffolds or 2D nanofibrous vs. solid films) were made from the same polymer and had the same macro-structures for cell accommodation, it appears that the nanofibrous architecture provides a more favourable environment for cell attachment and proliferation. Similar results were reported when comparing smooth muscle cells (SMCs) on electrospun nanofibrous films and plain polymer films. The adhesion and proliferation of SMCs on the nanofibrous films were significantly improved over those on the plain polymer films.

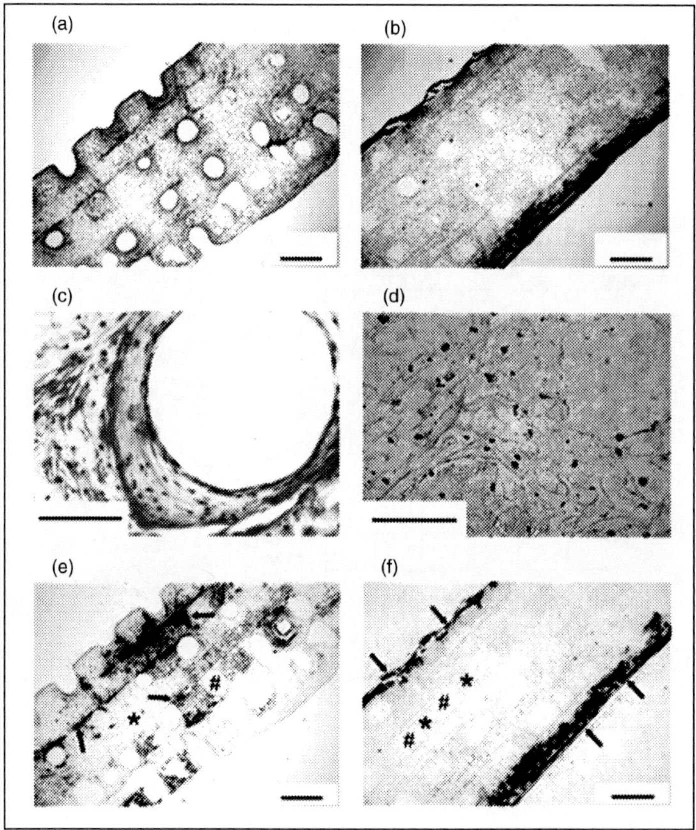

Figure: Responses of MC3T3-E1 Cells on Nanofibrous (A, C, E) and Solid-walled (B, D, F) PLLA scaffolds. Scaffolds were Prepared using a Combined Solid Freeform Fabrication and Phase Separation Technique, having the same Macropore Structures. Cells were Cultured in Vitro for 6 Weeks. (A-D): H&E Staining, (E, F): Von Kossa staining.

When bone tissue is engineered, cells and mineralized extracellular matrix (ECM) are distributed throughout nanofibrous scaffolds while cells and mineralized ECM are primarily deposited in the outer regions of the solid-walled scaffolds. Fast biomineralization is also demonstrated using a calcium assay, which shows 13-fold greater amount of mineral deposition on nanofibrous scaffolds than on solid-walled scaffolds. At the molecular level, cells on nanofibrous scaffolds express significantly higher levels of bone markers such as osteocalcin (OCN) and bone sialoprotein (BSP). In contrast, cells on solid-walled scaffolds express a substantially higher level of collagen (COL) mRNA. The results suggest that bone cells differentiate more quickly on nanofibrous scaffolds than on solid-walled scaffolds, resulting in decreased COL mRNA expression and the increased late-stage bone marker expression (OCN and BSP) on nanofibrous scaffolds. The enhanced osteoblastic differentiation by the nanofibrous scaffolds has recently been found to be associated with the RhoA-ROCK signaling pathway. In addition, our unpublished data show that nanofibrous architectures also facilitate the differentiation of both adult and embryonic stem cells. In a rat calvarial bone defect regeneration model, the amount of new bone formation was about two times greater in the nanofibrous scaffolds than in the solid-walled scaffolds (data to be published). These advantageous properties of the nano-fibrous scaffolds are also corroborated by data of cartilage tissue engineering using electrospun nanofibrous materials.

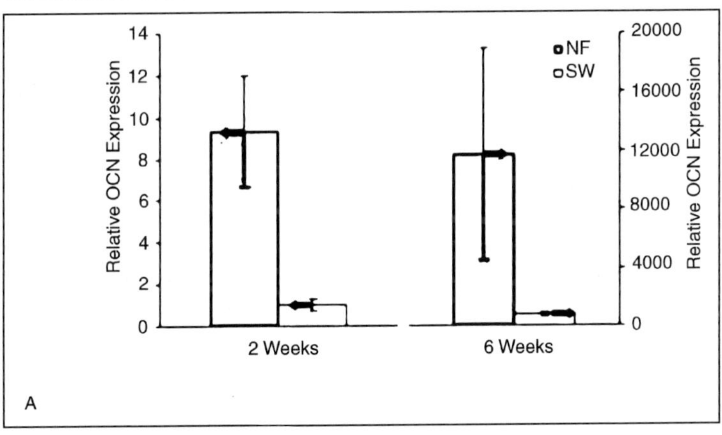

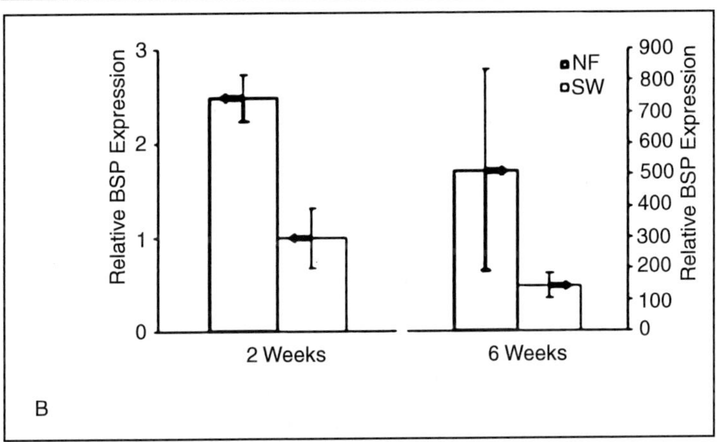

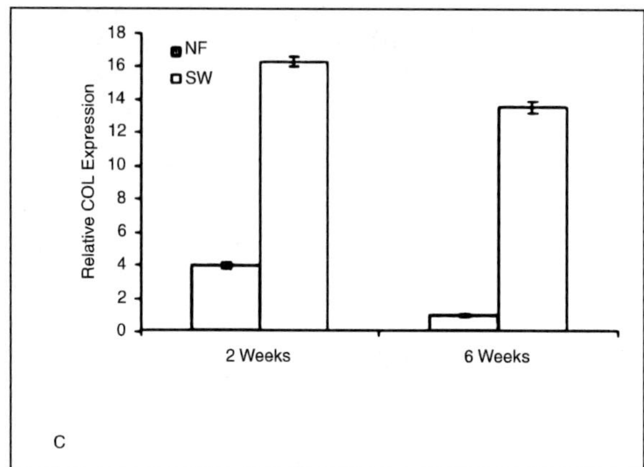

Figure: Relative Levels of Bone Marker Expression in Nanofibrous and Solid-walled Scaffolds after 2 and 6 weeks of Culture under Differentiation Conditions. (A) Osteocalcin (OCN) Expression; (B) Bone Sialoprotein (BSP) Expression; and (C) type I Collagen (COL) Expression.

The nanofibrous architecture, the high surface area, the microporosity between nanofibres (several μms), and the selective adsorption of ECM proteins, all likely contribute to the enhanced cellular responses and tissue regeneration in the nanofibrous scaffolds.

CLASSIFICATIONS AND BEHAVIOUR OF DIFFERENT TYPES OF BIOMATERIALS

BIOMATERIALS

Biomaterial by definition is "a non-drug substance suitable for inclusion in systems which augment or replace the function of bodily tissues or organs". From as early as a century ago artificial materials and devices have been developed to a point where they can replace various components of the human body. These materials are capable of being in contact with bodily fluids and tissues for prolonged periods of time, whilst eliciting little if any adverse reactions.

Historical Development of Biomaterials

Some of the earliest biomaterial applications were as far back as ancient Phoenicia where loose teeth were bound together with gold wires for tying artificial ones to neighbouring teeth.

In the early 1900's bone plates were successfully implemented to stabilise bone fractures and to accelerate their healing. While by the time of the 1950's to 60's, blood vessel replacement were in clinical trials and artificial heart valves and hip joints were in development.

Design Factors for Biomaterials

Even in the preliminary stages of this field, surgeons and engineers identified materials and design problems that resulted in premature loss of implant function through mechanical failure, corrosion or inadequate biocompatibility of the component. Key factors in a biomaterial usage are its biocompatibility, biofunctionality, and availability to a lesser extent. Ceramics are ideal candidates with respect to all the above functions, except for their brittle behaviour.

Implant Materials

It has been accepted that no foreign material placed within a living body is completely compatible. The only substances that conform completely are those manufactured by the body itself (autogenous) and any other substance that is recognized as foreign, initiates some type of reaction (host-tissue response). The four types of responses, which allow different means of achieving attachment of implants to the muscular skeletal system.

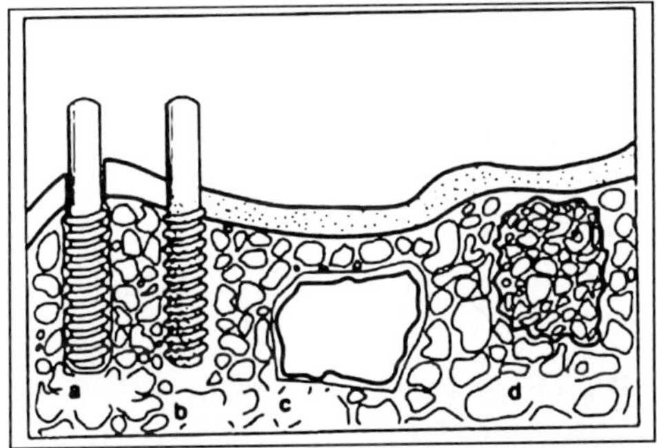

Figure : Classification of Biomaterials according to their Bioactivity (a) Bioinert Alumina Dental Implant, (b) Bioactive Hydroxyapatite [$Ca_{10}(PO_4)_6(OH)_2$] Coating on a Metallic Dental Implant, (c) Surface Active Bioglass and (d) Bioresorbable Tricalcium Phosphate ([$Ca_3(PO_4)_2$] Impant.

BIOMATERIALS CLASSIFICATIONS

When a synthetic material is placed within the human body, tissue reacts towards the implant in a variety of ways depending on the material type. The mechanism of tissue interaction (if any) depends on the tissue response to the implant surface. In general, there are three terms in which a biomaterial may be described in or classified into representing the tissues responses. These are bioinert, bioresorbable, and bioactive, which are well covered in range of excellent review papers.

Bioinert Biomaterials

The term bioinert refers to any material that once placed in the human body has minimal interaction with its surrounding tissue, examples of these are stainless steel, titanium, alumina, partially stabilised zirconia, and ultra high molecular weight polyethylene. Generally a fibrous capsule might form around bioinert implants hence its biofunctionality relies on tissue integration through the implant.

Bioactive Biomaterials

Bioactive refers to a material, which upon being placed within the human body interacts with the surrounding bone and in some cases, even soft tissue. This occurs through a time – dependent kinetic modification of the surface, triggered by their implantation within the living bone. An ion – exchange reaction between the bioactive implant and surrounding body fluids – results in the formation of a biologically active carbonate apatite (CHAp) layer on the implant that is chemically and crystallographically equivalent to the mineral phase in bone. Prime examples of these materials are synthetic hydroxyapatite [$Ca_{10}(PO_4)_6(OH)_2$], glass ceramic A-W and bioglass®.

Bioresorbable Biomaterials

Bioresorbable refers to a material that upon placement within the human body starts to dissolve (resorbed) and slowly replaced by advancing tissue (such as bone). Common examples of bioresorbable materials are tricalcium phosphate [$Ca_3(PO_4)_2$] and polylactic–polyglycolic acid copolymers. Calcium oxide, calcium carbonate and gypsum are other common materials that have been utilised during the last three decades.

BIOCOMPATIBILITY *IN VITRO* OF BIOMEDICAL-GRADE CHITOSAN DERIVATIVES

Biocompatibility of a biomaterial refers to the extent to which the material does not have toxic or injurious effects on biological systems. In the pre-commercialization steps of a biomedical-grade chitosan derivative as wound dressing, it is of great importance that the dressing be submitted to a number of pre-toxicity tests, *in vitro* or *in vivo*. Chitosan is a cationic polymer having an amino group in its chemical structure. Anti-tumor, anti-microbial and anti-inflammatory can be largely attributed to the cationic structure. Lee *et al.* performed trypan blue dye exclusion cytotoxicity assays on tumor cell lines, after increased positive charge density of chitosan by dialkylaminoalkylation and reductive amination followed by quaternization. Cytotoxicity of chitosan with various degrees of *N*-deacetylation to form aminoethyl-chitosan, dimetylaminoethyl-chitosan and diethylaminoethyl-chitosan, was examined only with 3-(4,5-dimethylthiazol-2-yl)-2,5-diphenyl tetrazolium bromide (MTT) on tumor cell lines and concluded that cytotoxic effect may be caused by cationic charge of the chitosan derivatives. Amino groups of chitosan attached with biologically active moieties such as vanillin, *p*-hydroxybenzaldehyde, *p*-chlorobenzaldehyde, anisaldehyde, methyl 4-hydroxybenzoate, methyl 2,4-dihydroxybenzoate, propyl 3,4, 5-trihydroxybenzoate and 2-hydroxymethylbenzoate can promote anti-microbial ability.

A thorough understanding of cell and proteins interactions with artificial surfaces is of importance to design suitable implant surfaces and substrates. The surface properties of newly synthesized biomedical-grade chitosan derivatives, including surface composition, wettability, domain composition, surface oxidation, surface charge and morphology, may influence protein adsorption and subsequently, the cellular responses to biomaterial implants. Though it has once been suggested that surfaces modified with polyethylene glycol (PEG) may limit protein adsorption and cell adhesion, the results for PEG blended with chitosan increased both the adsorption and the cell adhesion. Zhang *et al.* examined the biocompatibility of newly synthesized chitosan-PEG by growing and observing the morphological changes of 3T3 embryonic mouse fibroblasts and osteogenesis nodule cell suspension in culture. Cytotoxicity of chitosan in three dimensional fibre meshes was well determined with only scanning electron microscopy (SEM) and cell morphological observation on the use of mouse lung fibroblast (L929). Recent studies have been focused on the development of antibacterial surfaces of biomedical-grade chitosan derivatives to attain high functionality and high-value products. Huh *et al.* prepared chitosan-grafted poly(ethylene terepthalate) (PET) and quaternized chitosan-grafted PET, both showed high growth inhibition in the range of 75% to 86% and still retained 48% to 58% bacterial growth inhibition after laundering.

But, higher antibacterial activity of a wound dressing can retard the cell growth in culture. Laminin is known to be involved in metastasis of tumor cells. A peptide containing the Tyr-Ile-Gly-Ser-Arg (YIGSR) sequence, corresponding to a partial sequence of laminin, inhibited angiogenesis and therefore depressed tumor growth. Nishiyama *et al.* prepared YIGSR-chitosan conjugate and proved to have higher inhibitory activity against experimental lung metastasis of B16BL6 melanoma cells in mice than did the parent peptide. According to Koo *et al.*, water-soluble chitosan (WSC) significantly protected from serum starvation-induced cellular rounding up and from serum starvation-induced cell death as tested by flow-cytometry. WSC also protected from serum starvation-induced p53 activation as determined by the Western blot technique. From these results, it appears that WSC may prevent serum starvation-induced apoptosis of human astrocytes *via* p53 inactivation. Cross-linking will increase the toxicity of chitosan scaffold leachate to fibroblast cells. However, the chitosan crosslinked with dimethyl 3-3,dithio-bis-propionimidate (DTBP) permitted cell proliferation after three days in culture. Denuziere *et al.* examined the cytocompatibility of films of chitosan and chitosan associated with glycosaminoglycans (GAGs) with human skin keratinocytes and found that the cell growth was consistent approximately 60% of that of controls. Highly deacetylated chitosans can modulate human skin cell mitogenesis *in vitro*. Analysis of their effects on cells in culture may be useful as a screen for their potential activity *in vivo* as wound healing agents. But, it is important to select appropriate strains of cells for use in the *in vitro* biocompatibility screening. Still, most of the biocompatibility *in vitro* screenings to date for chitosan derivatives confined at cellular responses and reactions, namely at the cytocompatibility level.

BIOCOMPATIBILITY *IN VITRO* EXPERIMENTAL TOOLS

In vitro models for testing the biocompatibility of biodegradable chitosan and their derivatives are very useful to characterize the hidden toxic effects of leachable materials or derivatives, such as residual monomers, catalysts as well as other polymer erosion related properties, such as chemical compositions, molecular weight, polydispersity, degradation rate and so forth. Cytotoxicity assays measure only finite effects on cells during the first 12 to 24 hours after exposure to toxic substances. The host cells either recover from or succumb to their chemical injury. However, many biological reactions *in vivo* are not simply cytotoxic and are propagated beyond 24 hours. Examples are improper cells programmed apoptosis and necrosis due to DNA damage; inflammatory and immune reactions that all resulted from non-biocompatible chitosan derivatives.

In vitro toxicity examinations are preferable to *in vivo* tests since the newly developed dressing materials can be examined outside the body and the results are more reproducible, facilitating clinical evaluation of a medicinal product that can be replicated and estimated. It is necessary to determine the biocompatibility of any biomaterial in order to evaluate its strength, esthetics and feasibility for clinical manipulation. Thus, preliminary *in vitro* tests are continually being carried out to screen and characterize the potentially harmful effects of a dressing material before it is commercially used on humans. However, *in vitro* tests may not fully capture the *in vivo* situation; differences in sensitivities have been observed between primary and established cell lines. Current regulations in accordance with the United States Food and Drug Administration (FDA), the International Organization for Standardization (ISO) and the Japanese Ministry of Health and Welfare (JMHW) require that manufacturers conduct adequate safety testing of their finished devices through pre-clinical and clinical trials as part of the regulatory clearance process. Toxicity induced by a chemical substance or drug can limit the dose and duration of treatment, adversely affect patients' quality of life or even be life threatening. Side effects together with tissue toxicity will lead to long-term effects such as alteration to the immune system and development of malignancies, probably due to the genetic damage induced by a certain kind of drug treatment. Hence, biocompatibility assessment *in vitro* is necessary to evaluate cellular and molecular responses to the newly developed biomedical-grade chitosan derivatives.

Cytocompatibility Assessment in Vitro

There are a number of assays which are capable of evaluating the activation of biological processes (*e.g.* inflammation, immune reactions and mutagenesis), which require longer periods of tissue reaction to materials, as compared with cytotoxic reactions. Traditional assays have measured cytotoxicity using either an end-stage event (*e.g.* permeability of cytoplasmic membranes of dead and dying cells, or some metabolic parameters such as cell division or enzymatic reaction). Cytotoxicity assays are commonly used to measure the effects on cells during exposure to toxic substances at multiple concentrations and time course parameters. The colourimetric 3-(4,5-dimethylthiazol-2-yl)-2,5-diphenyl tetrazolium bromide (MTT) assay is a simple assay which is used to identify detrimental intracellular effects on mitochondria and metabolic activity, based on the selective ability of viable cells to reduce tetrazolium bromide into purple formazan crystals, which are only soluble in organic solvents. The extent of formazan crystal formation depends on intact metabolic activity and is frequently used for cytotoxicity screening. The MTT assessment to the cytocompatibility was adequately good and the results were easily interpreted, besides being cost effective.

In research paper, *in vitro* biocompatibility evaluation of biomedical-grade oligochitosan (O-C) and *N,O*-carboxymethyl-chitosan (NO-CMC) derivatives (O-C 1%, O-C 5%, NO-CMC 1% and NO-CMC 5%) correlated well with *in vivo* results in which O-C 1% being the most cytocompatible. The experiment was carried out as described by the International Standards Organization (ISO), with some modifications. One of the most basic and important parameters in the *in vitro* measurement of a medicinal product's biocompatibility is cytocompatibility, which measures the qualitative and quantitative aspects of the impact of a medicinal product with regard to the viability of target cultured cells. Biomedical-grade chitosan derivatives are to be used primarily in human wound management. Thus, the target cultured cells in our research are primary normal human skin epidermal keratinocytes (pNHEK) and primary normal human skin dermal fibroblasts (pNHDF), which are primarily found in the skin epidermis and dermis respectively.

Several transformed keratinocyte culture models have been used to study carcinogenesis, differentiation, apoptosis and cell cycle regulation. Our focus here is on normal (non-transformed) keratinocyte culture models because cutaneous toxicity testing of new drugs, cosmetic products and other chemicals requires phenotypically normal cell systems. This is because the loss of differentiated function in transformed cell types undoubtedly confounds interpretation of the results. Some have suggested that a better correlation of *in vitro* toxicity testing with irritation potential *in vivo* may be obtained by using normal keratinocyte culture models as opposed to transformed cell lines. Furthermore, various *in vitro* systems are used for cutaneous toxicity testing, ranging from submerged keratinocyte and liquid-air interface skin cultures to three-dimensional reconstructed skin models. However, our focus here is on submerged monolayer epidermal keratinocyte culture models. The primary reason for this focus is a key concept in support of incorporating keratinocyte culture models into a battery of cutaneous toxicity tests, namely keratinocytes make up the bulk (greater than 90%) of epidermal tissue mass, and viable keratinocytes (not exposed directly to air) participate in the conversion of environmental stimuli into biological signals that activate the inflammatory and hyper-proliferative responses of the skin. Therefore, submerged cultures of epidermal keratinocytes model a major component of the target epidermal tissue, allowing an assessment of the effects of drugs or chemicals on the skin. The use of low-density polyethylene (LDPE) as negative control is also recommended in the ISO cytotoxicity guidelines. LDPE is widely used in the manufacture of petri dishes and culture flasks, where LDPE is proven to be nontoxic to cultured cells. The comparison between control without any treatment and negative control (with LDPE) indicated in all tests that the cells in the negative control condition were slightly lesser than the untreated cells. This effect might be due to the weight impact of LDPE on cultured cells. LDPE might slightly inhibit the circulation of the enriched medium from the upper level to the lower level of a culture well. The net cytotoxicity effect of the biomedical-grade chitosan on cultured cells can be evaluated more accurately if they are compared to the negative control in order to eliminate the impact on physical weight from the calculation.

Genotoxicity Assessment in Vitro

Although the measures of a medical product's biocompatibility have largely been reported in terms of irritation, sensitization, and systemic toxicity, there is a growing concern that devices or biomaterial extracts may exert some genotoxic effects, even though they may be cytocompatible in cytocompatibility *in vitro* assessment. Thus, any attempts to assess the safety of a device intended for intimate body contact or permanent implantation would be incomplete without testing for the presence of toxins that might exert an effect on the genetic material of cells. Genotoxic effects will eventually lead to abnormal and reduced cell growth, even if the cells initially appear cytocompatible. In its set of standards for the biological evaluation of medical devices, the ISO has outlined the need for such genotoxicity testing in ISO 10993-3. A number of techniques for detecting deoxyribonucleic acid (DNA) damage as opposed to the biological results (*e.g.* Ames test, micronuclei, mutations, structural chromosomal aberrations) of DNA damage have been used long time ago to identify substances with genotoxic activity. According to Muramatsu *et al.*, they did not detect any genotoxicity performed *via* Ames test and chromosome aberration assay, on Chinese hamster lung fibroblasts with 5mg/mL (maximum concentration) of beta-tricalcium phosphate/ carboxymethyl-chitin. Abou Sereih *et al.* reported that their chitosan with concentrations 3.00 and 4.50 mg/mL have induced mutation on *Trichoderma*. However, those techniques have some shortcomings (*e.g.* time investment, requirement for proliferating cell populations).

Until recently, a more useful approach for assessing DNA damage has been the single-cell gel electrophoresis (SCGE) or Comet assay. The term "Comet" is used to denote the individual cell DNA migration patterns produced. Ostling and Johanson were the first to develop a microgel electrophoresis technique for detecting DNA damage at the single-cell level. For this technique, cells were embedded in agarose and placed on a microscope slide. The cells were lysed by detergents and high salt concentration to liberate the DNA. The liberated DNA was then electrophoresed under neutral conditions. Therefore, cells with an increased frequency of DNA double strand breaks (DSB) displayed increased migration of DNA towards the anode. The migrating DNA was then assessed quantitatively by staining with ethidium bromide using a microscope photometer. The neutral conditions used greatly limited the general utility of the assay for detecting the DSB of a cell. Subsequently, the microgel technique was introduced, which involved the use of electrophoresis under alkaline (pH >13) conditions for detecting DNA single strand breaks (SSB) and alkali-labile sites (ALS) in a single cell. Consequently, alkaline electrophoresis offers enhanced sensitivity for identifying various unexpected genotoxic agents that tend to induce more SSB and ALS than DSB. The Comet assay requires only a few cells per assay, and it is also cost effective. Many excellent reviews on the use of the Comet assay used aquatic animals to assess DNA damage caused by genotoxicants. In addition, production of DNA strand breaks correlates well with the mutagenic and carcinogenic properties of environmental pollutants. Therefore, application of comet assay to evaluate the genotoxicity of various newly modified chitosan derivatives can be another approach.

ASSESSMENT OF HUMAN SKIN PRO-INFLAMMATORY CYTOKINES

There are limited studies on the use of skin pro-inflammatory cytokines to assess the biocompatibility of various chitosan derivatives to date. Cytokines are low molecular weight glycoproteins produced by immune as well as non-immune cells. Their roles in regulating immunologic responses, hematopoietic development, and cell-to-cell communication, as well as host responses to infectious agents and inflammatory stimuli, have been widely investigated. Cytokines mediate interactions between various cells and their dysregulated production may lead to disease pathogenesis. Under normal circumstances, cytokines are not detectable at low levels in body fluids or in tissues. Therefore, their elevated levels of expression could indicate activation of cytokine pathways associated with inflammation or disease progression. All cytokines are pleiotropic: they are able to interact with a variety of cellular targets via specific receptors expressed on the surface of target cells. The binding of cytokines to cell surface receptors will initiate RNA signalling and protein synthesis. Cytokines function as networks or cascades of interacting cytokines, where they are able to induce the expression of other cytokines by regulating their own production through autocrine, juxtacrine or paracrine pathways in response to micro-environmental stimuli. Keratinocytes are able to secrete a number of inflammatory cytokines [*e.g.* interleukin 1α(IL-1α), tumor necrosis factor-α (TNF-α)] and chemotactic cytokines [*e.g.* interleukin-8 (IL-8)] when they come in contact with irritants. The changes in cytokine levels could potentially be used as biomarkers of cutaneous vesicant injury. Research on toxicity and inflammation in normal human epidermal keratinocyte (NHEK) cultures in accordance with the use of jet fuel showed similarities between porcine keratinocytes and NHEK in terms of TNF-α and IL-8 expression. Another experiment that demonstrated the chemical irritants induced the production of IL-1, TNF-α and IL-8. These findings indicate that the secretion of pro-inflammatory and chemotactic cytokines may provide a useful biomarker for determining the toxicity level of irritants on human skin.

One study has suggested that environmental pollutants such as diesel exhaust particles and formaldehyde affect the ability of human skin keratinocytes to release pro-inflammatory cytokines (IL-1α, IL-8 and TNF-α). It has been reported that IL-8 production by keratinocytes is under the control of protein kinase C and that protein kinase C signalling pathways are altered in keratinocytes from patients with psoriasis. Therefore, by quantitatively examining the amount (pg/ml) of human IL-1α, IL-8 and TNF-α secreted from treated pNHEK cultures, using the enzyme-linked immunosorbent assay (ELISA), the *in vitro* biocompatibility of chitosan derivatives can be evaluated in more detail.

PATHOLOGICAL CALCIFICATION AND REPLICATING CALCIFYING-NANOPARTICLES

PATHOLOGIC CALCIFICATION RELATED DISEASES

The mineral phase of many kinds of hard tissue in organisms is called biologic apatite (BA). Pure hydroxyapatite (HA) has the formula $Ca_{10}(PO_4)_6(OH)_2$. BA additionally contains several other ions, mainly carbonate but also trace amounts of other anions such as HPO_4^{2-}, Cl^-, and F^-. Other cation elements are also present in minor amounts including Mg^{2+}, Na^+, and Fe^{2+}. BA is the primary mineral of normal bone and teeth. By definition, pathologic calcification refers to the deposition of calcium phosphates (CaP) or other calcific salts at sites, which would not normally have become mineralized. Abnormal accumulation can occur in areas of tissue damage (dystrophic calcification), in hypercalcemic or hyperparathyroid states. Apatite deposition is not a rare phenomenon in pediatry. Calcification is most often seen in children who had prenatal infections. It is also found in association with or as a consequence of toxic or hypoxic damage, intracranial bleed, metabolic, and hereditary diseases like mitochondrial encephalopathy, biotinidase deficiency, and in certain conditions of unknown cause.

Kidney and bladder stones, dental pulp stones, some gall stones, salivary gland stones, chronic calculous prostatitis, testicular microliths, calcification in hemodialysis patients, atherosclerosis, malacoplakia, scleroderma (systemic sclerosis), calcinosis cutis, calcific aortic stenosis, several malignancies, some dementias, calcific tenditis, synovitis and arthritis, diffuse interstitial skeletal hyperostosis, juvenile dermatomyositis, systemic lupus erythematosus are the most common diseases involving extraskeletal calcification. BA deposits also occur in association with inflammation in a variety of other tissues including the eye after implantation of ocular lenses or the breast after breast implants. In each case, these deposits contribute to the morbidity and mortality of the underlying condition. Although the sizes of the mineral crystals in pathologic calcification are similar to those in bone, there is much more mineral in the deposits than there is in bone. Bone matrix proteins also accumulate associated with the deposits. The reasons for the formation of these deposits are not known. It was hypothesized that persistent inflammation is a component of BA and other HA deposition diseases. Other contributing factors are genetic, environmental, and physical chemical. Since inflammation has been the proposed trigger for the calcification, what is the underlying cause of the triggering inflammation?

We first consider the formation of kidney stones (KS), a major and increasing health problem. Approximately 5% of American women and 12% of men will have a KS at some time in life, at an annual cost of $2.1 billion. The prevalence of KS in the USA rose by 37% between 1976–1980 and 1988–1994 in both genders. Studies on the geographic variation in the USA in the prevalence of KS disease have shown a 50% higher prevalence in the southeast than the northwest, possibly associated with a changing state of dehydration related to high summertime temperatures and resulting in a low urine volume. Given the temperature rise worldwide due to the effects of global warming, it has been predicted that there could be an increase of 1.6–2.2 million lifetime cases of KS by 2050, particularly in the southeast regions of the USA. Rule *et al.* have shown that KS are a risk factor for chronic kidney disease, and studies are warranted to assess screening and preventive measures for chronic kidney disease in stone formers. Urinary macromolecules in children show stronger inhibition of Ca oxalate (CaOx) crystal growth, aggregation and adhesion than in adults. Furthermore osteopontin and calgranulin B expression is down-regulated in children due to this inhibitory effect and, thus, stone nidus formation is controlled. What happens in adults to make them more suseptable to KS formation? Could it be the acquisition of an infectious agent?

Another observation may be important. A strong association has been identified between the numbers of sessions of extracorporeal shock wave lithotripsy (SWL) the patient received with conversion of CaOx stones to increasing levels of CaP in recurring stones. The presence of CaP stones raises clinical concern, especially those stones containing brushite (calcium monohydrogen phosphate dihydrate, $CaHPO_4$), because of their increased hardness and, thus, greater resistance to communication by SWL. Why does SWL increase CaP KS formation? Is it possible that an infectious agent acquired over time enhances the formation of CaP rather than CaOx in recurring stones?

Another example of pathologic calcification occurs in cardiovascular disease. Surprisingly, cardiovascular disease is the leading cause of death in patients with chronic kidney disease. A growing body of data points to nontraditional risk factors, including disturbances in mineral metabolism, as important determinants of the extremely high cardiovascular morbidity and mortality rates in these patients. Disturbances in mineral metabolism, especially elevated calcium and phosphate levels, have been linked to vascular and valvular calcification, both of which are associated with poor prognosis in chronic kidney disease patients. New studies indicate that not only vascular intimal calcification (associated with atherosclerosis) but also vascular medial calcification are correlated with decreased survival in chronic kidney disease patients.

Calcification of cardiovascular prosthetic implants is a common and important problem. A variety of cardiovascular prostheses are affected clinically by calcification, including bioprosthetic heart valves, aortic homografts, and trileaflet polymeric valve prostheses. In addition, experimental studies have demonstrated calcification of artificial heart devices in ventricular assist systems in long-term calf studies. The pathophysiology of this disease process is incompletely understood. Is calcification on those devices form by the same mechanism of pathologic calcification in the soft tissue? Is the association of calcification-releated cardiovascular disease and KS formation closely related and perhaps caused by the same agent acquired like a bacterium or virus?

NUCLEATION AND GROWTH OF CAP IN HUMAN BODY

Pathologic calcification is a complicated, actively regulated process of mineralization that is similar to bone formation and remodeling. Mineralogists explain that all that is needed for crystal formation to start is nidi (nucleus) and an environment of available dissolved components at or near saturation concentrations, along with the absence of inhibitors for crystal formation. Bacteria or other agents producing such nidi, if present in blood and in urine, are very likely candidates to launch and accelerate pathologic calcification *in vivo*. This is clinically important since blood contains phosphate near its saturation level.

Matrix vesicles, which are membranous structures derived from the surfaces of hypertrophic chondrocytes, are thought to initiate calcification at the mineralization front by focally concentrating calcium within their already phosphate-rich structures. These vesicles have high levels of alkalinephosphatase and have the ability to concentrate calcium and increase phosphorus during mineralization. Electron microscope studies have indicated that biologic calcification occurs in the MVs in two steps, the first related to the initial deposition of HA within the lumen of the MVs and the second to the propagation of mineral outside the vesicles.

Acidic phospholipids are present at high levels in cells at the mineralizing front in bone and in pathologic calcifications. Detergent pretreatment of bioprosthetic heart valves, shown to inhibit mineralization, likely acts by the detergent-mediated extraction of phospholipids and other proteolipids, which takes place under these conditions.

ROLE OF NANOPARTICLES

Human exposure to nanoparticles is inevitable as nanoparticles become more widely used and, as a result, nanotoxicology research is now gaining attention. Although the number of nanoparticle types and applications continues to increase, studies to characterize their effects after exposure and to address their potential toxicity are few in comparison. In the medical field in particular, nanoparticles are being used in diagnostic and therapeutic tools to better understand, detect, and treat human diseases. When size-dependent cytotoxicity is studied, it has been shown that the smaller the particle is more cytotoxic it gets, and they may act as a nidus for pathologic calcification.

Apatite is generally not toxic to human cells. However, when apatite is present somewhere other than in bones and teeth, it may cause adverse effects. Other studies have concluded that *in vivo* apatite applications are generally safe. Although these disagreements have not been completely resolved, both BA and non-BA materials have been continuously used in drug delivery and transplantation. What has not been well studied, however, is the cytotoxcity of nanophase apatite and related particles.

EFFECT ON MAMMALIAN CELL CULTURES

Methods to diagnose CNP in biologicals, cells, tissues, blood, and urine include immunodetection with CNP-specific MAb, electron microscopy, and culture techniques. Because CNP pass through 0.22-μm pore size filters, which exclude most common microbes, filtration is often used to clean up fluid specimens before culture for CNP. Replication can be measured by particle counting and OD at 650 nm. It has been also shown that growth of the CNP could be detected by specific methods, such as ELISA and turbidity. Calcified and noncalcified forms of CNP have been observed as free, cell-attached, and internalized particles in mammalian cell cultures *in vitro*. In our experiments, we have chosen six different fibroblast lines as experimental models, because they are the most ubiquitous cells in the animal body and might be most accessible in wound tissue for invading pathogens with the exception of professional phagocytic leukocytes. We have shown that CNP were bound as clusters on the cell surfaces within 15 min. It is concluded that CNP are internalized either by receptor-mediated endocytosis or by a closely related pathway within 12 h. We showed that cytotoxicity was dependent on CNP concentration and exposure time. Dying cells always contained numerous ingested CNP. Hybridomas and many lymphocytes were found to be affected by CNP, but considerably higher doses were needed.

In Vivo Effects of CNP (Animal Experiments)

Åkerman *et al.* reported that radiolabeled (^{99m}Tc) viable CNP accumulated in the kidney and appeared in urine after 15 min of their *i.v.* injection into rabbits. This could be due to the fact that kidneys are the preferred sites for this agent, unlike other known nanoparticles, and the presence of injured epithelium or a nucleus in the kidney/urinary tract provides a preferable niche for CNP to adhere and grow, resulting in biocrystallization. A control study in rabbit was performed administering a similar dose of ^{99m}Tc-labeled albumin cross-linked particles or tin-technetium nanoparticles as nanocolloids. Nanocolloids were not targeted to kidneys as CNP were. Shiekh *et al.* too observed that CNP, when injected *i.v.* into rats, were localized in kidneys. In their research they observed regions of chronic inflammation infiltrated in the cortex and medulla, which could be due to damage induced by CNP. The research team has also shown CNP adhering to the surface of the epithelium and their penetration into the epithelial cells.

In a small study, Garcia Cuerpo *et al.* found that translumbar, percutaneous intrarenal injection of CNP (isolated from KS) into rats resulted in KS formation. In addition, Shiekh *et al.* has examined CNP's role in biocrystallization and *in vivo* effects on kidney pathology. CaOx monohydrate assay was carried out in the presence of CNP to study biocrystallization. Wistar rats were given an *i.v.* injection of CNP and the kidneys were examined for pathologic changes. The assay showed accelerated biocrystallization of CaOx in the presence of CNP, indicating them to be efficient candidates for biomineralization. Histopathological studies revealed bacteria induced renal tubular calcifications and various manifestations of infection. Their studies confirm that CNP may be involved in the pathogenesis of renal tubular calcification. Such findings are required to prove Koch's postulates linking CNP to other pathologic calcification related diseases. Schwartz *et al.* designed an experiment to test the hypothesis that the systemic delivery of planktonic forms of CNP derived from calcified, diseased human tissue, or bovine blood are transmissible particles that exacerbate arterial response to injury. New Zealand White rabbits in which the endothelium was mechanically removed from one carotid artery were injected *i.v.* with either saline (control), lipopolysaccharide, HA crystals. HA crystals exposed to culture media, or planktonic forms of bovine- or human-derived CNP. They have shown that the systemic administration of planktonic forms of human-derived CNP exacerbated arterial response to injury distinct from that of bovine-derived CNP and other inflammatory agents.

PATHOLOGIC CALCIFICATION

Most pathologic calcifications throughout the body contain mixtures of carbonate-substituted HA and octacalcium phosphate. According to the Merck Manuel, these ultramicroscopic crystals occur in snowball-like clumps. Those clumps can cause severe inflammation. Structures similar to these nanometre-sized snowballs were discovered almost two decades ago in blood and blood products. These structures, called calcifying nanoparticles (CNP), were detected in numerous pathologic calcification related diseases. CNP are the first CaP mineral containing particles isolated from human blood. CNP have the following unusual properties: bacteria-like, pleomorphic, infectious, self replicating, capable of passing sterilization filters because of their small size (80–500 nm), resistance to heat and γ irradiation at doses typically fatal for conventional bacteria, capable of forming a calcific coating at physiologic pH and mineral concentrations, lacking inherent DNA, sensitive to certain antibiotics, and capable of causing the formation of lipid compounds. CNP have now been linked to pathologic calcification related diseases such as arteriosclerosis, KS gall stone, dental pulp stone formation, prostatitis, Alzheimer's (Kajander *et al.* Do Autonomously replicating sterile-filterable particles have an association with amyloid accumulation? Viruses and virus-like agents in disease. Polycystic kidney (PKD), and cancer. CNP exert cytotoxic effects on some mammalian cells *in vitro* and on living organisms *in vivo*.

The incidence of KS formation and recurrences may continue to rise every year. Thus, new approaches in treatment and prevention could have a huge economic effect apart from benefits in terms of reduced morbidity. Over seventy years ago, Randall examined the papillae of cadaveric renal units and demonstrated that interstitial crystal plaques in the papillary tip were common in stone formers. These crystals were composed not of CaOx, the most common solid phase found in patients with nephrolithiasis, but of CaP. He believed that the CaP crystals serve as a nucleation surface for CaOx. Recently, Matlaga *et al.* analysed Randall's plaques (RP) and confirmed that they are formed of spherical CaP deposits with a multilaminated internal morphology. With the CNP detection methodologies we developed, important features of CNP and their triggering effect on nephrolithiasis have been suggested. In study, there was evidence of a link between detection of CNP and presence of RP. Although causality was not demonstrated, the results obtained suggest that further studies with negative control samples should be made to explore the etiology of RP formation, thus leading to a better understanding of the pathogenesis of stone formation. Additionally, this hypothesis may bring an explanation to the reason of the increase on CaP KS formation after SWL.

Another urinary disease, PKD, is the most common autosomal dominant lethal disease in humans. Interestingly, a higher prevalence of kidney calcifications is observed in PKD than in the normal population. It has been proposed that the currently known cellular toxicities, tissue distribution, and pharmacology of CNP are plausibly related to the known pathology and pharmacology of PKD. Hjelle *et al.* evaluated 13 PKD cyst fluids and detected CNP antigen positivity in each sample as well as in liver cystic fluid from affected individuals. This may explain the reason of high rate calcification problem in patients with PKD.

The presence of prostatic calculi in younger men is associated with both inflammation and symptoms of chronic pelvic pain syndrome. Biopsies, urine, and prostatic secretion cultures fail to demonstrate bacterial pathogens, however inflammation is often detected in chronic prostatitis. The core of prostatic calculi is typically CaP, which is the hallmark of CNP action. Wood and Shoskes proposed that there is a potential role of CNP in chronic prostatitis. Recent clinical research targeting these agents has proven effective in treating some patients with refractory category III prostatitis (chronic pelvic pain syndrome). In that research, CNP antigen or antibody was found in 60% of serum and 40% of urine samples. In 10 patients who underwent transrectal ultrasound after therapy, prostatic stones were decreased in size or resolved in 50%.

Testicular microlithiasis (TM) is an uncommon pathologic condition of unclear etiology, which is characterized by calcium deposits within the seminiferous tubules. Zhang *et al.* has studied infertility patients with and without TM. Their results have showed that CNP may be linked to the development of TM, which may provide a potential target for the diagnosis and treatment of infertility with TM.

Mechanisms mediating vascular calcification remain incompletely understood. Some have hypothesized the potential role of CNP in arterial calcification. Miller *et al.* cultured nano-sized particles from calcified but not from noncalcified aneurysms. These particles were stained with CNP-specific MAb, recognized by a DNA-specific dye and incorporated radiolabeled uridine, and, after decalcification, they appeared via electron microscopy to contain cell walls. Puskas *et al.* propagated CNP- like spherical particles from 26 of 42 sclerotic aorta and carotid samples and confirmed their nature by dot immunoblot by using CNP-specific MAb, light microscopy, and transmission electronmicroscopy (TEM). [3H]l-aspartic acid was incorporated into high molecular weight compounds of demineralized particles. PCR amplification of 16S rDNA sequences from the particles was unsuccessful using traditional protocols. Identification of CNP-like particles at the lesion supports but does not by itself prove the hypothesis that these agents contribute to the pathogenesis of atherosclerosis, especially vascular calcifications. Specific therapies targeting these particles has demonstrated reduced plaque formation, regression of plaques, and improved lipid profiles. The potential of anti-CNP treatments are controversial and await larger clinical trials. Epidemiologic studies have implicated antibodies made by the body against CNP to be a strong independent risk factor for coronary artery calcification. Also Hu *et al.* has cultured CNP-like material from calcified cardiac valves with rheumatic heart disease.

Apatite deposition is not a rare phenomena in pediatry. In an infant with idiopathic arterial calcification of infancy, prenatal diagnosis of arterial calcification was made by ultrasonography and allowed initiation of therapy *in utero*. Etidronate therapy produced apparent radiographic and ultrasonographic improvement in the degree of vascular calcification. Etidronate is a compound, which inhibits CNP replication.

Originally, Sedivy and Battistutti reported that CNP promoted crystallization of psammoma bodies in ovarian cancer. Hudelist *et al.* soon verified the 100% concordance between the expression of CNP and the presence of psammoma bodies in malignant ovarian tumours. In their research, several lines of evidence suggest the involvement of these organisms in the process of biomineralization. Therefore they have concluded that CNP infection of malignant ovarian tissue contributes to mechanisms leading to the formation of calcified deposits known as psammoma bodies.

In addition, CNP have been shown to be detected at higher rates in serum of patients with gallstone disease, and mitral valve calcification, dental calculus, and periodontitis. Others have suggested that CNP may contribute to the development of peripheral neutropathy in HIV-positive patients and even osteoporosis. All these hypothetical approaches require further investigation.

BIOLOGIC CLASSIFICATION OF CNP

Despite their potential role in major medical health problems, CNP have not been classified in any taxonomic group due to limited information on their biologic characteristics. Although the biologic characterization of CNP is yet to be fully understood, the precipitation and growth of CaP readily occurs in systems containing trace amounts of CNP, but not in identical control systems lacking CNP. Chemical analysis using energy-dispersive x-ray microanalysis (EDX) of these mineral layers shows Ca and P peaks. The effectiveness of CNP biomineralization is remarkable: apatite formation *in vitro* stopped only when the calcium level decreased by 50% from 1.8 to 0.9 mM, and the phosphate levels fell to near zero. Our results indicate that the CNP CaP phase can be formed at pH 7.4 consistent with human physiologic phosphate and calcium concentrations. This can be also monitored by ^{85}Sr incorporation and provides a unique model for *in vitro* studies on calcification.

EVIDENCE FOR SELF-REPLICATION OF CNP

To enhance our understanding of the self-replicating nature of CNP, we investigated their growth using inverted light microscope and BioStation IM time-lapse imaging system. The results are in conjunction with previous findings of metabolic activity antibiotic sensitivity, antibody specificity, morphologic aspects, and infectivity, all concomitantly validate CNP as live-like self-replicators.

We found a systematic increase in both the cell-like nanoparticles and the apatite coatings and apatite "igloos". The preferred formation of igloos over apatite-coated cell-like particles occurred when the serum or protein in the nutrient solution was depleted. With sufficient protein, the igloos did not readily form but the cell-like nanoparticles rapidly increased along a flattening growth curve. Although the mechanism for the formation of both the cell-like nanoparticles and the apatite crystals and igloos may be entirely nonbiologic, it mimics in many ways the life cycle of typical bacteria. As long as serum proteins, calcium, and phorphorous are present, new cell-like nanoparticles continue to form and apatite continues to precipitate as coatings on these nanoparticles.

Igloos continue to form in serum protein depleted media. Controls lacking an initial charge of CNP do not form either the cell-like nanoparticles or the apatite precipitates, even under identical conditions of chemistry and environmental conditions. Controls salted with inorganic apatite dust particles or albumin addition also do not produce the formation and growth of either the CNP or new apatite precipitates of any type. The "life cycle" of CNP is more complex than previously realised, and includes the formation of pleomorphic cell-like nanoparticles, their acquisition of precipitated apatite coatings built up from extremely small particles of apatite, the formation under a depleted protein environment of much larger apatite structures or igloos, the production of new cell-like nanoparticles within the igloos, the release of these nanoparticles, and a repeat of this cycle. In the case of abundant protein in the media, the igloo phase may be bypassed and the cell-like nanoparticles may be reproduced by budding or by fission. This "life cycle" is reproducible, relatively invariant, and predictable and has been observed over more than three decades in "cultures" inoculated with a small amount of initial CNP identical to those on file with the German cell bank, although some of the details are still being elucidated.

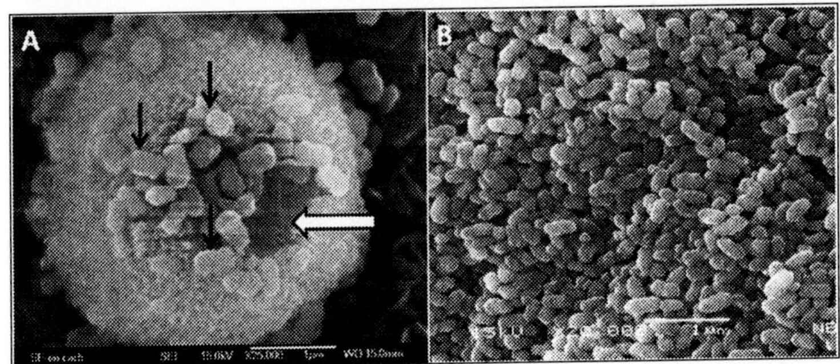

Figure : Scanning Electron Microscopic Image of CNP Cultured in Serum-free Condition (*A*), and in Serum Containing Condition (*B*). On the Left (*A*), an "Igloo" is Seen. White Arrow Shows the Igloo Opening, Small CNP Released from the Igloo. On the *B*, Coccobacillar Shaped Small CNP "Cells" are Seen. Bars, 1 μm.

CNP BIOFILM

CNP also produce a slime-like biofilm, which may contribute to the "stickiness" of the CNP complex (cell-like vesicles and associated apatite precipitates) in human tissue, circulatory systems, the obstruction of blood flow in arteries, and on catheter or stent-like artificial prosthesis (unpublished data). This biofilms links together the cell-like nanoparticles and provides anchoring to fixed structures such as arterial walls or kidney tissue. Detailed TEM imaging shows that this biofilms is usually present along with the cell-like vesicles and the precipating apatite.

The formation of this induced apatitic biofilm formation is dependent on the presence of oxygen. Additional studies have shown that the formation and growth of this biofilm can be prevented with several antibiotics and antimetabolites, and by high gamma irradiation at sterilizing doses. Schwartz *et al.* has shown that CNP biofilm retains some characteristics of conventional bacterial biofilm and requires protein-calcium interactions, although extracellular RNA is not required.

Formation of Apatite

Previous studies have shown that CNP form two distinct forms of apatite. The early-formed apatite consists of extremely small (10's of nanometre) flakes, plates, and crystals, which coat the cell-like organic nanoparticles. Under conditions of serum or protein depletion, these apatite crystals coalesce to form much larger structures (1–10 μm) which are usually concentrically zoned in texture (igloos).

These larger structures often contain a central cavity which typically contain multiple cell-like nanoparticles. In some cases, the igloo is partially open allow the egress of these cell-like particles which can then repeat the cycle. These subspherical units or igloos were identified in most human KS examined. Fourier Transform Infrared Spectroscopy of CNP revealed the mineral as almost identical to bone and apatite KS mineral.

In one interesting study, we examined CNP cultures in High Aspect Rotating Vessels designed at the National Aeronautics and Space Administration's Johnson Space Centre, which are designed to simulate some aspects of microgravity. CNP cultured in that system multiplied 4.6 times faster than under stationary conditions and 3.2 times faster than in shaker flask incubation.

Interestingly, the results demonstrated that the degree of apatite crystal formation on the cell-like nanoparticles, and the properties of the apatite are strongly affected by the gravity and other specific culture conditions used. Although some researchers believe that microgravity does not affect crystal formation and biomineralization, it has been shown that long periods in a microgravity environment does cause loss of bone, and enhance KS formation-like biomineralization disorders in astronauts. Our data from the experiments would support the hypothesis that reduced gravity may enhance the development of KS by enhancing the growth of CNP.

The Controversy

CNP were originally called "nanobacteria" because they behaved like known bacteria yet were very small. That terminology prompted a search for nucleic acid. (RNA or DNA). DNA fragments identified early subsequently were shown to be contamination or DNA absorbed by the particles from their environment. As of the present time, no undisputed DNA has been associated with these particles.

The original term nanobacteria was therefore dropped by our group as misleading and controversial. We have subsequently used the terms CNP. As we have learned more about their life cycle, we now realise that they consist of a membrane-enclosed vesicle or cell-like particle which has the ability to cause or catalyse precipitation of extremely small HA particles or crystals on the membrane surfaces, and to combine the precipitated apatite into more complex layered concentric structures, which we call igloos because they are often hollow containing newly formed membrane-enclosed vesicles. In one version of the life cycle, the hollow igloo is breached and releases the cell-like nanoparticles, which continue to reproduce and precipitate new apatite. In another version of the life cycle, the cell-like nanoparticles precipitate fresh apatite on their membrane surfaces but do not form igloos.

In both versions, a biofilm may link the nanoparticles together, link them to the igloo, or form an anchor against fluid movement and dispersion. The term CNP refers only to the membrane-enclosed cell-like particle. The associated apatite (both membrane coatings and igloos) and biofilms are separate, but the whole association is termed by us the CNP complex. No previous definition made clear that the membrane-enclosed cell-like particles are not apatite but can exist alone, but are closely associated with both a precipitated apatite coating and with a much larger apatite igloo. Previous definition did not always include the biofilms as separate and distinct part of the complex.

The very small size (50–200 nm) raises the critical question of whether these particles can contain sufficient DNA to operate as a true replicating and metabolizing cell. Maniloff's work suggests that to contain the DNA and proteins needed to function, a cell must be at least 140 nm across. However, recently it has been shown that a genome constructed to encode 387 protein coding and 43 structural RNA genes could sustain a viable synthetic cell, a *Mycoplasma laboratorium,* which can shrink its size below that limitation. CNP are also incredibly resistant to heat and other methods that would normally kill bacteria, which makes some scientists wonder if they might be an unusual form of crystal rather than organisms. Cisar *et al.* and Martel *et al.* presented alternative theories for the experimental findings of earlier CNP studies. They stated that biomineralization previously attributed to CNP may be initiated by nonliving macromolecules and transferred on "subculture" by self-propagating microcrystalline apatite. We are unaware of any report showing the nucleation and growth of inorganic apatite under physiologic conditions. In addition, for inorganic crystallization to continue over prolonged periods of time, conditions of nonequilibrium must exist and be maintained.

Martel *et al.* have investigated a purported similar material which they termed nanobacteria-like particles (NLP) in their research. It is not clear what the relationship of these NLP is to the original culture deposited by us in the German cell bank, so it is not possible to directly compare results. Cisar *et al.* also cultured similar particles from serum and human saliva sources and named them nanobacteria with no test comparing them to the original material deposited by the in the cell bank.

His team verified our findings, and also confirmed the extreme difficulties in performing PCR, but finally suggested his opinion that the culturable particles cannot be bacteria, since they were too small, were not inhibited with a respiratory poison, nucleic acids could not be detected with standard procedures and their protein patterns revealed only few proteins, much less than one would expect from a common bacterium. CNP could not be lysed with lysozyme, proteinase K, several other proteinases, lipases, amylases, alkali, ultrasound, X-press, detergents or solvents. In some experiments we needed to use acid or EDTA-like chelators before the analysis, which were the factors for structural change in nucleic acid. It has been our experience that CNP actually inhibit the amplification of added exogeneous classical bacterial DNA by polymerase chain reaction methods. We currently conclude that no convincing evidence exists for the presence of inherent nucleic acid within CNP. PCR analysis using universal primers practically impossible and worthless.

Cisar *et al.* has not sequenced any proteins. They did not do any DNA work besides staining with Hoechst 33258, where they got the same weakly positive result than we did. Contrary to Dr. Cranton's claims, Cisar *et al.* did not do a PCR phylogenetic analysis using 16S rRNA sequences.

As pointed out by Dr. Cranton, apatite can be formed under super-saturating concentrations of calcium and phosphate *via* several mechanisms. To our knowledge, CNP-mediated calcification is the only mechanism to make apatite at nonsaturating levels of calcium and phosphate.

Cisar *et al.* did not follow saturation degree analysis in his studies although saliva is known to be highly super-saturated with calcium and phosphate.

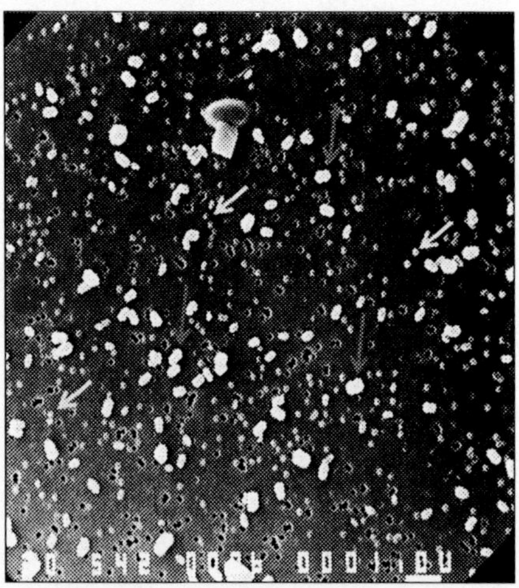

Figure : Scanning Electron Microscopic Image of a Filter (with 100 nm pore size) having CNP and Protein Precipitation on it.

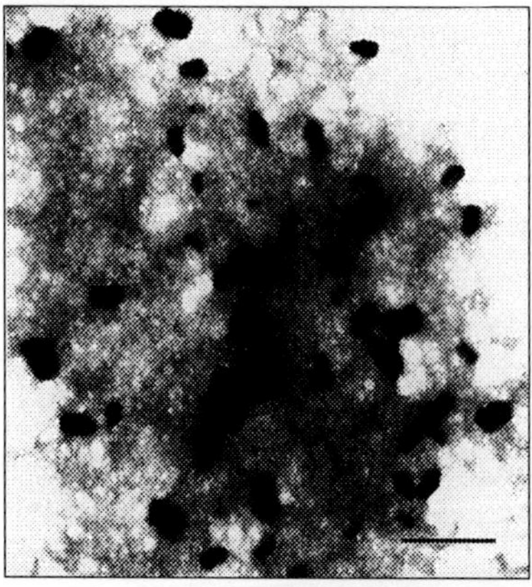

Figure : Transmission Electron Microscopic Image of CNP and Protein Precipitation Mixture.

In the study by Martel *et al.*, electron microscopic images from so called NLP morphologically look like the serum protein precipitation formed in long-term serum cultures and are not comparable with the earlier findings. Therefore, we find their entire report not comparable with our findings, and the immunostaining tests that they have pursued in their research cannot be comparable with the results we and other laboratories obtained in earlier years. As they have claimed, if the MAb raised for CNP would cross react with a major component of serum, in every experiment containing serum component, the results would be positive.

We would recommend using negative controls with no primary antibody and nonspecific MAb in every immunostaining experiments to eliminate false-positive possibilities. Vali *et al.* have also shown that nanoforms contain apatite-protein complexes and immuno electron microscopy reveals protein antigens in proximity to apatite suggesting a novel form of protein-associated mineralization.

The claims that CNP must be a nonbiogenic (albumin, fetuin-A, and apolipoprotein A1) crystallization still cannot explain the biogenic-like properties of CNP. Some examples of those properties are

- CNP cannot be cultured from every serum sample although each serum contains albumin, fetuin, and apolipopretein.
- TEM examination of CNP cultures have always shown a close association of apatite with submicrometer vesicles (cell-like bodies) enclosed within membraneous structures. We and others have always found the presence of these membranous vesicles remaining after the apatite has been chelated and dissolved away with acid or EDTA. Nonbiogenic apatite has larger crystals that are easily dissolved in acidic solution but CNP dissolves very slowly.
- When inorganic apatite is added to culture medium (DMEM), it dissolves but CNP introduced to identical media begins and continues to self propagate.
- CNP can be cultured in serum-free conditions and be passaged for years with no serum addition. Therefore, CNP cannot be simple serum protein-derived vesicles with associated crystal formation.
- Inorganic apatite crystals do not stimulate immune response and cause pathologic calcification as CNP do.
- CNP has specific staining properties. Other components of the serum or growth media do not stain using the same stain.
- CNP are cytotoxic and they have a specific route *in vivo*. Serum proteins do not have these properties. Serum protein coated inorganic apatite crystals have been cultured and injected to animals but their route in animals was different.
- CNP have and release endotoxin and thereby stimulate and mediate chronic local inflammatory reactions in atherosclerotic plaque. Other

suggested explanations (albumin, fetuin-A, and apolipoprotein A1, inorganically precipitated apatite particles) lack this property.

- CNP growth could be prevented with tetracycline, high doses of aminoglycoside antibiotics, EDTA, cytosine arabinoside, 5-fluorouracil, and gamma-irradiation.
- CNP structures have been isolated from many diseased tissues and when found, are always similar to other structures from other unrelated patients.
- The metabolic potential of CNP was confirmed using a tetrazolium salt detecting dehydrogenase activity, and S-methionine incorporation. Also, β-mercaptoethanol, known to enhance growth of certain microorganisms and mammalian cells, promoted CNP metabolism and growth.
- Polarized light was shown to reduce CNP biofilm formation indicating a light induced metabolic process within the CNP complex. Apparently, CNP have metabolic activity, which clearly differentiates them from inorganic crystal formation. Alternatively, the behaviour of these nanoparticles closely mimics metabolic activity using a mechanism not understood despite decades of research.

CNP Live or Dead

We summarized the properties of CNP that resemble the characteristics of living bacteria and also point out the differences that set them aside as a different and distinct type of small cell-like objects. We raise the question of whether they should be considered living entities. We conclude that CNPs do not fit the typical definition of life, but we suggest that perhaps the typical definition is too restrictive, and a broader definition is required, which encompasses the properties of CNPs.

One property of living systems is their ability to pass on to subsequent generations some information that restricts the new structure and properties to those of the earlier generation. Mutations are allowed, but basic properties should be passed on to subsequent generations. Bullard *et al.* put to test the "crystals-as-genes" hypothesis.

For crystals to resemble genes, there must be more inheritance than mutation in successive generations. However, despite the greatest of care taken to not expose the fresh crystal seeds to atmosphere, and even in the absence of cleavage, new hillocks, "mutations" proliferated so that the detailed structure and morphology of the new crystals differed considerably from the original structure. An antibody analog that recognized external structural morphology of the parent crystals would not recognize morphology of the subsequent crystals. By contrast, for more than a decade, CNP have been cultured and passaged under physiologic conditions similar to mammalian culture conditions, without any change in their growth characteristics and specific MAb recognizing epitopes, a property analogous to the morphology of the crystal surfaces of Bullard *et al.*

There is truly no universally agreed definition of life. Although the theme of DNA has permeated so deeply in the scientific world, lately, there have also been some contrary publications questioning the concept of the gene as the unit of life.

The theory that life could have started with very simple heterotrophic primordial cells is currently gaining recognition. Such could be the case with CNP. These particles consist of an organic membrane surrounding a vesicle. This membrane seems to have the ability to induce apatite precipitation and the formation of complex external structures embodied by the concentric layered larger precipitates including the igloos.

The membrane-encased cell-like nanoparticles seem to have the ability to self-replicate while maintaining a complex set of characteristics which are passed on to subsequent generations. One hypothesis for the origin of life is that it started with a simple membrane-enclosed vesicle, which allowed or promoted simple chemical reactions across the or through a membrane so that the interior chemistry was somewhat different from the exterior chemistry. The membrane may also have had catalytic properties so that it promoted chemical reactions not normally found in its surrounding environment. At some point, the membrane-enclosed vesicle split or budded and a new vesicle was formed.

If one of the chemical reactions induced at the membrane surface consisted of precipitation of a salt such as CaP, the system might resemble what we now call CNP. Continuation of this process would lead to growth, increase in complexity, and eventually to the formation of simple DNA. If life began in such a system, the early cell-like version would lack DNA, but would still be considered a living system. Perhaps the CNP can be considered a modern-day analog of such a primitive system, and could be considered a primitive form of life lacking nucleic acid, but still possessing self-replication, membrane-induced chemical reactions, the ability to maintain distinctive characteristics over multiple generations, and the ability to exist and grow over a range of environmental conditions. It is evident that CNP involved in pathologic biomineralization with their distinctive characteristics defy the old standard scientific expectations and definitions of life. Perhaps, an expanded definition of life is appropriate.

TYPES OF BIOLOGICAL WEATHERING

Weathering is the process wherein natural occurrences and reactions break down materials over time. One type of weathering is mechanical, caused by physical action.

Figure: Termites are one Cause of Biological Weathering.

The other type of weathering is chemical, caused by chemical reactions. Both of these types of weathering can at times fall under another category known as biological weathering. While there are several specific types of biological weathering, in general it is characterized as being caused by organic, or living, substances and organisms.

ACID SECRETION

A number of different organisms produce and secrete acids, including molds, fungi and lichens. According to jersey.uoregon. edu, these acids can chemically weather rocks and minerals and essentially "etch" irregularities into their surfaces. These irregularities can in turn serve as footholds or anchors for other organisms, which can further biologically weather the rocks or minerals.

ORGANIC DEBRIS

In addition to organisms producing and secreting acid directly, the decomposition of organic debris such as animal excrement, animal carcasses, and dead leaves and plants can also release this corrosive substance into the environment. According to jersey.uoregon.edu, when organic debris decomposes, carbon escapes into the atmosphere. This carbon then interacts with water molecules and forms acids, which can then fall as rain and weather rocks, minerals and other substances. When rain is especially acidic, it is referred to as "acid rain" and can be harmful to organisms and drinking water supplies. However, the majority of acid rain cases are associated with man-made pollution and are not "natural" occurrences.

ROOT DAMAGE

One of the most commonly occurring and easy-to-observe types of biological weathering is root damage. If you live in an area that features trees butting up against paved streets or sidewalks, you have likely seen this phenomenon. Root damage occurs when large, sturdy tree roots run out of growing space and break through the surfaces that are restricting them. This can produce cracks and holes in concrete. Root damage is a mechanical, as opposed to chemical, process.

DIGGING OR BORING

Another type of biological/mechanical weathering is the digging or boring of certain organisms into rock and other materials. According to mc2.vicnet.net.au, one example of this weathering can be seen in termites, which are small insects that eat complex carbohydrate cellulose, or plant matter, and commonly tunnel through wood in the process. However, termites also use chemicals to break down the cellulose, so the weathering in not entirely mechanical. According to geolsoc.org.uk, another example of biological/mechanical weathering caused by digging or boring can be observed in the piddock, a bivalve mollusk, which scrapes away bits of rock until it has bored out a suitable home.

BIO-CHEMICAL PROCESS OF WEATHERING

The processes of weathering which are mainly related to the activities of plants, animals and organisms like bacteria etc. are known as 'Biological weathering. It involves the role of plants and animals in the breaking down of rocks through mechanical ways as well as in the decomposition of rocks.

The role of organisms in chemical weathering depends much on their ability to assimilate various elements from the rocks and to evolve various organic acids etc. Plant roots dissolve various constituents of the rocks with the acids evolved by the tips of their roots. Besides, certain groups of bacteria algae and mosses breakup rock-forming silicates directly, removing from them elements like silica, potassium phosphorous, calcium, magnesium etc. that they need as nutrients. This facilitates rock-weathering.

Bird droppings have been stated to be capable of weathering limestone's. With the decay and degeneration of the dead remains of plants and animals, chemically active substances are produced like carbon dioxide, humic acid, together with traces of ammonia, nitric acid etc. These substances are capable of bringing about rock-weathering. Man himself is an important agent of destruction. All the processes of weathering, physical, chemical and biological, are interconnected and act simultaneously.

BLOOD INTERACTIONS WITH BACTERIAL CELLULOSE

Cardiovascular disease (CVD) is the main cause of death or invalidism in high-income countries today. Moreover, worldwide demographic changes are aiding CVD's rapid progression towards the number one killer in middle- and low-income countries. The World Health Organisation estimates that if current trends are allowed to continue, about 20 million people will die from CVD by 2015. This group of disorders, which affect the heart and blood vessels, includes coronary heart disease, cerebrovascular disease and peripherial arterial disease, deep vein thrombosis and pulmonary embolism.

The main cause of these acute life-threatening conditions is atherosclerosis. Atherosclerotic plaques and restenosis can result in severe occlusions of peripheral and coronary arteries. Current treatments include drug therapy and bypass surgery, and depend on the severity of the disease. All treatments require molecular understanding of the processes that govern atherosclerosis. This is especially important when introducing artificial graft materials in vivo.

Generally, the first choice for vascular replacement graft material is the patient's own vessels, *i.e.*, autologous vessels. If these are in shortage supply or do not exhibit sufficient quality due to, *e.g.*, other diseases or previous surgery, artificial alternatives become necessary. Today, clinics use biomaterials such as expanded polytetrafluorethylene and polyethylene terephtalate fibre (Dacron®) as prosthetic grafts for reconstructive vascular surgery. However, their performance is dismal in small diameter vessels (>6 mm) like coronary arteries and peripheral arteries below the knee, resulting in early thrombosis and intimal hyperplasia. Therefore, about 10% of patients with CVD are left untreated due to the lack of replacement material for small vessels.

Considering the large number of patients who need replacement vessels, the substantial demand for alternative small-caliber grafts is urgent, driving scientists to search for and develop new materials. Recently, this has even led to the use of completely biological vessels. However, the growth of such requires months, rendering them unsuitable for acute situations such as heart infarction, which demand a substitute vessel immediately.

IDEAL OF VASCULAR GRAFT

Several issues demand consideration when constructing a vascular graft: the mechanical properties of the graft must resemble those of a native blood vessel, and the graft must be biocompatible with its host. One important factor is compliance, *i.e.*, how well the vessel withstands pressure from the bloodstream and whether it can maintain systemic pressure in the vascular system. Vascular grafts should also be "invisible" to the immune system and possess non-thrombogenic properties.

One interesting option is bacterial cellulose (BC), whose unique properties (strength, good integration into host tissue and flexibility that allows production in various shapes and sizes) make it an exciting candidate for vascular graft material. The most abundant biopolymer on earth, cellulose is insoluble in water and degradable by microbial enzymes. Several organisms such as plants, algae and bacteria can produce BC. Some members of the bacterial genus Acetobacter, especially Gluconacetobacter xylinum, synthesize and secrete cellulose extracellularly. The network structure of cellulose fibrils resembles that of collagen in the extracellular matrix (ECM).

This chapter describes how BC interacts with human endothelial cells (EC) and blood. Specifically, we will evaluate whether surface modifications could promote adhesion of EC and also whether BC's thrombogenic properties compare favourably with conventional graft materials. These properties are critical because materials intended as vascular grafts must satisfy many important features, including blood compatibility, cell interactions and mechanical properties.

TISSUE ENGINEERING OF BLOOD VESSELS

Tissue engineering is a relatively new scientific discipline that combines cells, engineering and materials to improve or replace biological functions. Langer and Vacanti, two pioneers in the field, describe tissue engineering as an interdisciplinary area that applies the principles of engineering and life sciences to the development of biological substitutes that restore, maintain or improve tissue formation.

The basic concept of tissue engineering includes a physical support (3D-scaffold) composed of synthetic polymers or natural materials (collagen, elastin or fibrin). This support mimicks ECM and initially serves as a scaffold or template on which cells can organize and mature in vitro prior to implantation at the appropriate location.

Initial research in the mid-20th century focused on developing bioinert materials, eliciting a minimal host response characterized by passive blood transport and minimal interactions with blood and tissues. Although widely available, these industrial materials, including Teflon and silicone, were not developed specifically for medical applications. Later, the production of completely non-reactive substances became unrealistic.

Today, other biomaterials are being developed to stimulate reactions between proteins and cells at the molecular level in a highly precise and controllable manner. The key concept underpinning development of such biomaterials is that the scaffold should contain chemical or structural information that mimicks cell-cell communication and controls tissue formation, such as growth factors, the adhesion peptide Arg–Gly–Asp (RGD) and other molecules that mimic ECM components. RGD is the minimal sequence in basement membrane proteins such as fibronectin, fibrinogen and von Willebrand Factor, all required for cell adhesion.

A successful tissue engineered blood vessel must: be biocompatible, *i.e.*, noninflammatory, nontoxic, nonimmunogenic and noncarcinogenic; infection-resistant and nonthrombogenic. It also must have appropriate mechanical properties, *e.g.*, tensile strength, burst strength, good suture retention and compliance, and possess appropriate vasoactive physiological properties, including contraction or relaxation in response to neural or chemical stimuli and more.

The feasibility of constructing and using tissue engineered blood vessels was first demonstrated in landmark studies by L'Heureux and Niklason. The vessels, which were produced using different in vitro techniques, had very good mechanical properties and functioned well in experimental animals. Although in vitro and experimental techniques have been developed since then, no clinical implantations have been made until now.

Biomaterials/biomaterial Scaffolds

Williams defines biomaterial as any natural or man-made material that comprises the whole or part of a living structure or biomedical device that performs, augments, or replaces a natural function.

Many different materials have been investigated for biomaterial applications. They can be divided into natural materials, *i.e.*, collagen; fibrin; hyaluronic acid; silk fibroin and BC and synthetic polymers, *i.e.*, polyglycolic acid (PGA), polyethylene terephthalate (PET) and ePTFE. The required properties for biomaterials vary with cell type, implantation site and strategy for tissue formation. Common demands for all biomaterials include biocompatibility, *e.g.*, avoiding foreign body reactions, capsule formation and chronic inflammatory reactions. Additionally, materials intended to be in contact with blood require evaluation for thrombogenicity. Mechanical properties are important and depend on the target tissue. Since biomaterials used as vascular grafts must withstand blood pressure, they must be investigated for burst pressure, compliance, suture strength and fatigue before using them as implants.

A recent and popular approach involves electrospinning different materials to create nanofibre constructs. Both electrospun synthetic polymers and native ECM proteins have been used for cell seeding to construct vascular grafts.

Materials for Vascular Grafts

Jaboulay and Briau performed the first arterial transplantation in 1896, but imperfect anastomoses resulted in thrombosis. Since then, more sophisticated techniques have been developed. The search for arterial vascular grafts began in 1952, when Voorhees discovered Vinyon N (nylon), the first fabric graft. A few years later, DeBakey discovered Dacron® in 1958. Today, arterial and even especially venous autografts are used routinely in surgery, creating bypasses for patients with peripheral or coronary occlusive vascular diseases. However, autograft availability is limited, particularly for arteries.

Dacron® and ePTFE are still widely used as arterial replacements. Despite their success in replacing large diameter (>6 mm) high-flow vessels, these materials show thrombogenicity and compliance mismatch in low-flow or small-diameter vessels. Sophisticated techniques have been evaluated to enhance patency, including chemical modifications, coatings and seeding of the surface with different cells. In contrast to natural materials, synthetics often lack adhesion sites. Although passive materials can reproduce sufficient physiological mechanical strength, proper metabolic function and cellular signalling requires intact cellular machinery.

BC is an attractive material for biomaterial applications. Its structure resembles that of collagen, the component in arteries and veins that gives the blood vessel its strength. BC's manufacturing process allows versatility in shape and size, including tubes. Studies have shown successful growth of cardiac rat-derived myocytes and fibroblasts, rat-derived hepatocytes and osteoprogenitor cells from mice on cellulose-based materials. However, these matrices are based not on natural cellulose but rather on derivatives such as cellulose acetate and regenerative cellulose.

Although BC is biocompatible, it generally does not promote cell growth. Thus, BC must be modified to support EC adherence. Modification of wet state BC is challenging because fibre structure and strength must be maintained. This is especially important for vascular grafts, which must withstand blood pressure.

COLLAGEN-BASED BLOOD VESSEL MODEL

Weinberg and colleagues developed a collagen-based blood vessel model. Improvement of the construct's mechanical properties is ongoing due to poor mechanical integrity.

CELL SELF-ASSEMBLY MODEL

The cell self-assembly model is made using intact layers of human vascular cells grown to overconfluence to form visible sheets of cells and ECM. A sheet of smooth muscle cells (SMCs) is rolled around a mandrel to form the medial layer. Similarly, a sheet of fibroblasts is rolled over the SMC sheet media, forming an adventitial layer. Finally, ECs seeded onto the lumen of the matured vessel form a confluent monolayer. These constructs withstood more than 2000 mmHg pressure before bursting.

Cell-seeded Polymeric Scaffold–hybrid Graft

This graft was developed by Niklason. Vascular cells were seeded into a biodegradable scaffold (PGA) and cultured for 8 weeks under pulsatile radial stress (165 beats per minute and 5% radial strain). ECs seeded onto the lumen of the construct formed a confluent monolayer. Cultured under pulsatile conditions, the histological structure of the constructs resembled that of native arteries. Due to high collagen content, the constructs had a burst pressure greater than 2000 mmHg and they could contract. In vivo studies of tissue engineered blood vessels in Yucatan miniature pigs were promising.

With degradable material such as PGA, it is critical to ensure adequate strength of ECM (collagen, elastin) produced by the vascular cells. Degradation products may be toxic to the cells. Non-degradable scaffolds offer durable support, but tissue acceptance is mandatory. Ideally, the scaffold should be compliant, similar to native vessels.

Acellularized Construct

A rolled small intestinal submucosa (SIS) has been used as a small diameter vascular graft. A cell-free, 100-µm-thick collagen layer derived from small intestine, SIS is compliant, making it an interesting candidate for vascular implantation and requiring investigation.

Artificial Artery Generated in the Peritoneal Cavity

In the peritoneal cavity, artificial arteries are generated on silastic tubes. The arteries are lined by nonthrombogenic, mesothelial (endothelial-like) cells. The feasibility of this approach in humans is a matter of debate.

BACTERIAL SYNTHESIZED CELLULOSE AS BIOMATERIAL

Cellulose, the most abundant biopolymer on earth, is insoluble in water and degradable by microbial enzymes. Several organisms, *e.g.*, plants, algae and bacteria can produce cellulose, and the bacteria Gluconacetobacter xylinum can synthesize and secret cellulose extracellularly. BC is composed of linear nanosized fibrils of D-glucose molecules. The network structure of cellulose fibrils resembles that of collagen in the ECM of native connective tissue.

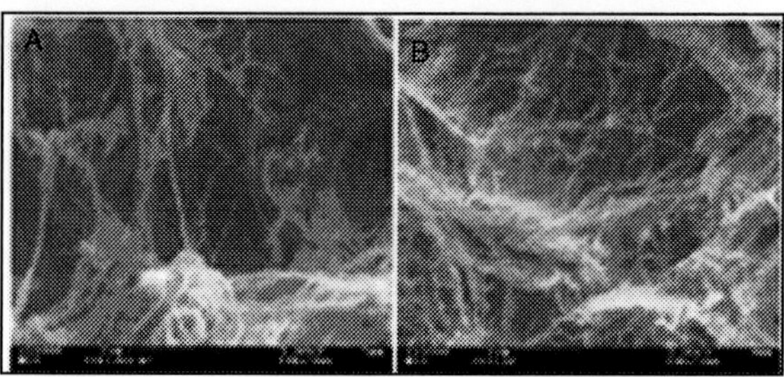

Figure: SEM Images of (A) Collagen and (B) BC.

Although BC is not a hydrogel in the true sense of the meaning, it is often referred to as such because of its high water content, insolubility in water and highly hydrophilic nature. Since BC consists of a highly entangled network of fibrils, it also provides strong mechanical properties that ensure the ability of tissue engineered blood vessels to withstand mechanical forces and prevent rupture. BC can be designed and shaped into three dimensional structures such as tubes or sheets. A major advantage of using BC rather than cellulose produced by any other organism is that BC is completely free of biogenic compounds such as lignin, pectin and arabinan found in, *e.g.*, plant cellulose. During the production process, it is also possible to modify several other properties including pore size, surface properties and layering of the material.

BC is used in various areas including food matrix (nata de coco), dietary fibres, acoustic or filter membranes and ultra-strength paper. In addition, BC has been suggested as a potential material for tissue engineering in several areas, *e.g.*, as scaffold for tissue engineering of cartilage, blood vessels (BASYC?) and successful treatment of second and third degree burns, stomach ulcers and other situations that require a temporary skin substitute (Biofill?, Gengiflex?, XCell?) or to recover periodontal tissue (Gengiflex?).

STRUCTURE AND MORPHOLOGY

Beginning with the water-soluble monosaccharide D-glucose, cellulose synthesis is produced extracellularly as pellicles at the air/liquid interface. Glucan chains of BC are extruded from several enzyme complexes and aggregated by van der Waals forces to form sub-fibrils, approximately 1.5 nm wide. BC sub-fibrils crystallise into microfibrils and then into bundles, which form a dense reticulated structure stabilized by hydrogen bonding. In culture medium, the bundles assemble into ribbons, forming a network of cellulose. This network of cellulose nanofibrils provides BC with high mechanical strength and a water retention capacity of about 99%.

The macroscopic morphology of BC varies with different culture conditions. In static conditions, BC accumulates on the surface of nutrient-rich culture medium, at the oxygen- rich air-liquid interface. Statically cultured BC has lower crystallinity than BC fermented during agitation.

Our molecular studies on EC-blood interactions with BC used BC synthesized by Gluconacetobacter xylinum. Cellulose tubes were grown in corn steep liquid media at 30°C for 7 days. The cellulose was then purified by boiling, first in 0.1 M NaOH at 60°C for 4h and then in Millipore water. Finally, the cellulose was sterilised by autoclaving for 20 minutes. Due to the production process, BC consists of two distinctly different layers: one side has a compact network of fibrils with few if any pores, and the other side has a porous network structure. A density gradient arises between the sides.

Mechanical Properties

The optimal scaffold is a biocompatible biomaterial that provides proper mechanical and physical properties, thus promoting cell adhesion and tissue formation. Prior to implantation into animals, BC tubes undergo extensive mechanical testing (burst strength, compliance and tensile strength). Films or sheets of BC show remarkable mechanical strength, due to high crystallinity, high planar orientation of the ribbons, ultrafine structure and a complex network. The mechanical properties of BC tubes are similar to those of pig carotid arteries. BC's compliance curve resembles that of a native artery more than any other synthetic material on the market, which is advantageous. Material density can be altered by varying the culture conditions or by post-culture modifications.

Biocompatibility

Integration of a material with the host tissue is essential for the success of tissue engineered blood vessels. According to Williams, the biocompatibility of a material is defined as "the ability of a material to perform with an appropriate host response in a specific application". Therefore, an appropriate host response would involve a biomaterial that induces a very low inflammatory and foreign body response in the host tissue.

A study by Helenius *et al.* showed that BC is well integrated into the host tissue and does not induce inflammatory or foreign body responses. They implanted BC pieces subcutaneously in rats and explanted them after 1, 4 and 12 weeks. Incorporation of the implant in the host tissue made it difficult to distinguish a clear interface between the implant and the host tissue. These results are supported by another in vivo study, where BC tubes were implanted into the carotid arteries in pigs. Therefore, BC clearly has good biocompatibility and shows promising potential as scaffold material.

Surface Modification

One challenge in the field of vascular grafts involves promoting EC attachment and spreading, since many biomaterials similar to BC exhibit limited support for cellular adhesion. Over the years, many strategies have been developed to modify material surfaces.

To optimize cell-biomaterial interactions, manufacturers coat synthetic scaffolds with cell adhesive proteins such as collagen, fibronectin or laminin. However, varying protein composition results in a biofilm with passive protein adsorption, and that composition can modify over time. Additionally, protein adsorption to BC is very low.

Much attention has focused on cell adhesion peptide RGD and its derivatives as possible alternative for stimulating reproducible and predictable cell adhesion. Most RGD modifications occur via covalent binding to the material. Although cellulose contains reactive hydroxyl groups that can be chemically modified, these very same hydroxyl groups participate in hydrogen bonding, which holds the cellulose fibre network together. Disruption of these bonds associates with loss of fibre ultrastructure. Dry films of BC have been modified with carboxymethyl and acetyl groups. However, surface modifications to wet state BC remain incompletely understood. Thus, modification of a BC hydrogel is especially challenging, since solvent exchange and cellulose modification typically destroy the hydrogel morphology.

Modification of a BC hydrogel is especially challenging because solvent exchange and cellulose modification typically destroy hydrogel morphology.

Thus far, most modifications have been performed on dried BC. Consequently, a new method is needed to increase cell attachment without altering the structure of the BC network.

Xyloglucan

Xyloglucan (XG), the most abundant hemicellulose, is present in the primary wall of many plants. In contrast to cellulose, XG is water-soluble and interacts strongly with cellulose fibres. We have taken advantage of these properties, which provide an elegant means of introducing cell adhesion peptide RGD with XG as a carrier molecule to BC. BC and cotton linters, as reference material, were modified with XG and XG bearing a GRGDS pentapeptide. Compared with organic solvents, modification in the water phase was clearly advantageous for preserving the morphology, as observed with SEM. XG adsorption increased the wettability only to a minor extent, possibly explaining the decreased or undetectable adsorption of adhesive proteins shown by QCM-D. QCM-D studies further revealed that fibrinogen antibodies do not bind to BC, leading to the conclusion that cell enhancement would result from the presence of RGD epitopes, not from unspecific protein adsorption, *e.g.*, fibronectin, from the cell culture medium. XG also enhances hepatocyte adhesion, and modification of BC with XG does not adversely affect ECs.

BIOLOGICAL RESPONSES TO MATERIALS

The goal of this heading is to provide material scientists and engineers with an appreciation of the fundamental aspects of tissue responses to materials, as well as the in vivo evaluation of tissue responses to materials. Fundamental aspects of tissue responses to materials include the tissue response continuum, which is initiated when a material (biomaterial), medical device, or prosthesis is implanted in living tissue. The tissue response continuum is the series of responses that are initiated by the implantation procedure, as well as by the presence of the biomaterial, medical device, or prosthesis. The fundamental aspects of the tissue response continuum are viewed from the classical medical perspective of the pathologist. It includes our current understanding of inflammatory and wound healing responses, foreign body reactions, and ultimately fibrous encapsulation (scar formation) of the biomaterial, medical device, or prosthesis.

The second part of this review addresses the in vivo evaluation of tissue responses to materials. From a practical perspective, *i.e.* manufacturing, clinical, and regulatory, the in vivo evaluation of prostheses and medical devices, *i.e.* biomaterials in their ready-to-use form, is necessary to determine their biocompatibility. Biocompatibility is generally defined as the ability of a biomaterial, prosthesis, or medical device to perform with an appropriate host response in a specific application, and biocompatibility assessment, *i.e.* evaluation of biological responses, is a measure of the magnitude and duration of the adverse alterations in homeostatic mechanisms that determine the host response. Practically speaking, the evaluation of biological responses to a medical device is carried out to determine that the medical device performs as intended and presents no significant harm to the patient or user. Thus the goal of biological response evaluation is to predict whether a biomaterial, medical device, or prosthesis presents potential harm to the patient or user by evaluating conditions that simulate clinical use.

BLOOD-MATERIAL INTERACTIONS AND INITIATION OF THE INFLAMMATORY RESPONSE

Blood-material interactions and the inflammatory response are intimately linked and, in fact, early responses to injury involve mainly blood and the vasculature. Regardless of the tissue or organ into which a biomaterial is implanted, the initial inflammatory response is activated by injury to vascularized connective tissue. Because blood and its components are involved in the initial inflammatory responses, thrombi and/or blood clots also form. Thrombus formation involves activation of the extrinsic and intrinsic coagulation systems, the complement system, the fibrinolytic system, the kinin-generating system, and platelets. Thrombus or blood clot formation on the surface of a biomaterial is related to the well-known Vroman effect of protein adsorption. From a wound-healing perspective, blood protein deposition on a biomaterial surface is described as provisional matrix formation.

Immediately following injury, changes occur in vascular flow, caliber, and permeability. Fluid, proteins, and blood cells escape from the vascular system into the injured tissue in a process called exudation. Following changes in the vascular system, which also include changes induced in blood and its components, cellular events occur and characterize the inflammatory response. The effect of the injury and/or biomaterial in situ on plasma or cells can produce chemical factors that mediate many of the vascular and cellular responses of inflammation. Although injury initiates the inflammatory response, released chemicals from plasma, cells, and injured tissue mediate the response. Important classes of chemical mediators of inflammation. Several important points must be noted in order to understand the inflammatory response and how it relates to biomaterials. First, although chemical mediators are classified on a structural or functional basis, different mediator systems interact and provide a system of checks and balances regarding their respective activities and functions. Second, chemical mediators matrix consists of fibrin, produced by activation of the coagulative and thrombosis systems, and inflammatory products, released by the complement system, activated platelets, inflammatory cells, and endothelial cells. These events occur early, within minutes to hours following implantation of a medical device. Components within or released from the provisional matrix, *i.e.* fibrin network (thrombosis or clot), initiate the resolution, reorganization, and repair processes such as inflammatory cell and fibroblast recruitment. Platelets, activated during the fibrin network formation, release platelet factor 4, platelet-derived growth factor (PDGF), and transforming growth factor b (TGF-b), which contribute to fibroblast recruitment. Upon activation, monocytes and lymphocytes generate additional chemotactic factors, including LTB4, PDGF, and TGF- b, to recruit fibroblasts.

REFERENCES

- Chrystler, H Lari : *Encyclopaedia of Nanoinformatics*, Dominant, Delhi, 2008.
- Don Norman: *A Student Handbook of Nanobiotechnology*, Dominant, Delhi, 2008.
- Eric, K. : *Nanosystems: Molecular Machinery, Manufacturing, and Computation*, New York: Wiley, 1992.
- Herzberg, G. : *Molecular Spectra and Molecular Structure*, Van Nostrand Reinhold Company, New York, 1979.
- Kapoor, Manish : *BioMEMS and Biomedical Nanotechnology*, Oxford Book Company, Delhi, 2012.
- Kaushik, N. L. : *Nanotechnology and Micromachines*, Pearl Books, Delhi, 2007.
- Marc Madou: *Fundamentals of Microfabrication*, CRC Press, Delhi, 1997.

7

Nanofabrication and Nanomanufacturing

Nanofabrication is of interest to computer engineers because it opens the door to super-high-density microprocessor s and memory chip s. It has been suggested that each data bit could be stored in a single atom. Carrying this further, a single atom might even be able to represent a byte or word of data. Nanofabrication has also caught the attention of the medical industry, the military, and the aerospace industry.

There are several ways that nanofabrication might be done. One method involves scaling down integrated-circuit (IC) fabrication that has been standard since the 1970s, removing one atom at a time until the desired structure emerges. A more sophisticated hypothetical scheme involves the assembly of a chip atom-by-atom; this would resemble bricklaying. An extension of this is the notion that a chip might assemble itself atom-by-atom using programmable nanomachines. Finally, it has been suggested that a so-called biochip might be grown like a plant from a seed; the components would form by a process resembling cell division in living things.

CARBON NANOMATERIALS

Nanotechnology refers to engineering materials at the atomic scale: the size of nanometres. Richard Smalley, Robert Curl and Harold Kroto (currently a technical advisor to Luna) were awarded the Nobel prize for their discovery of a spherical crystal form of carbon they named a buckminsterfullerene for its resemblance to the geodesic shape that architect incorporated in many of his designs. The original fullerene sphere, sometimes called a "Buckyball," contained 60 carbon atoms bound by single and double bonds that form a three dimensional geodesic spheroidal crystal. Since that discovery other related forms have included larger spheres. Collectively these materials are called fullerenes. Fullerene structures have a number of unusual properties (chemical, physical and biological) that are significantly different from other organic materials. They are comprised of pure carbon, hence insoluble in water and in many organic solvents. Fullerenes also have an intrinsic thirst for absorbing free radicals, which are molecules that have an uneven number of electrons. Having an uneven number of electrons makes a molecule unstable because the electrons want to mate. The reactive molecule may attach to another molecule to make itself more stable. Sometimes this starts a chain reaction that can harm the cells. This chain reaction of free radicals may be responsible for numerous age-related diseases. Fullerenes are exceptional free radical scavengers, or antioxidants which can intercept free radicals and neutralize them before they cause harm. An area of development at Luna is supplementing natural anti-oxidants with fullerenes to treat conditions where free radicals cause chronic health problems.

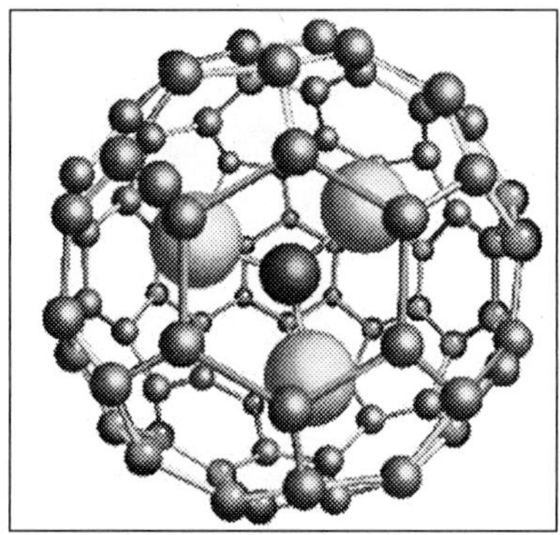

Luna's proprietary TRIMETASPHERE® carbon nanomaterials are hollow molecules of carbon that encapsulate nitride metal atoms. Luna is developing these unique molecules for applications in medical and alternative energy.

A class of novel fullerenes comprising 80 carbons (C_{80}) forming a sphere which encloses a complex of three metal atoms and one nitrogen atom were discovered at Virginia Tech. Luna licensed the patent for these nanomaterials and named them TRIMETASPHERE® carbon nanomaterials. Since these fullerenes encapsulate metals it puts them in the subset referred to as metallofullerenes. Carbon nanomaterials have the potential for use in diagnostics as safe imaging agents and in therapeutics targeting and treating diseases.

The entrapped metals of TRIMETASPHERE® carbon nanomaterials provide unique physical, chemical, thermal, magnetic, biological, optical and electronic properties that differentiate them from other carbon nanomaterials. TRIMETASPHERES have a unique electron configuration. The packed nitride cluster has a net charge of +6, and the fullerene cage has a compensating-6 charge. Thus while the overall charge is zero, the surface of the sphere presents the electronegative moiety of a potential dipole moment. This charge distribution affects the magnetic and chemical reactivity of the sphere, its ability to transfer electrons, and its biological behaviour.

For many applications TRIMETASPHERE® carbon nanomaterials must be derivatized. It is possible to attach one or more side chains to functionalize the TRIMETASPHERE for specific applications. By chemical modification of the surface of TRIMETASPHERE carbon nanomaterials with the metal gadolinium inside ($Gd_3N@C_{80}$), scientists at Luna have invented a new family of molecules named HYDROCHALARONE™ [hî-drô-kal-a-rône]. The name comes from the words Hydro, meaning water and Chalaro, the Greek word for relax. Relaxivity is a measurement of the relative amount of contrast provided by media used in Magnetic Resonance Imaging (MRI) that helps to provide a meaningful image.

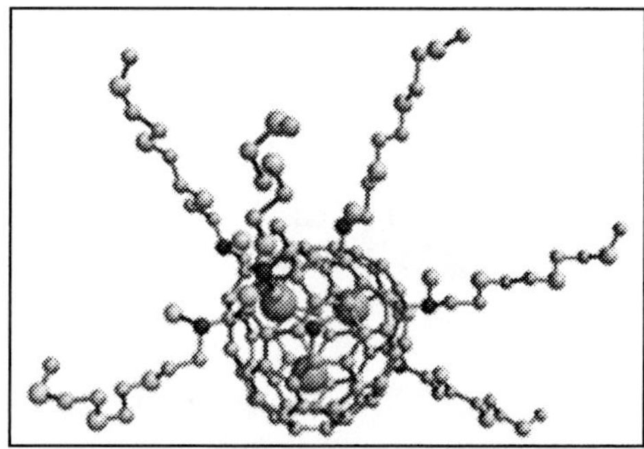

By chemical modification of the surface of a TRIMETASPHERE with the metal gadolinium inside, scientists at Luna have invented a new family of nanomaterials named HYDROCHALARONE. The image above is a hypothetical depiction of what this new variety of molecule might look like.

Luna's HYDROCHALARONE™ nanomaterials have unique properties which we plan to develop into exciting products in the pharmaceutical industry. An early commercial application is the next generation of medical diagnostic agents. Luna has demonstrated its contrast agent prototype in mouse studies which show the potential to provide a better quality image. Additionally, Luna's MRI contrast agent prototype has the potential to be safer for use in medical imaging versus today's gadolinium chelate-based agents which are the subject of class-action lawsuit concerns. Read more about the warnings of gadolinium-containing contrast agents from the U.S. Food and Drug Administration.

CARBON NANOTUBES IN MEDICINE

Carbon nanotubes (CNTs) are very prevalent in today's world of medical research and are being highly researched in the fields of efficient drug delivery and biosensing methods for disease treatment and health monitoring. Carbon nanotube technology has shown to have the potential to alter drug delivery and biosensing methods for the better, and thus, carbon nanotubes have recently garnered interest in the field of medicine. The use of CNTs in drug delivery and biosensing technology has the potential to revolutionalize medicine. Functionalization of SWNTs has proven to enhance solubility and allow for efficient tumor targeting/drug delivery. It prevents SWNTs from being cytotoxic and altering the function of immune cells.

Cancer, a group of diseases in which cells grow and divide abnormally, is one of the primary diseases being looked at with regards to how it responds to CNT drug delivery. Current cancer therapy primarily involves surgery, radiation therapy, and chemotherapy. These methods of treatment are usually painful and kill normal cells in addition to producing adverse side effects. CNTs as drug delivery vehicles have shown potential in targeting specific cancer cells with a dosage lower than conventional drugs used, that is just as effective in killing the cells, however does not harm healthy cells and significantly reduces side effects.

Current blood glucose monitoring methods by patients suffering from diabetes are normally invasive and often painful. For example, one method involves a continuous glucose sensor integrated into a small needle which must be inserted under the skin to monitor glucose levels every few days. Another method involves glucose monitoring strips to which blood must be applied. These methods are not only invasive but they can also yield inaccurate results. It was shown that 70 percent of glucose readings obtained by continuous glucose sensors differed by 10 percent or more and 7 percent differed by over 50 percent. The high electrochemically accessible surface area, high electrical conductivity and useful structural properties have demonstrated the potential use of single-walled nanotubes (SWNTs) and multi-walled nanotubes (MWNTs) in highly sensitive noninvasive glucose detectors.

CNT PROPERTIES

CNTs have several unique chemical, size, optical, electrical and structural properties that make them attractive as drug delivery and biosensing platforms for the treatment of various diseases and the noninvasive monitoring of blood levels and other chemical properties of the human body, respectively.

Electrical and Structural

Carbon nanotubes can be metallic or semiconducting depending on their structure. This is due to the symmetry and unique electronic structure of graphene. For a given (n,m) nanotube, if n = m, the nanotube is metallic; if n–m is a multiple of 3, then the nanotube is semiconducting with a very small band gap, otherwise the nanotube is a moderate semiconductor. Thus all armchair (n=m) nanotubes are metallic, and nanotubes, etc. are semiconducting. Thus, some nanotubes have conductivities higher than that of copper, while others behave more like silicon.

Dimensional

Due to their nanoscale dimensions, electron transport in carbon nanotubes will take place through quantum effects and will only propagate along the axis of the tube. These electrical and structural properties best serve CNTs as far as biosensing is concerned because current changes in the CNTs can signify specific biological entities they are designed to detect. The fact that CNTs are small (nm scale) allows them to deliver smaller doses of drugs to specific disease cells in the body thus reducing side effects and harm to healthy cells unlike conventional drugs, whilst improving disease cell targeting efficiency.

Chemical

CNTs have been observed to have enhanced solubility when functionalized with lipids which would make their movement through the human body easier and would also reduce the risk of blockage of vital body organ pathways. As far as optical properties are concerned CNTs have been shown to exhibit strong optical absorbance in certain spectral windows such as NIR (near-infrared) light and when functionalized with tumor cell specific binding entities have allowed the selective destruction of disease (*e.g.* cancer) cells with NIR in drug delivery applications.

CNTS IN DRUG DELIVERY AND CANCER THERAPY

Drug delivery is a rapidly growing area that is now taking advantage of nanotube technology. Systems being used currently for drug delivery include dendrimers, polymers, and liposomes, but carbon nanotubes present the opportunity to work with effective structures that have high drug loading capacities and good cell penetration qualities. These nanotubes function with a larger inner volume to be used as the drug container, large aspect ratios for numerous functionalization attachments, and the ability to be readily taken up by the cell. Because of their tube structure, carbon nanotubes can be made with or without end caps, meaning that without end caps the inside where the drug is held would be more accessible. Right now with carbon nanotube drug delivery systems, problems arise like the lack of solubility, clumping occurrences, and half-life.

However, these are all issues that are currently being addressed and altered for further advancements in the carbon nanotube field. The advantages of carbon nanotubes as nanovectors for drug delivery remain where cell uptake of these structures was demonstrated efficiently where the effects were prominent, showing the particular nanotubes can be less harmful as nenovehicles for drugs. Also, drug encapsulation has been shown to enhance water dispersibility, better bioavailability, and reduced toxicity. Encapsulation of molecules also provides a material storage application as well as protection and controlled release of loaded molecules. All of these result in a good drug delivery basis where further research and understanding could improve upon numerous other advancements, like increased water solubility, decreased toxicity, sustained half-life, increased cell penetration and uptake, all of which are currently novel but undeveloped ideas.

Boron Neutron Capture Therapy

Narayan Hosmane and his co-workers have recently developed a new approach to Boron Neutron Capture Therapy in the treatment of cancer using substituted Carborane-Appended Water-Soluble single-wall carbon nanotubes. Substituted C2B10 carborane cages were successfully attached to the side walls of single wall carbon nanotubes (SWCNTs) via nitrene cycloaddition. The decapitations of these C2B10 carborane cages, with the appended SWCNTs intact, were accomplished by the reaction with sodium hydroxide in refluxing ethanol. During base reflux, the three-membered ring formed by the nitrene and SWCNT was opened to produce water-soluble SWCNTs in which the side walls were functionalized by both substituted nido-C2B9 carborane units and ethoxide moieties. All new compounds were characterized by EA, SEM, TEM, UV, NMR, and IR spectra and chemical analyses. Selected tissue distribution studies on one of these nanotubes, {([Na+][1-Me-2-((CH2)4NH-)-1,2-C2B9H10][OEt])n(SWCNT)} (Va), showed that the boron atoms are concentrated more in tumours cells than in blood and other organs, making it an attractive nanovehicle for the delivery of boron to tumor cells for an effective boron neutron capture therapy in the treatment of cancer.

Selective Cancer Cell Destruction

Carbon nanotubes can be used as multifunctional biological transporters and near-infrared agents for selective cancer cell destruction. Biological systems are known to be highly transparent to 700- to 1,100-nm near-infrared (NIR) light. Researchers showed that the strong optical absorbance of single-walled carbon nanotubes (SWNTs) in this special spectral window, an intrinsic property of SWNTs, can be used for optical stimulation of nanotubes inside living cells to afford multifunctional nanotube biological transporters. They used oligonucleotides transported inside living Hela cells by nanotubes. The oligonucleotides translocated into the cell nucleus upon endosomal rupture triggered by NIR laser pulses. Continuous NIR radiation caused cell death because of excessive local heating of SWNT in vitro. Selective cancer cell destruction was achieved by functionalization of SWNT with a folate moiety, selective internalization of SWNTs inside cells labelled with folate receptor tumor markers, and NIR-triggered cell death, without harming receptor-free normal cells. Thus, the transporting capabilities of carbon nanotubes combined with suitable functionalization chemistry and their intrinsic optical properties can lead to new classes of novel nanomaterials for drug delivery and cancer therapy.

Tumor Targeting

Research has been conducted on in vivo biodistribution and highly efficient tumor targeting of carbon nanotubes in mice for cancer therapy. Investigations are being done on the biodistribution of radio-labelled SWNTs in mice by in vivo positron emission tomography (PET), ex vivo biodistribution and Raman spectroscopy. It was found that SWNTs that are functionalized with phospholipids bearing polyethylene glycol (PEG) are surprisingly stable *in vivo*. The effect of PEG chain length on the biodistribution and circulation of the SWNTs was studied. Effectively PEGylated SWNTs exhibited relatively long blood circulation times and low uptake by the reticuloendothelial system (RES). Efficient targeting of integrin positive tumor in mice was achieved with SWNTs coated with PEG chains linked to an arginine–glycine–aspartic acid (RGD) peptide. A high tumor accumulation was attributed to the multivalent effect of the SWNTs. The Raman signatures of SWNTs were used to directly probe the presence of nanotubes in mice tissues and confirm the radio-label-based results.

NANOFABRICATION WITH PULSED LASERS

When nanostructured, many materials start to exhibit new optical properties making them unique for a plethora of applications. In particular, despite small and indirect band gaps in the bulk state, the nanostructured IV group semiconductors (*e.g.* Si, Ge) become efficient size-dependent emitters in the visible light range, but also can work as photosensitizers to generate singlet oxygen under photoexcitation. Another prominent example relates to noble metal nanostructures, which provide a number of unique plasmonic effects, including size-dependent absorption peaks, drastic local electric field enhancement, resolution beyond the diffraction limit, nanotrapping etc. These new properties of emerging nanomaterials appear to be extremely promising for photovoltaics and optoelectronics, as well as for biological sensing, imaging and therapeutics.

The employment of pulsed lasers offers a novel unique tool for nanofabrication. When focused on the surface of a solid target, pulsed laser radiation causes a variety of effects, including heating, melting and finally ablation of the target and such processes can lead to an efficient material nanostructuring. First, the laser-assisted removal of material from the irradiated spot can result in a spontaneous formation of variety of periodic micro- and nanoarchitectures within this spot. Second, laser ablation of material from a solid target leads to the production of nanoclusters. When produced in gaseous environment or in vacuum, these nanoclusters can then be deposited on a substrate yielding to the formation of a nanostructured film. When produced in liquid environment, the nanoclusters can be released into the liquid forming a colloidal nanoparticle solution. In all cases, properties of formed nanostructures can be unique and not reproducible by any other route. As an example, the fabrication of nanoparticles in aqueous solutions does not require any chemical reducing agent, which conditions unique surface chemistry and purity of produced nanomaterials. Furthermore, when synthesized in clean, biocompatible environment, laser-synthesized nanomaterials are exempt of any residual toxicity that is typical for chemically synthesized nanoparticles.

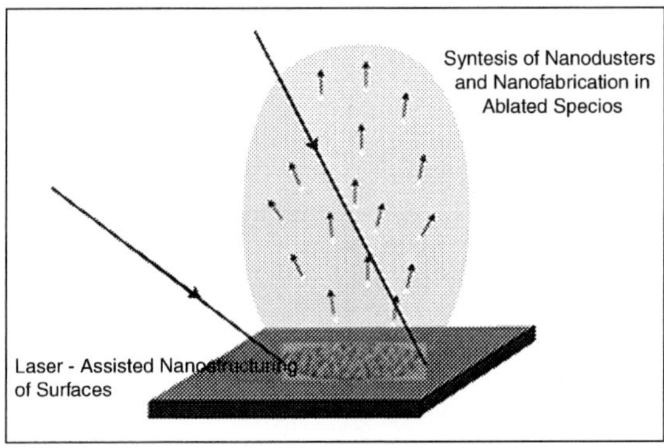

Figure : Schematics of Laser–target Interaction and Material Nanostructuring

LASER-ASSISTED NANOSTRUCTURING OF SURFACES

Properties of laser-materials interaction are known to be strongly dependent on parameters of laser radiation. Among these parameters, the wavelength and pulse length are especially important to determine the efficiency of radiation absorption, dynamics of plasma plume expansion and nanoclustering. Ultraviolet, visible and near-infrared lasers are normally considered as most adequate for laser ablative tasks. Indeed, most industrially important materials efficiently absorb laser radiation in the spectral range of 200–1,000 nm, yielding to material ablation and nanostructuring. In addition, radiation of UV and visible lasers is relatively transparent to laser plasma, which minimizes laser beam distortions and power losses before reaching the target. In contrast, the interaction of infrared radiation (1–11 μm) with materials is characterized by a strong energy absorption by plasma itself, which completely changes conditions of nanostructuring. The pulse length is another important parameter in laser–matter interaction. Micro- and nanosecond laser–matter interactions are typically associated with long ablation regime, in which the ablation process takes place during the laser pulse itself. In contrast, pico- and femtosecond lasers offer short ablation regimes, in which moments of radiation absorption and material ablation are temporally separated. Although multi-pulse laser ablation from the target surface is always accompanied by the formation of micro- and nanoscale features and periodic structures, properties of these features strongly depend on the wavelength and pulse length. Therefore, depending on the task, one can select appropriate radiation parameters to condition the ablation regime and obtain prescribed properties of laser-synthesized nanostructures. Below, we give several examples of laser-assisted methodologies to fabricate nanostructures within the irradiation spot.

Formation of Black Si for Photovoltaics Applications

It is accepted that femtosecond pulses give at least two major advantages to micromachining compared to nanosecond and longer pulses : (1) the reduction of the pulse energy which is necessary to induce ablation for fixed laser wavelength and focussing conditions and (2) a significant reduction or complete removal of heat-affected zone (HAZ) and, as consequence, the improvement of the contour sharpness for the laser-processed structures. The second advantage is a direct consequence of the pulse being shorter than the heat diffusion time, given by the phonon transport.

As in the case of nanosecond- and microsecond laser ablation, multi-pulse femtosecond ablation leads to a spontaneous formation of nanoarchitectures. However, in contrast to long ablation, the fs regime is characterized by the absence of target melting effects, yielding to the formation of clean micro- and nanoscale features.

In particular, using multi-pulse fs ablation of Si in the presence of SF_6 reactive gas, Mazur *et al.* managed to fabricate extremely narrow micro-spikes within the irradiation spot, which are capable of efficiently absorbing light in the visible and infrared ranges. The efficient absorption of light exceeding 95% in the visible was attributed to a geometric multi-reflection effect offered by a unique spike-based structure, while the enhanced absorption in the infrared was explained by a sulphur doping.

Due to the wide-range absorption effect, the produced spike-based structure was called "black silicon" and was later used for the development of Si-based photodetectors. It is worth noting that such absorptive features cannot be reproduced by any alternative non-laser route. Other studies reported the fabrication of nanostructured metal films exhibiting colours ("coloured metals") using similar fs ablation approach.

This sub-project is devoted to the fabrication of "black silicon" structures for photovoltaics solar panel applications. The choice of photovoltaics as target application imposes new criteria on nanostructuring conditions. First, these applications require a high absorption mostly in the visible—near-IR range (300–1,000 nm), which enables us to exclude the necessity of using sulphur-based doping species. Second, these applications require uniform high-quality doping of nanostructured layers to maximize the photovoltage response. People succeeded in developing of a novel methodology to produce "black silicon" with such parameters, employing a Ti:Sapphire laser (wavelength 800 nm, pulse energy 5 mJ, repetition rate 1 kHz). In contrast to, we carry out multi-pulse laser processing in vacuum under the residual pressure of (1–5) × 10^{-5} mBar. In addition, we avoid the doping procedure during the laser processing process and do it afterwards.

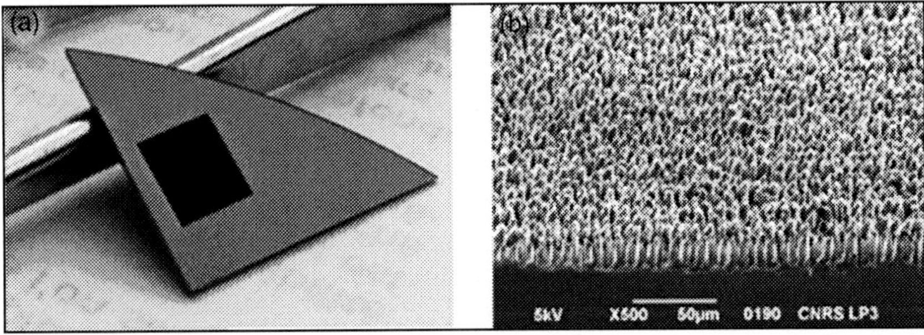

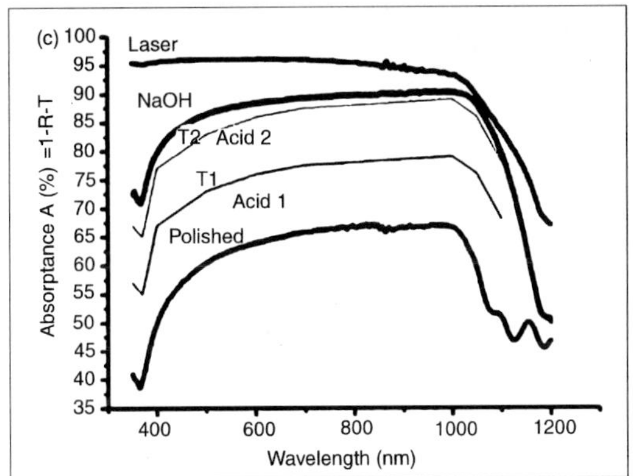

Figure : (*a*) Typical Image of "Black Silicon" Spot Fabricated on a Si Wafer by Multi-pulse fs Laser Ablation in Vacuum; (*b*) Typical Scanning Electron Microscopy (*SEM*) of Penguin-like Structure of Black Silicon; (*c*) Typical Absorption Spectra from "Black Silicon" and Silicon Treated by Different Methods

To achieve a high quality doping of deep layers, the laser-structured samples are boron implanted by Plasma immersion (PULSION, BF3, 2 kV, 900°C, 30 min) and thermally annealed (TA). The junction depth obtained by this method is estimated to be about 150 nm, which is much shallower than the 3D laser structures; therefore, the junction follows the topography of the structures.

The demonstrates a silicon wafer surface after the laser processing and boron implantation procedure. Here, a rectangular area of 3 × 2 cm^2 is written by a programmed displacement with the speed of 150 μm/s of a femtosecond laser beam having the spot size of 35 × 35 μm^2. One can clearly see a black area on the silicon wafer associated with "black silicon". The treated surface contains penguin-like nanospikes with the length of up to 10 μm and sub-μm lateral dimensions. Although the morphology of femtosecond laser-treated surface is rather different compared to narrow spike-like structures in, it is also characterized by an enhanced absorption in the visible range exceeding 90%. Depositing grating-like contacts on the top on the treated area, they were able to obtain the amplification of photocurrent by 50% compared to the untreated surface area. Such result was attributed to an enhanced absorption granted by the penguin-like structures, much larger surface of nanostructured silicon used for signal collection, and high quality of boron implantation offered by the post-ablation plasma implantation procedure. The fabricated structures are now actively tested as photovoltaics solar cells.

Laser Plasma-assisted Nanostructuring of Surfaces

UV or ultrashort lasers contribute to a good radiation absorption by the target itself, while plasma remains relatively transparent to the incoming radiation. Such parameters ensure good quality of surface treatment in laser processing tasks. The plasma effect can be further minimized by reducing the pressure of the ambient gas. Depending on plasma plume size conditioned by the ambient gas pressure, the material can be re-deposited either within the irradiation spot (for high atmospheric pressure) or into the environment (for reduced pressures). In particular, for atmospheric pressure, the ablation process results in the formation of a deep crater, containing microscale spikes, covered by re-deposited nanoparticles. In this case, chemical transformations in ablated species are minimal, since the ablated material rapidly cools down while interacting with the environment.

A novel method introduced recently for surface nanostructuring, which is characterized by radically different nanofabrication conditions. The method may look paradoxical, since it disaccords with main principles of laser processing requiring the minimization of plasma-related effects as one of main conditions to achieve high quality of laser treatment. In contrast, in this method, plasma-associated effects are amplified by all possible means. Basically, we use infrared radiation from CO_2 laser, which is strongly absorbed by the plasma itself. When focused in air and any other gas having atmospheric pressure, infrared radiation is capable of efficiently igniting the gas breakdown and this phenomenon is called the "laser spark".

The presence of a target decreases the breakdown threshold by 2–3 orders of magnitude. In the latter case, the target serves to provide first electrons. Then, an avalanche plasma discharge develops in ambient air moving towards the focusing lens. Absorbing main radiation power through the inverse Bremsstrahlung mechanism, the plasma accumulates an enormous amount of energy and is supposed to radically change conditions of nanocluster production and growth. Indeed, in contrast to conventional laser ablation, the ablated species find themselves in a plasma "reactor" with extremely high temperatures (10^4 K) and strong electromagnetic fields, yielding to a deep chemical transformation of properties of ablated clusters. The clusters then move back to the irradiated spot forming a film of clearly separated and

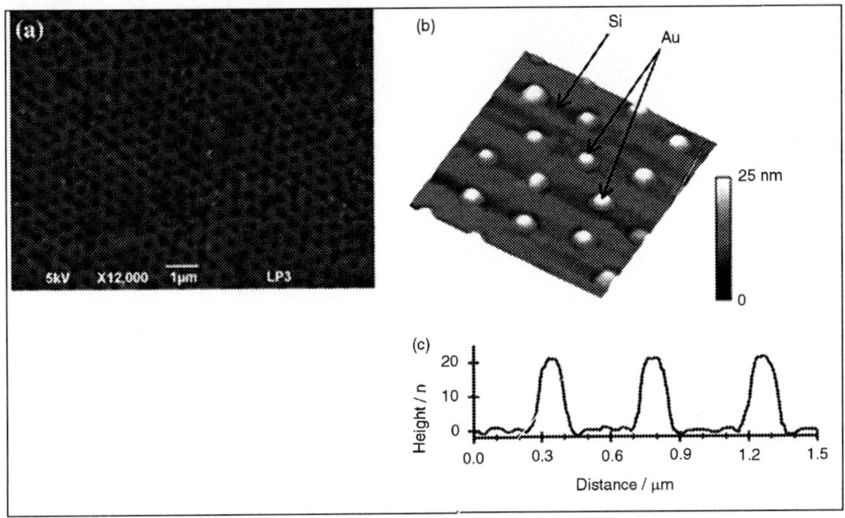

Figure : (*a*) SEM of an Al_2O_3 thin Film (20 nm) deposited on a Si Substrate and Simultaneously Drilled by the Near-field Enhancement of a Single Nanosecond Laser Pulse, which is Produced by a Lattice of SiO_2 Spheres (R = 250 nm); (*b*) AFM and (*c*) Depth Profile of Gold Nanodots Created on Silicon Substrates by the LF-PAM-based Process.

LASER ABLATIVE SYNTHESIS OF NANOCLUSTERS

The evolution of properties of formed nanostructures after the ablation process is mainly determined by the interaction of the species with the environment. In fact, in the first approximation, the nanocluster formation process can be described by the classical theory of condensation and nucleation in a vapour layer. However, the growth of nanoclusters strongly depends on initial parameters of ablated species (energy, angular distribution, density) and by laser interaction with the environment, which are conditioned by characteristics of pumping radiation. In particular, the radiation can directly affect the nucleation process if the pumping laser pulse is long enough. Indeed, the energy of photons can be sufficient to produce nucleation centres, change the dynamics of the nuclei growth and modify the diffusion of species in the vapour phase.

Laser Ablation in Residual Gases and Deposition of Nanostructured Films

The first possibility is related to the ablation of material in gaseous environment and the deposition of ablated species onto a substrate in pulsed laser deposition (PLD) geometry. In this geometry, the employment of a UV radiation from an excimer laser is normally preferable. UV radiation is well absorbed by most industrially important materials, while the formed plasma plume is relatively transparent to it. The material can be in general ablated in vacuum; but due to a low probability of nanocluster coalescence in vacuum, it is difficult to control their growth and the nanoclusters normally deposit as ablated, forming a dense film with a significant amorphous fraction. Therefore, it is important to have a residual neutral light gas to affect the growth procedure. In this case, nanoclusters cool down under collisions with gas molecules or atoms, which contribute to their coalescence in the vapour phase. Under such conditions, the nanocluster growth process can be efficiently controlled by varying the pressure of the ambient gas.

To fabricate Si-based nanostructured films, we normally use radiation of a pulsed ArF or KrF lasers (193 or 248 nm, respectively, 15 ns FWHM, repetition rate 30 Hz) to ablate material from a rotating Si target. The radiation is focused at the incident angle of 45° to the surface. A substrate is placed on a rotating holder in front of the target. The experimental chamber is filled with helium for a deposition at a constant pressure ranging between 0.05 and 10 Torr. The film thickness after several thousands laser shots is 100–700 nm.

A transmission electron microscopy (TEM) image of several isolated nanoparticles, synthesized by laser ablation from a Si target and deposited on a carbon-coated TEM grid (a) and corresponding nanoparticle size distribution (b). One can see that the produced nanoclusters are very small with the size in the range of 1–4 nm.

The mean size of nanoparticles slightly increases under the increase of He pressure. For example, the increase of the pressure from 0.5 to 8 Torr results in the increase of the nanoparticle size from 1.5 to 4 nm. Another important feature relates to essentially porous texture of the films prepared by pulsed laser deposition, and the porosity of films increases with the increase of the ambient gas pressure. Indeed, while the deposition under 1 Torr results only in some germs of roughness, the experiment under 2 Torr provides a developed porous structure with pore size of about 50–100 nm. A further pressure increase up to 4 Torr leads to a formation of web-like aggregations of particles.

The deposition of films at high pressures (>4 Torr) leads to porosities exceeding 90%, corresponding to the formation of powders on the substrate. Thus, the pressure of the ambient gas appears to be one of key parameters, which determines both the size of synthesized nanocrystals and the porosity of deposited nanostructured layers.

Such a strong impact of the gas pressure on nanoclustering process suggests the importance of cooling of ablated species under their collisions with gas atoms. If the pressure of the gas is high enough, the nanoclusters experience a sufficient number of collisions to rapidly cool down and crystallize in the vapour phase. As a result, they arrive on the substrate in the form of a powder. In contrast, low collision regime at low pressures advantages the formation of dense and low-porous films.

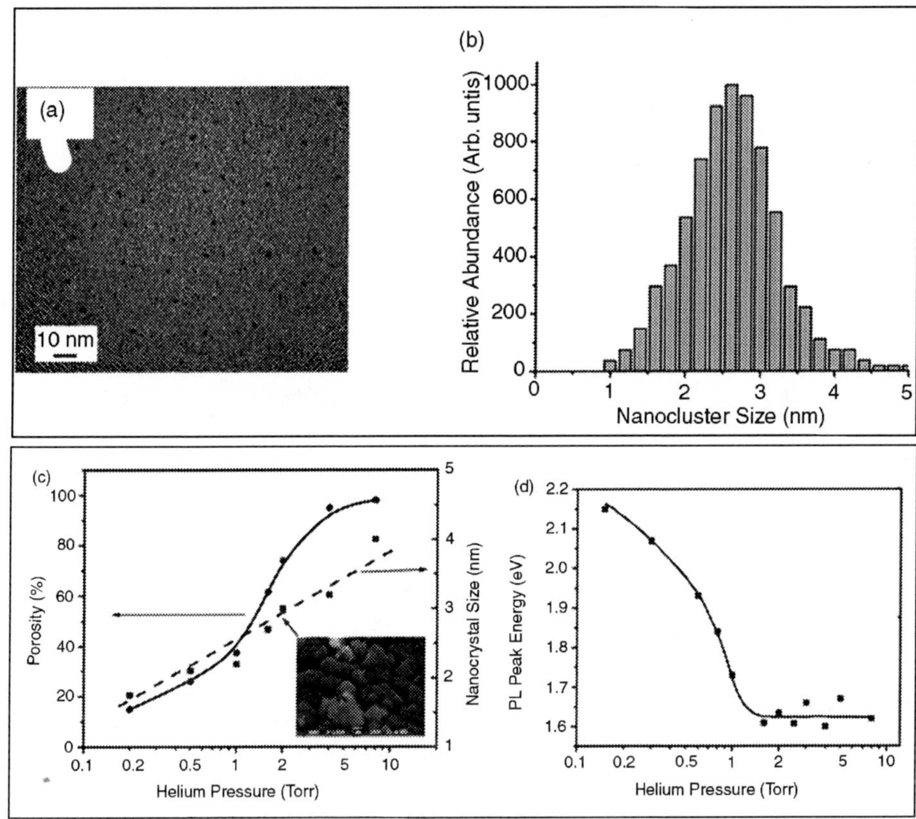

It is important that all Si-based films prepared by the PLD method exhibit strong visible emission (PL) under photoexcitation, while PL properties of the films are quite different for films of different porosities. Low-porous films $P < 40\%$, deposited at reduced pressures $P < 1.5$ Torr, exhibit relatively weak PL with peak energy strongly depending on the gas pressure. In particular, the pressure decrease from 1.5 to 0.15 Torr in different depositions causes a blue shift of the peak from 1.6 to 2.15 eV. In contrast, films with an enhanced porosity $P > 40\%$, deposited at higher pressures, provide only spectra with fixed peaks. The first band (1.6–1.7 eV) is independent of the pressure and can be seen just after the fabrication process. An additional 2.2–2.3 eV band can appear under the oxidation of samples in humid air. Here, PL properties of low-porous films (porosity <40%) are almost unchangeable under these conditions, while the integral PL intensity from highly porous films significantly increases with the prolonged oxidation. We attribute such a difference of PL properties to the impact of post-fabrication natural oxidation, controlled by the level of porosity. Dense and self-coagulated structures of the films fabricated under $P < 1$ Torr minimize the impact of ambient atmosphere on the film properties; and for these films, mechanisms related to core silicon crystals became predominant. Since the blue shift of the spectra under the decrease of helium residual pressure is accompanied by a certain decrease of the nanocrystal size, the quantum confinement mechanism can be considered as one of main opportunities. In contrast, a high porosity enhances the surface area, which is subjected to surface chemistry modifications due to interactions of nanocrystallites with oxygen. This can drastically enhance the role of oxidation in the formation of PL centres and the domination of oxygen-related PL mechanisms connected either to defects in the SiO_2 structure (usually, this mechanism provides 2–2.4 eV PL) or to the interfacial layer (1.65 eV).

Liquid Environment to Form Colloidal Nanoparticles

In the case of liquid ambience, laser ablation process leads to the release of nanoclusters into the liquid and the formation of a colloidal nanoparticle solution. In contrast to conventional chemical reduction methods, this method enables avoidance of the use of toxic chemical reduction agents to control the growth of the nanoparticles. As an example, the laser ablation-based synthesis can be implemented in pure deionized water. The independence of laser-based synthesis of dirty colloidal chemistry makes it unique for the fabrication of markers of bioanalytes for sensing and imaging applications. In pure water, however, or any other aqueous solution exempted of additional chemically active components, the size of nanoparticles produced tends to be relatively large, since a natural coalescence of hot ablated nanoclusters cannot be easily overcome. In particular, nanosecond laser-based ablation used in most works, generally gives relatively large (10–300 nm) and strongly dispersed (50–300 nm) particles. Although certain size control can be achieved by decreasing the wavelength of pumping radiation or varying the laser fluence, the range of size variations stays rather moderate in the nanosecond pulse case. Mafune showed that size characteristics of nanoparticles can be improved by adding some reactive surfactants such as sodium dodecyl sulphate (SDS) or CTAB during the ablation. As an example, thiol-containing SDS covered gold nanoclusters just after their production and thus protected them from further coalescence. However, bioimaging applications of so produced nanoparticles are hardly possible since the surfactants are not biocompatible.

We recently studied a femtosecond laser ablation-based method for nanoparticle synthesis, which makes possible an efficient control of size of prepared nanoparticles by varying physical parameters. The experiments are normally carried out with an Ytterbium (400 fs FWHM, 1,025 nm, 1 kHz) or a Ti/Sapphire laser (110 fs FWHM, 800 nm, 1 μJ/pulse, 1 kHz). The radiation is focused onto a target of different materials (Au, Ag, Si, Ti, Cu), which is placed on the bottom of a glass vessel filled with aqueous solutions. The vessel is placed on a horizontal moving platform to avoid the ablation of material from the same area. The ablation experiments are carried out in pure deionized water and in aqueous solutions (biopolymers, cyclodextrins).

When performed in relatively neutral environment such as deionized water, fs laser ablation normally leads to the formation of two nanoparticle populations, independently on the material of the target. The first population is characterized by a small mean nanoparticle size and narrow size dispersion, whereas the second one has a much larger mean size and broader size dispersion. The presence of the two populations suggests the involvement of two different mechanisms of nanoparticle growth.

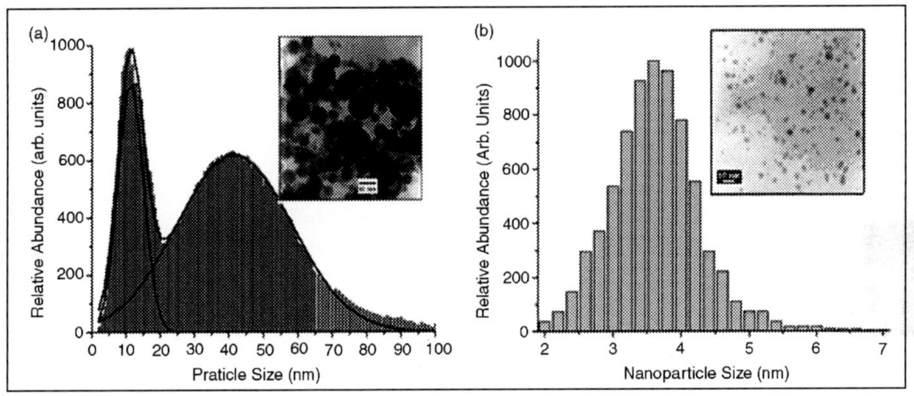

Figure : Transmission Electron Microscopy and Corresponding Size Distributions of a TiO_x Nanoparticles Prepared by fs Laser Ablation from a Ti Target in Deionized Water; (*b*) Gold Nanoparticles Prepared in Aqueous Solution of Polyethylene Glycol.

The production of the first, less dispersed population is characterized by the absence of target melting effects, suggesting direct radiation-related ablation of material. In contrast, the production of the highly dispersed population is accompanied by a strong melting of material inside the ablated crater. This melting is usually attributed to the explosion of a cavitation bubble formed as a result of energy transfer from laser plasma to the liquid. It is important that in the fs laser ablation regime, the nanoparticle size can be efficiently controlled by varying the intensity of radiation during the nanosynthesis process. In particular, the mean size of gold nanoparticles can be reduced from 120 to 4 nm by the decrease of laser fluence down to the threshold values. Similar effect can be achieved by changing the radiation focusing on the target surface. Such efficient method of size control, not possible with nanosecond or microsecond pulses, is granted by specific conditions of fs laser–materials interaction. In the fs ablation regime, much less radiation energy is transferred to the cavitation bubble (15% compared to 80% in the nanosecond pulse regime). We believe that the decrease of the laser fluence down to near-threshold values enables one to completely avoid cavitation phenomena and thus eliminate the second highly size-dispersed nanoparticle population. In addition, the decrease of the laser fluence changes parameters of ablated nanoclusters (energy, angular distribution etc.), which can in turn affect the final size of nanoparticles of the first population. In particular, using this method in the case of gold, we managed to vary the size of synthesized nanoparticles between 4 and 20 nm.

Another important issue is related to chemical properties of laser-synthesized nanoparticles. Since the nanoparticles are produced by pure physical ablation from a target without the involvement of any specific chemicals, surface chemistry of these nanoparticles can drastically differ from that of counterparts prepared by conventional colloidal chemistry. In particular, laser-synthesized gold becomes susceptible to oxidation and, in contrast to chemically prepared gold, the surface of these nanoparticles is partially covered by a layer of oxide. Furthermore, the oxidized nanoparticle surface can have different termination, depending on the pH of the environment.

The oxidized portion of the gold surface normally has Au–O– groups at $pH > 5.8$ and increasing numbers of Au–OH groups at $pH < 5.8$. This oxide-related surface termination makes possible interactions of gold nanoparticles with groups, for which conventional gold is normally inert. A prominent example of new gold chemistry is related to the use of biopolymers and oligosaccharides. Although these substances do not contain gold-reactive thiol group, they react with laser-synthesized gold nanoparticles, yielding to a drastic reduction of the nanoparticle size. The effect of the reduction of nanoparticle mean size under the use of a biopolymer polyethylene glycol (PEG). One can see that the nanoparticles' mean size and size dispersion can be reduced down to 3 nm with the size dispersion not exceeding 1.5 nm FWHM. We believe that such reduction of nanoparticle size is a result of the hydrogen bonding of the –OH groups of these compounds and the $-O^-$ at the gold surface.

The molecules of PEG cover gold nanoclusters just after ablation and act like "bumpers", limiting contact between particles, preventing their coalescence (when the particles are still "hot") and aggregation (when the particles are "cold"). Similar mechanism takes place during laser ablation in aqueous solutions of other polymers (dextran etc.) and oligosaccharides (cyclodextrin). Thus, OH groups of different biocompatible compounds can efficiently react with oxidized gold surface leading to the reduction of the nanoparticle size. It is important that in contrast to SDS or other surfactants previously used to control the nanoparticle size, biopolymers and oligosaccharides are essentially biocompatible. Moreover, the ultra-pure laser-ablated nanoparticles can be functionalized by a proper chemical modification of chemicals. We believe that this gives a huge advantage over the chemically produced nanoparticles for nano-engineering and functionalization of nanoparticles produced, as well as for a solution of toxicity problems. In particular, PEG is known as one of best materials to avoid the immune response in "in vivo" applications of inorganic nanoparticles. When covered by PEG, the nanoparticles become invisible for the immune system. In the case of laser-synthesized nanoparticles, one does not need to use any intermediate chemical group to link nanoparticles to PEG, as it takes place in the case of chemically synthesized nanoparticles.

NANOMANUFACTURING

Nanomanufacturing is a term used to describe either the production of nanoscaled materials, which can be powders or fluids, or to describe the manufacturing of parts "bottom up" from nanoscaled materials or "top down" in smallest steps for high precision, used in several technologies such as laser ablation, etching and others. Nanomanufacturing should not be confused with molecular manufacturing, which refers specifically to the manufacture of complex, nanoscale structures by means of nonbiological mechanosynthesis (and subsequent assembly).

The term "nanomanufacturing" is widely used, *e.g.* by the European Technology Platform MINAM and the U.S. National Nanotechnology Initiative (NNI). The NNI refers to the sub-domain of nanotechnology as one of its five "priority areas." There is also a nanomanufacturing programme at the U.S. National Science Foundation, through which the National Nanomanufacturing Network (NNN) has been established. The NNN is an organization that works to expedite the transition of nanotechnologies from laboratory research to production manufacturing and it does so through information exchange, strategic workshops, and roadmap development.

The NNI has defined nanotechnology very broadly, to include a wide range of tiny structures, including those created by large and imprecise tools. However, nanomanufacturing is not defined in the NNI's recent report, *Instrumentation and Metrology for Nanotechnology*. In contrast, another "priority area," nanofabrication, is defined as "the ability to fabricate, by directed or self-assembly methods, functional structures or devices at the atomic or molecular level". Nanomanufacturing appears to be the near-term, industrial-scale manufacture of nanotechnology-based objects, with emphasis on low cost and reliability.

HOW WE MAKE SEMICONDUCTORS

The raw material for a 'chip', semiconductor product or Integrated Circuit (IC) is silicon- the basic constituent of sand, and one of the earth's most abundant materials. Silicon is a 'semiconductor'- under some conditions it conducts electricity and under other conditions it doesn't. That can make it act as an on/off switch, and the basic circuit that does this is a transistor. Transistors are the basic building blocks of electronic circuits, and the on/off action is what we understand as the I and 0 of the digital equipment. IC processing combines millions of transistors on a chip, allowing us to make phenomenally complex circuits.

The first essential process is purifying the silicon, since any flaws or unwanted impurities will make circuits unusable. The near 100% purity silicon is melted, and a perfect silicon 'seed' crystal is lowered into it. The result is a long rod (ingot) of single-crystal silicon with a diameter (for the latest equipment generations) of 300 mm. This work is done by specialist manufacturers, who then accurately saw the ingot bare wafers (substrates) and deliver them to us, the semiconductor manufacturer.

A chip is manufactured in layers, with perhaps as many as two dozen layers containing millions and sometimes over a billion transistors in a single microchip. The basic manufacturing process is simple in concept, but very complex in its details. Circuits must be built with incredible precision to switch quickly and exactly each and every time. Getting the signals to the next transistors in line is itself a major issue, needing metal tracks across the chip that must be insulated from each other and from other circuits.

In the Front-End operation, a blank wafer, typically silicon, is the start of the process. Essentially, semiconductor manufacturing uses chemicals to create circuits by adding and removing layers on the wafer. In this way, the interconnected layers that form the circuit within the microchip are built and replicated across the wafer. Although the exact processing depends on the technology being used and the complexity of circuits, there are several major processing stages.

An insulating silicon dioxide (sand) layer is first grown onto the silicon wafer. Then, the electrical characteristics of the individual transistors are altered by diffusing or ion implanting precise amounts of impurities (dopants) into the silicon.

These impurities are what make silicon so useful, and they can be placed wherever desired on the wafer using photolithography 'masks'. The principle is the same as printing a photograph from a negative. The image of the chip is patterned in several layers. For each layer, the image is exposed on a photo resist across the wafer, and developed. The areas not covered by the photo resist can then be etched away using a plasma or chemical etchant, and/or electrically modified by adding dopants.

The interconnections between transistors are made by metallization. This process also makes the bonding pads that connect the chip itself to the package leads, and so to the rest of the world. The whole wafer is then- if necessary- planarized by Chemical Mechanical Polishing (CMP), given a protective layer, and tested to ensure the circuits work as they should.

In the Back-End operation, devices are then assembled using four steps. 1) 'Die preparation' cuts the wafer into individual integrated circuits or dice. 2) 'Die attach' attaches the die to the support structure (*e.g.* the lead frame) of the package. 3) 'Bonding' connects the circuit to the electrical contacts of the package, and so to the outside world. 4) 'Encapsulation' (usually by plastic moulding) give physical and chemical protection to the circuit.

Subsequent steps that give the package its final form and appearance vary from package to package. Steps like marking and lead finish give the product its own identity and improve reliability.

Once assembled, the integrated circuit is ready to use. However, many things can go wrong and make a device fail: the die may have wafer fab-related defects or may crack during assembly, or the bonds may be poorly connected or not connected at all. So, before being shipped to the customer, assembled devices must first be electrically tested.

Surface mounted devices are then normally packed in tapes with pockets, while the tape is wound around a reel. Reels, trays or tubes are boxed and labelled to customer requirements for shipping.

SEMICONDUCTOR MANUFACTURING STOCKS

The semiconductor stocks, as represented by the SPDR S&P Semiconductor ETF (XSD) have underperformed the market, down by about 9% YTD compared to the 8% rise for the general market. The under-performance comes on the back of a miserable 2011, when the average stock in the group corrected by about 20%, compared to the flattish performance for the general market.

The weak global economies and sagging consumer confidence that have plagued the sector continue to be issues, and the sector may not be out of the doldrums this year, perhaps preparing for a resurgence in 2013. In this section, via an analysis based on the latest available Q1 institutional 13-F filings, we identify the semiconductor manufacturing companies that are being accumulated and those being distributed by the world's largest fund managers.

These mega fund managers, such as Fidelity Investments, Goldman Sachs, BlackRock Inc., Vanguard Group, and 22 others, manage between $100 billion and over $1 trillion each, and together control about 40% of the assets invested in the U.S. equity markets. Together, these mega fund managers are bearish on the semiconductor manufacturing group, cutting a net $90 million in Q1 from their $93.66 billion prior quarter position in the group.

The investing activities of these mega fund managers in the group in the prior quarter can be found here. The following are the semiconductor manufacturing companies that these mega fund managers are most bullish about :

TAIWAN SEMICONDUCTOR ADR OR TSMC (TSM)

Taiwan Semiconductor is the world's largest dedicated IC foundry manufacturing logic and mixed-signal ICs for fabless semiconductor companies and integrated device manufacturers. Mega funds together added a net 19.90 million shares in Q1 to their 449.98 million share prior quarter position in the company, and taken together mega funds held $6.09 billion or 9.1% of the outstanding shares. The top buyer was Vanguard Group, with $1.7 trillion in assets under management, that purchased 8.05 million shares. Other large mega fund purchasers included Boston-based MFS Investment Management (7.20 million shares), with over $278 billion in assets under management, and JPMorgan Chase & Co. (5.05 million shares), with $1.4 trillion in assets under management.

In its latest Q1 (March), TSM beat analyst revenue and earnings estimates. Its shares trade at a current 15.5 P/E and 3.0 P/B compared to averages of 31.1 and 3.8 for its peers in the semiconductor foundries group, while earnings are projected to rise from 87c in 2011 to $1.12 in 2013, at a respectable annual growth rate of 13.5%. The shares are off about 20% from the multi-year $16.15 high set in late-April, and besides being attractively valued, the stock also sports a dividend yield of 3.2% compared to the 1.4% average for its peers in the group.

Other semiconductor manufacturing companies that mega fund managers are bullish about include:

- SanDisk Corp. (SNDK), a manufacturer of non-volatile removable memory cards used in various computing and electronic devices, in which mega funds together added a net 6.87 million shares to their 101.17 million share prior quarter position in the company;
- International Rectifier Corp. (IRF), that is a manufacturer of power management semiconductors, in which mega funds together added a net 1.76 million shares to their 27.51 million share prior quarter position in the company; and
- United Microelectronics ADR (UMC), a Taiwanese foundry for high performance semiconductor applications for telecom, consumer electronics, PC and memory markets, in which mega funds together added a net 10.42 million shares to their 72.15 million share prior quarter position in the company.

The following are semiconductor manufacturing companies that mega funds are bearish about :

- Micron Technology (MU), a leading manufacturer of semiconductor memory solution, including DRAM, NAND and NOR flash memory, phase change memory, and image sensors, in which mega funds together cut a net 55.67 million shares from their 425.02 million share prior quarter position in the company;
- RF Micro Devices Inc. (RFMD), that manufactures radio frequency systems and solutions that drive mobile communications. Its applications are used in mobile devices, base stations, wireless LANs and advanced metering devices, in which mega funds together cut a net 11.02 million shares from their 118.58 million share prior quarter position in the company; and

- STMicroelectronics (STM), that is a Swiss manufacturer of a broad range of ICs and discrete devices used in a wide variety of microelectronic applications, including telecommunications systems, computer systems, consumer products, automotive products and industrial automation and control systems, in which mega funds together cut a net 5.73 million shares from their 10.46 million share prior quarter position in the company.

Furthermore, the following are other notable semiconductor manufacturing companies that are among the top holdings of mega funds in the group, relative to their market-cap :

- Texas Instruments (TXN), that is a global manufacturer of analog ICs and digital signal processors used in cell phones, navigation systems and PCs, in which mega funds together hold 475.78 million or 41.6% of the outstanding shares; and
- ON Semiconductor Corp. (ONNN), that is a supplier of broadband and power management ICs and standard semiconductors used in fibre-optic networking equipment and portable electronics, in which mega funds together hold 220.99 million or 48.7% of the outstanding shares.

General Methodology and Background Information: The latest available institutional 13-F filings of the largest 25 mega hedge fund and mutual fund managers were analysed to determine their capital allocation among different industry groupings, and to determine their favourite picks and pans in each group. These mega fund managers number less than one percent of all funds and yet they control almost half of the U.S. equity discretionary fund assets. The argument is that mega institutional investors have the resources and the access to information, knowledge and expertise to conduct extensive due diligence in informing their investment decisions. When mega Institutional Investors invest and maybe even converge on a specific investment idea, the idea deserves consideration for further investigation. The savvy investor may then leverage this information either as a starting point to conduct his own due diligence.

SOFT LITHOGRAPHY

SCIENTIFIC DRIVERS

Fabrication of new types of electronic, magnetic, and optical devices currently consists of redesigning existing structures in down-sized versions. We are exploring an alternate strategy that uses "soft lithography" for the fabrication and manufacture of nanostructures. With these techniques we are able to generate patterns with critical dimensions as small as 30 nm. These techniques use transparent, elastomeric polydimethylsiloxane (PDMS) "stamps" with patterned relief on the surface to generate features. The stamps can be prepared by casting prepolymers against masters patterned by conventional lithographic techniques, as well as against other masters of interest.

Several different techniques are known collectively as soft lithography. They are as described below:

- *Near-Field Phase Shift Lithography.* A transparent PDMS phase mask with relief on its surface is placed in conformal contact with a layer of photoresist. Light passing through the stamp is modulated in the near-field. If the relief on the surface of the stamp shifts the phase of light by an odd multiple of (a node in the intensity is produced). Features with dimensions between 40 and 100 nm are produced in photoresist at each phase edge.
- *Replica Moulding.* A PDMS stamp is cast against a conventionally patterned master. Polyurethane is then molded against the secondary PDMS master. In this way, multiple copies can be made without damaging the original master. The technique can replicate features as small as 30 nm.
- *Micromolding in Capillaries (MIMIC).* Continuous channels are formed when a PDMS stamp is brought into conformal contact with a solid substrate. Capillary action fills the channels with a polymer precursor. The polymer is cured and the stamp is removed. MIMIC is able to generate features down to 1 μm in size.
- *Microtransfer Moulding ((TM).* A PDMS stamp is filled with a prepolymer or ceramic precursor and placed on a substrate. The material is cured and the stamp is removed. The technique generates features as small as 250 nm and is able to generate multilayer systems.
- *Solvent-assisted Microcontact Moulding (SAMIM).* A small amount of solvent is spread on a patterned PDMS stamp and the stamp is placed on a polymer, such as photoresist. The solvent swells the polymer and causes it to expand to fill the surface relief of the stamp. Features as small as 60 nm have been produced.
- *Microcontact Printing ((CP).* An "ink" of alkanethiols is spread on a patterned PDMS stamp. The stamp is then brought into contact with the substrate, which can range from coinage metals to oxide layers. The thiol ink is transferred to the substrate where it forms a self-assembled monolayer that can act as a resist against etching. Features as small as 300 nm have been made in this way.
- Techniques used in other groups include micromachining of silicon for microelectricalmechanical systems (MEMS), and embossing of thermoplastic with patterned quartz.

Our techniques require little in capital investment and most can be carried out in ambient laboratory conditions at low cost. Unlike conventional lithography, these techniques are able to generate features on both curved and reflective substrates and rapidly pattern large areas. A variety of materials have been patterned using the above techniques, including metals and polymers. The methods complement and extend existing nanolithographic techniques and provide new routes to high-quality patterns and structures with feature sizes (30 nm. There are, however, important unresolved problems such as distortions in the mask and issues such as runout and registration in multilevel fabrication.

Applications of soft lithography in the near future could include simple optical devices, such as polarizers, filters, wire grids, and surface acoustic wave (SAW) devices. Longer term goals include working towards optical data storage systems, flat panel displays, and quantum devices. Soft lithographic techniques are currently not competitive with conventional photolithography for multilayer fabrication where there are critical requirements for pattern regularity.

Critical Parameters

Soft lithography appears to be a promising route to nanostructures and nanosystems. Many issues remain to be solved, however, before the techniques can be brought to the market. Some critical parameters are applicable to all the techniques, while others apply only to specific techniques.

Multilevel structures and devices necessitate the ability to accurately place one layer on top of another. High-resolution registration is problematic with elastomeric, distortion-prone materials. Distortion currently limits soft lithographic techniques to the fabrication of single-layer structures and devices. Registration is possible on micron-sized features using a mask aligner. Studies are also underway in some group to quantify the distortion in elastomeric stamps and to limit this distortion by using thick stamps and rigid backings. The formation and distribution of defects also must be controlled if these techniques are to be used in device fabrication. Defects arise from dust particles, poor adhesion to the substrate and poor release from the stamp, bubbles in the precursor, and contamination from low molecular weight monomers in the stamp. Another problem is the presence of a thin film of polymer under the nanometre-sized features in certain of the soft lithographic methods; microtransfer moulding and solvent-assisted embossing often have problems with underlayer control. The presence of the underlayer makes etching and lift-off difficult. Although reactive ion etching can be used to remove thin layers, the time necessary to remove thick layers would damage small features.

The resist used has been a limiting factor in some cases. Phase shift lithography is able to generate features in photoresist more than an order of magnitude smaller than those for which the resist and mask aligner are designed. The resist is, however, only filtered to particles of 40 nm, suggesting that it will be difficult to achieve smaller features consistently with current materials. In microcontact printing where the SAM functions as a resist for etching, the resolution of the grain size of gold in the substrate produced by evaporation (30 nm) limits the edge resolution of the etched area.

Phase-shift lithography has its own set of problems. Features are formed at a phase edge: a mask with a 4 μm period generates two 90 nm lines with a distance of 2 μm between them. Increasing the density of the features is challenging. The fabrication of features such as T-junctions is complex because of the necessity of the phase edge. Integration of large and small features in phase shift lithography is also difficult. The compatibility of these patterning techniques with the production of microelectronic circuitry is starting to be addressed. The above difficulties are tractable; creativity, however, will be necessary to overcome them.

Bringing Soft Lithography to the Market

Nanostructures are fabricated ordinarily using advanced nanolithographic techniques such as e-beam writing, X-ray lithography, and proximal-probe lithography. These techniques are capable of providing very small features, but their development into methods for generating large areas of nanostructures rapidly and at low cost will require some ingenuity. Conventional techniques are also limited to the formation of two-dimensional structures in a limited number of materials on planar surfaces. The lack of techniques capable of generating and manufacturing nanostructures rapidly and economically represents a limiting step in the area of nanoscience and nanotechnology. Soft lithography suggests a new conceptual approach to nanomanufacturing: advanced nanolithographic techniques would be used to make masters, and these structures would then be transferred into organic polymers or other materials using procedures such as printing, moulding, embossing, or a combination thereof. The techniques could be automated. The ability to make both positive and negative replicas and to modify the dimensions and shapes of features present on elastomeric masters by mechanical deformation adds further flexibility to this methodology. Currently, the ability to manufacture is limited to devices where small feature size is important but lateral distortions are not: single-level structures such as polarizers and waveguides. The limitations are due to the yet unresolved issues of registration and distortion. Further practical technological uses of these techniques will require further development.

SEMICONDUCTOR NANOCRYSTALS

The silicon microelectronics industry is the technological driver of modern society. The industry is built upon two major inventions: the solid state transistor and the photolithographic integrated circuit. The present technology is robust and advancing rapidly, as shown by the amazing improvement in and decrease in cost of simple PCs. Much of this is due to continued miniaturization in both magnetic memory and Si device size. This engineering improvement will continue to an ultimate Si feature size of perhaps 3-6 nm, an order of magnitude smaller in size than the smallest production devices at present. At even smaller sizes, several factors slow progress: basic statistical fluctuations in the power necessary for reliable driving of one device by the next and in the performance of individual transistors, coupled with the optical problem of photolithography on such a small scale. It is likely to be only several decades into the future when Si circuits ultimately reach a practical size limit. However, in approaching the 3-6 nm size, manufacturing and design must evolve, as semiconductors in this size regime are not true bulk materials. Nanocrystal research will thus guide the "ultimate" design principles of photolithographic Si circuits.

Going much smaller towards single-electron and small nanocrystal ("molecule") devices will require an absolute revolution in all aspects of the industry. Such hypothetical circuits do not appear to violate the fundamental laws of physics and would offer huge performance advances. Today, with slow scanned probe techniques, we can move single atoms and assemble molecular-level devices; however, this is not a realistic manufacturing process. At present we are far from understanding how to carry out a practical engineered "chemical self-assembly" of diverse components to create a circuit approaching the complexity and especially the reliability of modern photolithographic circuits. Just as the photolithographic Si circuit of 1997 could not be imagined in the vacuum tube era of 1937, today we can not envision the circuit of 2057. Major, unexpected inventions and truly novel scientific insights are necessary; it is not simply a question of increasing money for development. The engineering of such circuits would represent a merging of synthetic chemistry and industrial technology. Indeed, academic chemistry is evolving towards fundamental study of very large molecules, self-assembly, and macroscopic matter. Solid state physics is evolving towards new materials with molecular level structure and very large unit cells. The nanometre range is being approached from both ends: top down and bottom up.

In this context, nanocrystals are basic building blocks. Even the most elementary questions are not quantitatively understood: How does charge transfer on and off nanocrystals? What is the influence of the environment? How can the internal electron-hole kinetics and the charge transfer be controlled by surface passivation, with both inorganic and organic species? Can we entirely eliminate surface states? Can we find ways to make crystalline nanocrystals of uniform shape, size, and surface stoichiometry? Can we find ways to assemble nanocrystals of different materials and electronegativities? What role does the difference in dielectric constants play? In a material with high internal interface area, can long-term stability while carrying current approach the stability of bulk semiconductors?

SOLAR CELLS, DISPLAYS, AND NEW MATERIALS

In the next century, the question of global warming due to carbon dioxide in the atmosphere is likely to become critical as increased industrialization in the third world causes accelerated warming. Direct conversion of sunlight into electricity would require large-area, inexpensive, environmentally benign solar cell designs that could be integrated into building construction. The O'Regan-Gratzel liquid junction, thin film solar cell is composed of intimate, percolating networks of liquid electrolyte and partially fused titanium dioxide 10 nm particles. Photoexcited charge separation occurs at the interface between the two phases. This cell represents a new design in solar cells and is very efficient, despite the fact it is made from inexpensive, poor electrical quality oxide semiconductor nanocrystals presently manufactured in huge quantities for the ceramic and paint industries. Can a liquid junction device be as inherently stable as an all solid state device? How do we understand charge transport through such as device when the diffusing carrier interacts with both the electrolyte and titanium dioxide? This device offers a glimpse of what novel designs might be possible; can we learn from this device to optimize other applications?

Nanocrystals have higher electron affinities than organic polymers such as PPV, which are used as hole conductors in organic displays. Moreover, nanocrystal affinities are size-tunable, and when used as luminescence centres for electron-hole recombination in electroluminescent displays, the emission colour is also size-tunable. Thus, in the construction of inexpensive flat panel displays, layers of neat passivated nanocrystals make good electron transport layers, and the resulting hybrid organic/nanocrystal device can emit a range of colours. Can we learn to optimize transport, and can we achieve a fundamental understanding of junctions between nanocrystal layers and organic layers? Monodisperse nanocrystals can self-assemble into three-dimensional "supercrystals" that structurally resemble naturally occurring gem opals. The optical and electrical properties of such organized nanocrystal solids could be tuned through nanocrystal size, through the inter-nanocrystal coupling as controlled by surface passivation, and by doping via material in the interstitial spaces.

PRESENT SCIENTIFIC UNDERSTANDING AND SYNTHETIC ABILITIES

Work over the past decade has lead to an understanding of the evolution from molecules to bulk solids, at least in outline form, for individual nanocrystals.

In spectroscopic properties, there are three physical regimes: molecular, quantum dot (or nanocrystal), and polariton. Small clusters of some tens of atoms are molecules — the bulk unit cell is not present. In the nanocrystal regime, the unit cell is present, yet the electronic states are strongly quantum-confined in three dimensions. The excited electronic states and the band gap variation with size can be quantitatively calculated with advanced theory. In the polariton regime, the interaction with the electromagnetic field becomes strong and can be described by Mie theory.

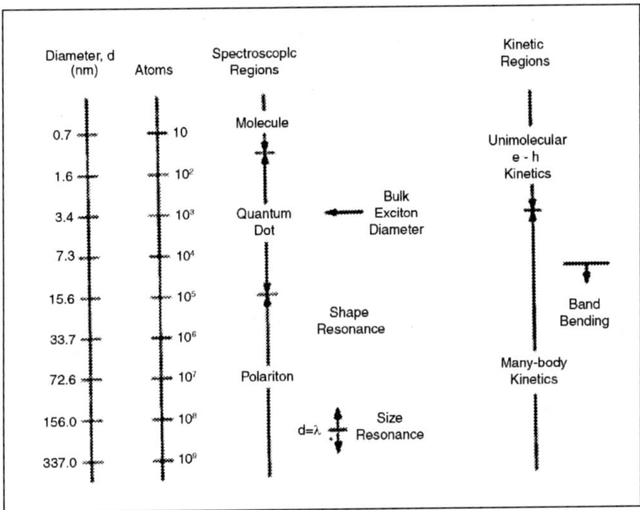

Figure: Schematic Size Regimes for Semiconductor Nanocrystals.

In kinetic properties, there are two limiting regimes: unimolecular kinetics in small nanocrystals, and many-body kinetics in large nanocrystals. In small nanocrystals, the electrical field of a moving charge exits the nanocrystal and interacts with the local environment: this is essentially a molecular property. In the absence of a moving charge, nanocrystals are intrinsic dielectric particles. (In bulk semiconductor transport, the physics is entirely internal; the fields of moving carriers terminate on ionized donor atoms.) Also, monolayer surface passivation with a higher band gap material can eliminate most nanocrystal surface states, while still allowing efficient electron transfer on and off the particle. In the CdSe system, organometallic synthesis has achieved the ability to make nearly uniform nanocrystals. An equivalently good synthesis for Si nanocrystals does not exist.

The ability to do single nanocrystal luminescence spectroscopy should accelerate our efforts to rigorously understand the physics. Nanocrystals are highly polarizable molecules, some thousands of atoms in size; in the electronic properties, every atom counts. Real samples have a wide distribution of shapes and surface structures; thus, the ability to examine properties one nanocrystal at a time is quite useful. The room temperature luminescence from single CdSe nanocrystals. The luminescence is intermittent; the nanocrystal appears to "blink" on and off on a time scale of minutes. Such blinking is incoherent from one nanocrystal to the next and was averaged out in prior hole-burning and size-selective photoluminescence ensemble studies. The blinking is slowed down by surface passivation with ZnS and has been tentatively attributed to photoionization. The single molecule method is quite powerful for such complex systems.

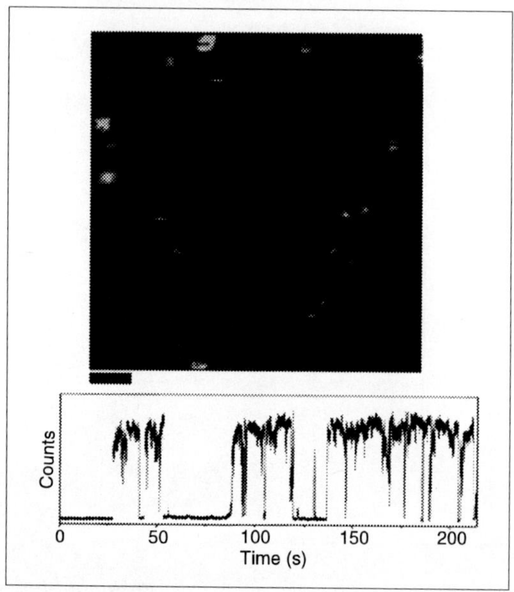

Figure : Scanning Luminescence of Single CdSe Nanocrystals (top); Time Trace of Single Nanocrystal Luminescence under Quasi-cw Excitation (bottom).

In the past five years, we have begun to understand the size dependence of indirect gap materials such as Si. The remarkably high photoluminescence quantum yield of Si nanocrystals and porous Si thin films can be traced to the size dependence of Auger nonradiative recombination. Small silicon nanocrystals remain indirect gap-like in all essential properties. They do not become partially direct gap-like, as was initially conjectured when the quantum properties of porous Si films were first discovered.

Just recently, systematic study of solid-solid phase transitions in nanocrystals has lead to the realization that nanocrystals in six-coordinate, dense structural phases, which are only theromodyanmically stable at high pressure, may be metastable at STP. These phases have completely different optical and electrical properties than the normal sp3 hybridized phases. This opens up the possibility of an expanded inventory of possible nanocrystal materials in the future; to achieve this, novel synthetic methods are required.

NANOIMPRINT LITHOGRAPHY

Nanoimprint lithography is a method of fabricating nanometre scale patterns. It is a simple nanolithography process with low cost, high throughput and high resolution. It creates patterns by mechanical deformation of imprint resist and subsequent processes. The imprint resist is typically a monomer or polymer formulation that is cured by heat or UV light during the imprinting. Adhesion between the resist and the template is controlled to allow proper release.

HISTORY

The term "Nanoimprint Lithography" (NIL) was coined in the scientific literature in 1996, when Prof. Stephen Chou and his students published a report in Science, although hot embossing (now taken as a synonym of NIL) of thermoplastics had been appearing in the patent literature for a few years already. Soon after the Science paper, many researchers developed many different variations and implementations. At this point, nanoimprint lithography has been added to the International Technology Roadmap for Semiconductors (ITRS) for the 32 and 22 nm nodes.

PROCESSES

There are many different types of nanoimprint lithography, but two of them are most important: thermoplastic nanoimprint lithography and photo nanoimprint lithography.

THERMOPLASTIC NANOIMPRINT LITHOGRAPHY

Thermoplastic nanoimprint lithography (T-NIL) is the earliest nanoimprint lithography developed by Prof. Stephen Chou's group. In a standard T-NIL process, a thin layer of imprint resist (thermoplastic polymer) is spin coated onto the sample substrate. Then the mold, which has predefined topological patterns, is brought into contact with the sample and they are pressed together under certain pressure. When heated up above the glass transition temperature of the polymer, the pattern on the mold is pressed into the softened polymer film. After being cooled down, the mold is separated from the sample and the pattern resist is left on the substrate. A pattern transfer process (reactive ion etching, normally) can be used to transfer the pattern in the resist to the underneath substrate. Alternatively, cold welding between two metal surfaces could also transfer low dimensional nanostructured metal without heating (especially for critical sizes less than ~10nm). Three-dimensional structures can be fabricated by repeating this procedure. The cold welding approach has the advantage of reducing surface contact contamination or defect due to no heating process, which is a main problem in the latest development and fabrication of organic electronic devices as well as novel solar cells.

Photo Nanoimprint Lithography

In photo nanoimprint lithography (P-NIL), a photo(UV) curable liquid resist is applied to the sample substrate and the mold is normally made of transparent material like fused silica. After the mold and the substrate are pressed together, the resist is cured in UV light and becomes solid. After mold separation, a similar pattern transfer process can be used to transfer the pattern in resist onto the underneath material. The use of a UV-transparent mold is difficult in a vacuum, because a vacuum chuck to hold the mold would not be possible.

Schemes

Full Wafer Nanoimprint

In a full wafer nanoimprint scheme, all the patterns are contained in a single nanoimprint field and will be transferred in a single imprint step. This allows a high throughput and uniformity. An at least 8-inch (203 mm) diameter full-wafer nanoimprint with high fidelity is possible.

To ensure the pressure and pattern uniformities of full wafer nanoimprint processes and prolong the mold lifetime, a pressing method utilizing isotropic fluid pressure, named Air Cushion Press (ACP) by its inventors, is developed and being used by commercial nanoimprint systems.

Step and Repeat Nanoimprint

Nanoimprint can be performed in a way similar to the step and repeat optical lithography. The imprint field (die) is typically much smaller than the full wafer nanoimprint field. The die is repeatedly imprinted to the substrate with certain step size. This scheme is good for nanoimprint mold creation.

Applications

Nanoimprint lithography has been used to fabricate devices for electrical, optical, photonic and biological applications. For electronics devices, NIL has been used to fabricate MOSFET, O-TFT, single electron memory. For optics and photonics, intensive study has been conducted in fabrication of subwavelength resonant grating filter, polarizers, waveplate, anti-reflective structures, integrated photonics circuit and plasmontic devices by NIL. sub-10 nm nanofluidic channels had been fabricated using NIL and used in DNA strenching experiment. Currently, NIL is used to shrink the size of biomolecular sorting device an order of magnitude smaller and more efficient.

Benefits

A key benefit of nanoimprint lithography is its sheer simplicity. The single greatest cost associated with chip fabrication is the optical lithography tool used to print the circuit patterns. Optical lithography requires high powered excimer lasers and immense stacks of precision ground lens elements to achieve nanometre scale resolution. There is no need for complex optics or high-energy radiation sources with a nanoimprint tool. There is no need for finely tailored photoresists designed for both resolution and sensitivity at a given wavelength. The simplified requirements of the technology lead to its low cost.

Imprint lithography is inherently a three-dimensional patterning process. Imprint molds can be fabricated with multiple layers of topography stacked vertically. Resulting imprints replicate both layers with a single imprint step, which allows chip manufactures to reduce chip fabrication costs and improve product throughput. The imprint material does not need to be finely tuned for high resolution and sensitivity. A broader range of materials with varying properties are available for use with imprint lithography. The increased material variability gives chemists the freedom to design new functional materials rather than sacrificial etch resistant polymers. A functional material may be imprinted directly to form a layer in a chip with no need for pattern transfer into underlying materials. The successful implementation of a functional imprint material would result in significant cost reductions and increased throughput by eliminating many difficult chip fabrication processing steps.

Concerns

The key concerns for nanoimprint lithography are overlay, defects, template patterning and template wear. However, recently Kumar *et al.* have shown that amorphous metals (metallic glasses) can be patterned on sub-100 nm scale, which can significantly reduce the template cost.

Overlay

The current overlay 3 sigma capability is 10 nm. Overlay has a better chance with step-and-scan approaches as opposed to full-wafer imprint.

Defects

As with immersion lithography, defect control is expected to improve as the technology matures. Defects from the template with size below the post-imprint process bias can be eliminated. Other defects would require effective template cleaning and/or the use of intermediate polymer stamps. When vacuum is not used during the imprint process, air can get trapped, resulting in bubble defects. This is because the imprint resist layer and the template or stamp features are not perfectly flat. There is an elevated risk when the intermediate or master stamp contains depressions (which are especially easy air traps), or when the imprint resist is dispensed as droplets just before imprinting, rather than pre-spun onto the substrate. Sufficient time must be allowed for the air to escape.

Template Patterning

High resolution template patterning can currently be performed by electron beam lithography or focused ion beam patterning; however at the smallest resolution, the throughput is very slow. As a result, optical patterning tools will be more helpful if they have sufficient resolution. Such an approach has been successfully demonstrated by Greener *et al.* whereby robust templates were rapidly fabricated by optical patterning of a photoresist-coated metal substrate through a photomask. Other patterning techniques (including even double patterning) may also be used. Kumar and Schroers at Yale developed the nanopatterning of amorphous metals which can be used as inexpensive templates for nanoimprinting.

Template Wear

The use of substantial pressure to not only contact but also penetrate a layer during imprinting accelerates the wear of imprint templates compared to other types of lithographic masks.

Other

Future applications of nanoimprint lithography may involve the use of porous low-k materials. These materials are not stiff and, as part of the substrate, are readily damaged mechanically by the pressure of the imprint process.

REMOVAL OF RESIDUAL LAYERS

A key characteristic of nanoimprint lithography (except for electrochemical nanoimprinting) is the residual layer following the imprint process. It is preferable to have thick enough residual layers to support alignment and throughput and low defects. However, this renders the nanoimprint lithography step less critical for critical dimension (CD) control than the etch step used to remove the residual layer. Hence, it is important to consider the residual layer removal an integrated part of the overall nanoimprint patterning process. In a sense, the residual layer etch is similar to the develop process in conventional lithography. It has been proposed to combine photolithography and nanoimprint lithography techniques in one step in order to eliminate the residual layer.

Proximity Effects

Nanoimprint lithography relies on displacing polymer. This could lead to systematic effects over long distances. For example, a large, dense array of protrusions will displace significantly more polymer than an isolated protrusion. Depending on the distance of this isolated protrusion from the array, the isolated feature may not imprint correctly due to polymer displacement and thickening. Resist holes can form in between groups of protrusions. Likewise, wider depressions in the template do not fill up with as much polymer as narrower depressions, resulting in misshapen wide lines. In addition, a depression at the edge of a large array fills up much earlier than one located in the centre of the array, resulting in within-array uniformity issues.

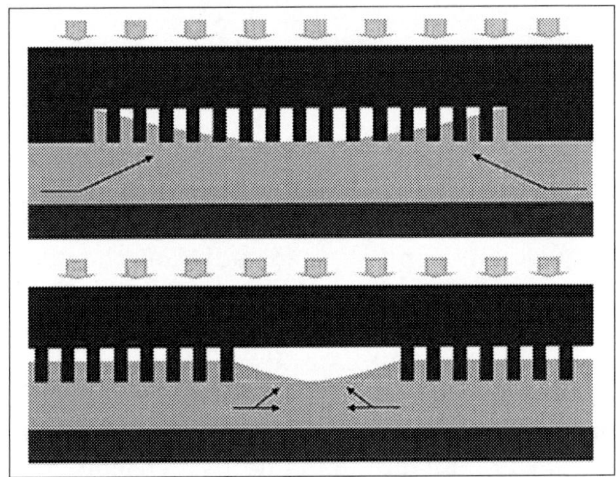

Figure: Nanoimprint Proximity Effect. *Top:* Array of Depressions is more Quickly Filled at the Edge than the Centre, Resulting in Less Imprinting at the Centre of the Array. *Bottom:* The Wide Space between Two Groups of Protrusions Tends to be Filled Slower than the Narrow Spaces between the Protrusions, Resulting in the Formation of Holes in the Unpatterned Area.

3D-patterning

A unique benefit of nanoimprint lithography is the ability to pattern 3D structures, such as damascene interconnects and T-gates, in fewer steps than required for conventional lithography. This is achieved by building the T-shape into the protrusion on the template.

ALTERNATIVE APPROACHES

Electrochemical Nanoimprinting

Electrochemical nanoimprinting can be achieved using a stamp made from a superionic conductor such as silver sulfide. When the stamp is contacted with metal, electrochemical etching can be carried out with an applied voltage. The electrochemical reaction generates metal ions which move from the original film into the stamp. Eventually all the metal is removed and the complementary stamp pattern is transferred to the remaining metal.

Laser Assisted Direct Imprint

Laser assisted direct imprint (LADI) is a rapid technique for patterning nanostructures in solid substrates and it does not require etching. A single or multiple excimer laser pulses melt a thin surface layer of substrate material, and a mold is embossed into the resulting liquid layer. A variety of structures with resolution better than 10 nm have been imprinted into silicon using LADI, and the embossing time is less than 250 ns. The high resolution and speed of LADI, attributed to molten silicon's low viscosity (one-third that of water), could open up a variety of applications and be extended to other materials and processing techniques.

THE FUTURE OF NANOIMPRINT

Nanoimprint lithography is a simple pattern transfer process that is neither limited by diffraction nor scattering effects nor secondary electrons, and does not require any sophisticated radiation chemistry. It is also a potentially simple and inexpensive technique. However, a lingering barrier to nanometre-scale patterning is the current reliance on other lithography techniques to generate the template. It is possible that self-assembled structures will provide the ultimate solution for templates of periodic patterns at scales of 10 nm and less. It is also possible to resolve the template generation issue by using a programmable template in a scheme based on double patterning. Toshiba is the only company to have validated nanoimprint lithography for 22 nm and beyond. What is more significant is that nanoimprint lithography is the first sub-30 nm lithography to be validated by an industrial user.

LITHOGRAPHICALLY INDUCED SELF-ASSEMBLY (LISA)

Lithographically Induced Self-Assembly (LISA) is a novel pattern formation phenomenon first discovered here at NSL.

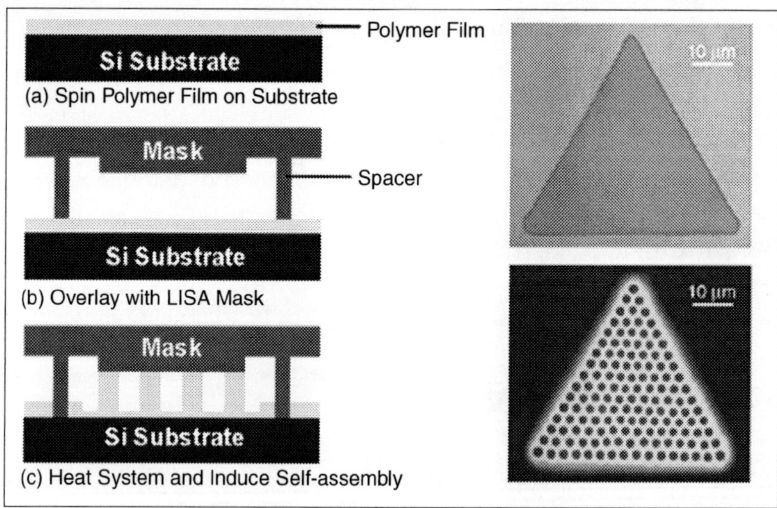

Figure : Schematic of LISA Process Showing how a Mask Held in Close Proximity to a Polymer Surface can Induce the Formation of Periodic Structures.

It is a simple process by which a mask is used to control the formation of micro and nanostructures in a thin polymer film. A polymer film is spun onto a flat substrate and is overlayed with a mask. The mask is held apart from the polymer surface by spacers. The system is heated above the glass transition temperature of the polymer. The result is that the polymer rises up against the forces of gravity and surface tension forming periodic structures that are aligned to any pattern on the mask.

LISA DYNAMICS

Not only is the LISA pattern ordered, the manner in which it forms is well ordered as well. To establish this, we used a transparent mask and directly monitored the pattern formation process through the top of the mask using a microscope. A series of snapshots showing the ordered growth of a square LISA array formed in a PMMA film. Patterns always start at the corners and gradually propagate towards the centre.

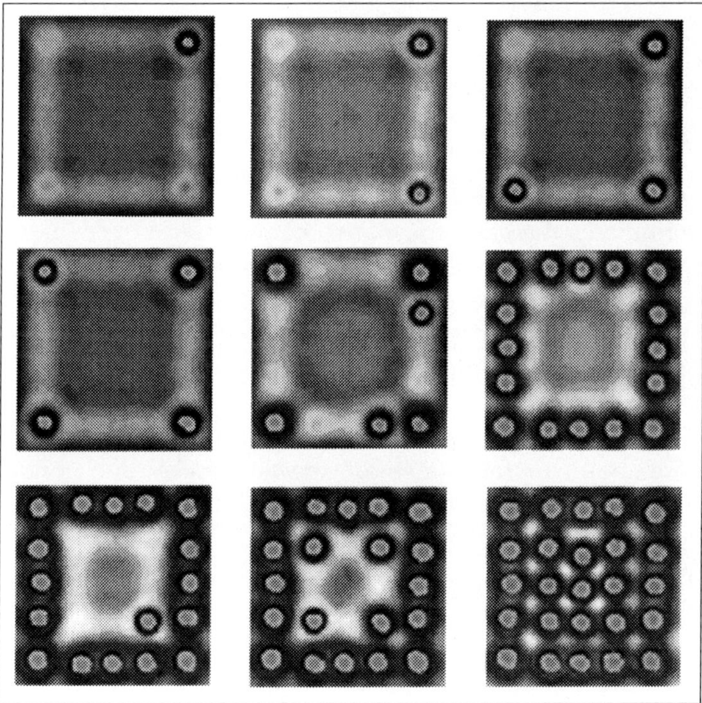

Figure : Optical Micrographs Showing the Sequence of LISA Formation under a Square Mask Pattern.

Other Novel Structures

Other novel patterns have also been observed using the LISA technique. These include well-controlled concentric ring patterns and mesa structures. Atomic force micrographs of such formations in PMMA. The rings are formed by first nucleating a central pillar using a small protrusion on the mask. Rings then form spontaneously around this central point. Mesa structures can be formed simply by increasing the surface energy of the mask allowing the polymer to easily flow on the mask surface. Rather than being fixed in the shape of a pillar, the polymer pillars flow together giving a mesa structure.

Figure : Atomic Force Micrographs Showing other Novel LISA Structures.

APPLICATIONS AND CURRENT PROJECTS

LISA lends itself to a variety of applications because of its simplicity and because it eliminates the need for many chemicals and exposure techniques commonly associated with lithography. This self-assembly process may be particularly useful for organic electronic and opto-electronic devices. The self-assembled nature of the periodic structures are also an interesting mean for patterning quantized magnetic disks or electronic memories. The regularity of LISA patterns could further be employed in photonic crystal applications.

In conjunction with work on these applications, we are interested in learning more about the fundamental aspects of LISA. Such understanding will inevitably help in accelerating the application of LISA in many areas.

DIP-PEN NANOLITHOGRAPHY

Dip Pen Nanolithography (DPN) began as a scanning probe lithography technique where an atomic force microscope tip was used to transfer alkane thiolates to a gold surface. This technique allows surface patterning on scales of under 100 nanometres. DPN is the nanotechnology analog of the dip pen (also called the quill pen), where the tip of an atomic force microscope cantilever acts as a "pen," which is coated with a chemical compound or mixture acting as an "ink," and put in contact with a substrate, the "paper."

DPN enables direct deposition of nanoscale materials onto a substrate in a flexible manner. Recent advances have demonstrated massively parallel patterning using two-dimensional arrays of 55,000 tips. Applications of this technology currently range through chemistry, materials science, and the life sciences, and include such work as ultra high density biological nanoarrays, and additive photomask repair.

The uncontrollable transfer of a molecular 'ink' from a coated AFM tip to a substrate was first reported by Jaschke and Butt in 1995, but they erroneously concluded that alkanethiols could not be transferred to gold substrates to form stable nanostructures. A research group at Northwestern University led by Chad Mirkin independently studied the process and determined that under the appropriate conditions, molecules could be transferred to a wide variety of surfaces to create stable chemically-adsorbed monolayers in a high resolution lithographic process they termed "DPN". Mirkin and his coworkers hold the patents on this process, and the patterning technique has expanded to include liquid "inks". It is important to note that "liquid inks" are governed by a very different deposition mechanism when compared to "molecular inks".

DEPOSITION MATERIALS

Molecular Inks

Figure: Classic DPN Mechanism: Molecular Ink Diffusing from a Nanoscale Tip to a Surface through a Water Meniscus.

Molecular inks are typically composed of small molecules that are coated onto a DPN tip and are delivered to the surface through a water meniscus. In order to coat the tips, one can either vapour coat the tip or dip the tips into a dilute solution containing the molecular ink. If one dip-coats the tips, the solvent must be removed prior to deposition. The deposition rate of a molecular ink is dependent on the diffusion rate of the molecule, which is different for each molecule. The size of the feature is controlled by the tip/surface dwell-time (ranging from milliseconds to seconds) and the size of the water meniscus, which is determined by the humidity conditions (assuming the tip's radius of curvature is much smaller than the meniscus).

- Water meniscus mediated (exceptions do exist)

- Nanoscale feature resolution (50 nm to 2000 nm)
- No multiplexed depositions
- Each molecular ink is limited to its corresponding substrate.

Examples:

- Alkane thiols written to gold
- Silanes written to glass or silicon.

Liquid Inks

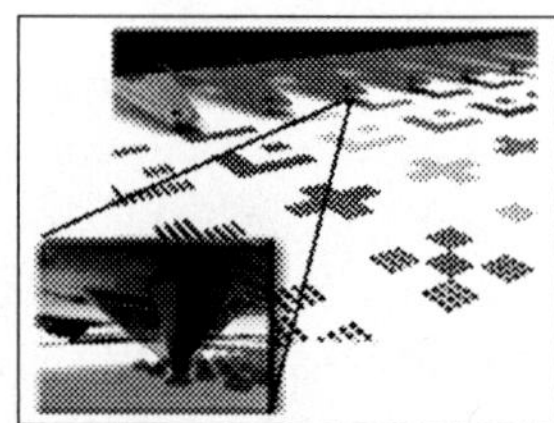

Figure: Liquid Ink Deposition Mechanism.

Liquid inks can be any material that is liquid at deposition conditions. The liquid deposition properties are determined by the interactions between the liquid and the tip, the liquid and the surface, and the viscosity of the liquid itself. These interactions limit the minimum feature size of the liquid ink to about 1 micrometre, depending on the contact angle of the liquid. Higher viscosities offer greater control over feature size and are desirable. Unlike molecular inks, it is possible to perform multiplexed depositions using a carrier liquid. For example, using a viscous buffer, it is possible to directly deposit multiple proteins simultaneously.

- 1-10 micrometre feature resolution
- Multiplexed depositions
- Less restrictive ink/surface requirements
- Direct deposition of high viscosity materials

Examples

- Protein, peptide, and DNA patterning
- Hydrogels
- Sol gels
- Conductive inks
- Lipids

Applications

In order to define a good DPN application, it is important to understand what DPN can do that other techniques can't. Direct-write techniques, like contact printing, can pattern multiple biological materials but it cannot create features with subcellular resolution. Many high-resolution lithography methods can pattern at sub-micrometre resolution, but these require high-cost equipment that were not designed for biomolecule deposition and cell culture. Micro contact printing can print biomolecules at ambient conditions, but it cannot pattern multiple materials with nanoscale registry.

Industrial Applications

The following are some examples of how DPN is being applied to potential products.

Figure: Cantilever Biosensor Functionalized with 4 Different Proteins.

- Biosensor Functionalization- Directly place multiple capture domains on a single biosensor device
- Nanoscale Sensor Fabrication- Small, high-value sensors that can detect multiple targets
- Nanoscale Protein Chips- High-density protein arrays with increased sensitivity.

EMERGING APPLICATIONS

Cell Engineering

DPN is emerging as a powerful research tool for manipulating cells at subcellular resolution

- Stem cell differentiation
- Subcellular drug delivery
- Cell sorting
- Surface gradients
- Subcellular ECM protein patterns
- Cell adhesion.

Rapid Prototyping

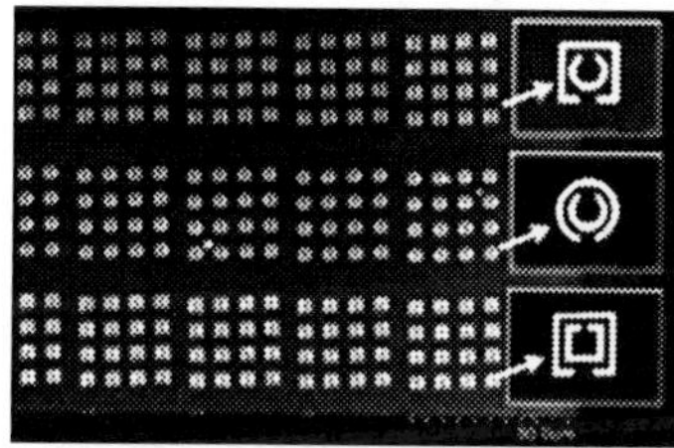

Figure: SEM Image of DPN Fabricated Gold Metastructure Arrays.

- Plasmonics and Metamaterials
- Cell and tissue screening.

DPN PROPERTIES

Direct Write

DPN is a direct write technique so it can be used for top-down and bottom-up lithography applications. In top-down work, the tips are used to deliver an etch resist to a surface, which is followed by a standard etching process. In bottom-up applications, the material of interest is delivered directly to the surface via the tips.

Figure: Gold on Silicon Metastructure Fabricated with Top-down DPN Methods.

Unique advantages:

- Directed Placement- Directly print various materials onto existing nano and microstructures with nanoscale registry
- Direct Write- Maskless creation of arbitrary patterns with feature resolutions from as small as 50 nm and as large as 10 micrometres
- Biocompatible- Subcellular to nanoscale resolution at ambient deposition conditions
- Scalable- Force independent, allowing for parallel depositions.

COMMON MISCONCEPTIONS

Direct Comparisons to other Techniques

Figure: Streptavidin (4 nm Thickness) Deposited using Micro Contact Printing.

The criticism most often directed at DPN is the patterning speed. The reason for this has more to do with how it is compared to other techniques rather than any inherent weaknesses. For example, the soft lithography method, micro contact printing (μCP), is the current standard for low cost, bench-top micro and nanoscale patterning, so it is easy to understand why

DPN is compared directly to micro contact printing. The problem is that the comparisons are usually based upon applications that are strongly suited to μCP, instead of comparing them to some neutral application. μCP has the ability to pattern one material over a large area in a single stamping step, just as photolithography can pattern over a large area in a single exposure. Of course DPN is slow when it is compared to the strength of another technique. DPN is a maskless direct write technique that can be used to create multiple patterns of varying size, shape, and feature resolution, all on a single substrate. No one would try to apply micro contact printing to such a project because it would never be worth the time and money required to fabricate each master stamp for each new pattern. Even if they did, micro contact printing would not be capable of aligning multiple materials from multiple stamps with nanoscale registry. The best way to understand this misconception is to think about the different ways to apply photolithography and e-beam lithography. No one would try to use e-beam to solve a photolithography problem and then claim e-beam to be "too slow". Directly compared to photolithography's large area patterning capabilities, e-beam lithography is slow and yet, e-beam instruments can be found in every lab and nanofab in the world. The reason for this is because e-beam has unique capabilities that cannot be matched by photolithography, just as DPN has unique capabilities that cannot be matched by micro contact printing.

REFERENCES

- Geckeler, K. E. : *Biomedical Applications of Nanostructured Materials,* Macmillan Publishers India, Delhi, 2011.
- Gupta, Sushmita: *Understanding of Nano Science and Technology,* Global Vision, Delhi, 2006.
- Jana, S.: *Surface Analysis Methods for Characterizing Polymeric Biomaterials,* New Delhi: Long Life Publication, 2003.
- Janes, D.B. : *The Handbook of Nanostructured Materials and Nanotechnology,* San Diego: Academic Press, 1998.
- Kuhn T.: *"The Structure of Scientific Revolutions."* University of Chicago Press Chicago, 1962.
- Kumar, H.D. : *Material Science: Nanotechnology and Applications,* I.K. International, Delhi, 2011.
- Reddy, K. : *Nanotechnology in Electronics,* Campus Books International, Delhi, 2007.
- Siegel, R.W.: *Physics of New Materials,* Berlin, Springer 1994.

8

Pharmaceutically Important of Nanomaterials

INTRODUCTION

Nanoparticles are having a great position in pharmaceutical field for drug delivery (targeted and sustained) as well as used for disease diagnosis. Following are the important types of nanoparticles used in pharmaceutical field.

MICELLES

Micelles are small, monolayer structures, defined as amphipathic lipid aggregates that naturally formed in water. Micelle can be formed using block copolymers that helps in combining the desired aspects of more than one homopolymer with a hydrophobic core. The micelles are measured about 10-100 nm, and the hydrophilic outer core offer good drug entrapment without significant risk of premature dissolution by the reticuloendothelial system (MPS) as observed in liposomes.

Micelle systems based upon poly (hexyl-substituted lactides), poly (lactide)-poly (ethylene glycol), DEX-g-PEO-C16 and self-assembling PEG-p (CL-co-TMC) copolymers are used as potential carrier systems for hydrophobic drugs.

LIPOSOMES

Liposomes are used as delivery systems for drugs, vaccines and nutrients and an ideal model for biological membranes. Liposomes comprise of lipid that form a bilayer membranes around an aqueous interior. The active drug compound can be positioned in the aqueous centre if it is water soluble or in the lipid layer if it is fat soluble. Although liposome technology is an older technology but still, it is having problems of stability, poor mechanical And integrity low incorporated drug concentrations.

SOLID LIPID NANOPARTICLES

Solid lipid nanoparticle is an alternative colloidal carrier system for emulsions, liposomes and polymeric nanoparticles in controlled drug delivery and is having advantages such as good biocompatibility, low toxicity and stability. Solid lipid nanoparticles are prepared by lipids and the surfactants that have an accepted GRAS status. Solid lipid nanoparticle can be produced in large scales by high-pressure homogenization without using organic solvents Mean particle size of prepared Solid lipid nanoparticles was ranged from 100 to 150 nm. It was found that the drug loading capacity was improved with the most lipophilic drug and low concentration of surfactant.

POLYMERIC NANOPARTICLES

Biodegradable polymers are mostly preferred for preparation of nanoparticles due to its efficiency and effectiveness in controlled delivery of drugs and biological products. Synthetic biodegradable polymers like polycyanoacrylate or poly (D, L-lactide) and poly (lactide-co-glycolide) (PLGA), and natural polymers like chitosan, gelatine and sodium alginate, are mostly used for nanoparticle preparation, But natural polymers are preferred over synthetic polmers to minimize toxicity. When the drug molecules are uniformly dispersed in the matrix of nanoparticle they are called as nanosphere whereas when the drug molecules are surrounded by a polymeric membrane it is called as nanocapsules. These polymeric nanoparticles are having more therapeutic potential, higher stability in the biological fluid and during storage. This kind of particles may have specificity, allowing them to deliver a higher concentration of pharmaceutical agent to a desired location due to the possible change in surface charge or other properties.

PEGYLATED NANOSTRUCTURES

Poly Ethylene Glycol a water-soluble compound with terminal hydroxyl groups can be easily converted into reactive compound by simple organic reactions. The method of attaching Polyethylene glycol to any drugs, protein, polymer or other compounds is called as Pegylation with a number of advantages and application in biological system. Pegylation improves the pharmacokinetics of protein by shielding them from the degrading proteolytic enzymes, resulting in proteins with long circulating time, better acceptability, targated drug delivery in the biological system.

Dendrimers Represent

Dendrimers represent a new class of controlled-structure polymers with nanometric dimensions. They are considered to be basic elements for large-scale synthesis of organic and inorganic nanostructures with dimensions of 1 to 100 nm, displaying unique properties. Compatible with organic structures such as DNA, they can also be fabricated to interact with metallic nanocrystals and nanotubes or to possess an encapsulation capacity. Dendrimers are three dimensional structures with a central core, internal branches and symmetrically distributed terminal groups. Mono-dispersed dendrimers are nanosized spheres carrying multiple attachment sites and a hydrophobic interior for binding and release of hydrophobic chemicals.

Nanotubes and Carbon Fullerenes

Carbon fullerenes and nanotubes are made of carbon 60 atoms and have numerous points of attachment. Nanotubes are among the most utilized particles because they offer strength and excellent electrical properties. They can be single-walled or multi-walled structures. This type of material has also been used for controlled release of testosterone in living rats.

Metallic Nanoparticles

Metallic particles include magnetic compounds, such as iron oxide gold- and silver-coated particles. Magnetic compounds are being used for targeting drugs. Gold-coated particles display excellent optical properties needed for diagnostic imaging. Silver-coated particles show outstanding anti-infection properties. There is a synergistic bactericidal action on Escherichia coli when amoxicillin and silver nanoparticles were administered together because silver nanoparticles act as a drug carrier. The cell membranes consist of hydrophobic groups like phospholipids/glycoprotein therefore silver nanoparticles but not amoxicillin (hydrophilic) are likely to approach the membrane of the target cells and facilitate the transport of amoxicillin to the cell surface.

TOXICITY OF NANOPARTICLES

Usually biodegradable nanoparticles are less toxic than nonbiodegradable ones, assuming that no toxic degradation products are involved. For example, biodegradable polyalkylcyanoacrylate nanoparticles show a reversible mild liver inflammatory reaction whereas non-biodegradable nanoparticles shows sever toxic reaction in experimental animals. When polyalkylcyanoacrylate and polystyrene nanoparticles are compared the biodegradable polyalkylcyanoacrylate nanoparticles removed from experimental animals whereas the polystyrene nanoparticles remain at relatively high levels at the end of the experiment.

At the micro-level there are indications that the chemical make-up of particles is at least as important in the distribution pattern as its size. Drug delivery nanoparticles are often designed to degrade to result in drug delivery thus resulting in relatively short exposures to the actual nano-sized particle. An extreme example is the increased bioavailability of the pH- sensitive nanoparticle delivery form for RR01 (a lipophilic drug). This is based on the nanoparticle dissolving in the upper part of the intestine and thus locally releasing the drug, but simultaneously it ensures there is only very limited and acute exposure to the nanopartic *per se*.

Importance of Shape

Shape is having immense importance in determining toxicity of nanoparticles. We can take two types of nanoparticles like nanotubes (rods) and fullerenes (spheres) to discuss how shape is influencing toxicity. Carbon nanotubes with an average diameter of 0.9 nm were shown to block a large variety of K+ -channels 2 to 3 fold more efficiently than carbon fullerenes (spheres) with an average diameter of 0.7 nm. Carbon nanotubes also increase platelet aggregation and in vivo thrombosis in experimental animals but not observed in case of fullerenes. The reason may be considered as nanotubes mimic the molecular bridges involved in platelet interactions while fullerenes do not.

Different Size

The biological system behaves differently to nanoparticles having different size. For example, cell entry mechanisms for paclitaxel in larger nanospheres appear to be different for vesicles of a few hundred nm diameter versus 1-2 microns. Micro-sized particles induced cell death mainly by necrosis where as larger nanospeheres by apoptosis. It is observed that larger nanospheres (200-300 nm) show different distribution profiles as compared to smaller nanospheres (25-50 nm) may be due to lower liver uptake of the smaller spheres.

Charge

Different experimental data shown that drug bioavailibity from nanoparticle is dependent on its zeta potential. Zeta potential is related to the surface potential of the particle and usually measured by electrophoretic mobility. Relative oral bioavailability compared to conventional drug formulations increases with increasing positive zeta potential, where decreases with decreasing in negative zeta potential. The anionic mucus layer of GI tract strongly affects cell-association opportunities of the nanoparticle and thus significantly affects in vivo oral bioavailability. On exposure to blood, nanoparticles of varying surface characteristics, size and morphology attract different opsonins as well as other plasma proteins. The adsorption patterns of nanoparticle also strongly influence the rate and site of its clearance from the vasculature For example; neutral liposomes with diameters less than 100 nm have circulation half lives up to 20 hours, whereas the anionic counter-parts have half-lives less than 1 hour. The reason may be neutral vesicles are poor activators of the complement system as compared to anionic liposomes. Interestingly, plasma protein adsorption of nanoparticle was shown to increase with increasing surface charge density, but there was no change in the detected protein species. This was found in a series of nanoparticle of different chemical compositions but with very comparable particle size, hydrophobicity and surface topography. Cellular uptake of chitosan nanoparticle is mainly depend on degree of deacetylation and hence more on zeta-potential of nanoparticle and less on the molecular weight of the polymer used for nanoparticle preparation. Cell associated chitosan nanoparticle were internalized readily compared to cell associated chitosan molecules. Despite this difference, chitosan molecules and nanoparticle demonstrated comparable toxicity against A549 cells.

Cellular Toxicity

The common mechanism of cellular toxicity appeared to be the induction of oxidative stress.

Iron Oxide Nanoparticle

Super paramagnetic iron oxide-based colloids (SPION, median particle size > 50 nm), coated with dextran (in ferumoxide) or siloxane (in ferumoxil) used intravenously as MRI contrast fluids. After injection the particles accumulate in the reticuloendothelial system (RES) of the liver and the spleen, but as RES activity is reduced or absent in liver tumours it helps in liver tumor visualization. The main disadvantage with these products in clinical practice is the risk of anaphylactic/anaphylactoid reactions. In clinical trials, anaphylactic and allergic adverse events occurred in around 0.5% of the patients who received FERIDEX I.V. It is not definite whether the anaphylactic/ anaphylactoid reactions are due to direct mediator releasing effects of iron (or dextran) or an IgE-mediated reaction.

Non-biodegradable Healthcare Nanoparticles

The quantum dots are made of a central core of inorganic semiconductor atoms covered with a monolayer of organic ligands and for biological applications this nanoparticle is then solubilised using amphiphilic polymers that shield the organic fluid and core. The polymer on the dots can be linked to specific targeting molecules. The size varies from 1nm to a maximum of 10nm. The polymer coating is biodegradable, but the core is not easily biodegradable but the toxicity information is extremely limited. PEG-phospholipid-coated dots have been injected into frog embryos and did not affect embryo development for up to 4 days. The blood clearance rates of quantum dots were significantly dependent on the coating material and size of the dots. Using four different coatings, all dots half-lives were less than 12 minutes, excepting the one coated with the longest chain PEG (5000 Da) where half-life was 70 minutes. It is concluded that such PEG-coated dots are small enough and sufficiently hydrophilic to slow down opsonisation and reticuloendothelial uptake, but large enough to avoid renal filtration. The same work demonstrated that quantum dots injected into mice were taken up by the lymph nodes and could still be observed there for more than 4 months. No long term toxicity studies or metabolism study could be found. However, cytotoxicity studies using primary hepatocytes indicated that cytotoxicity is correlated to the release of toxic cadmium ions from the core and identify the formation of reactive oxygen species as the likely mechanism of toxicity.

MEDICAL AND PHARMACEUTICAL APPLICATIONS FOR NANOMATERIALS AND NANOPARTICLES

BIOSENSORS AND BIOLABELS

Using Biosensors and Biolabels to Understand Living Cells

Nanotechnology has the potential to augment our ability to understand the fundamental working of living cells. Many potential applications for metal nanomaterials as biosensors and biolabels are under investigation. They have found use in cellular studies, enhanced spectroscopic techniques, biochips, and protein and enzyme analysis.

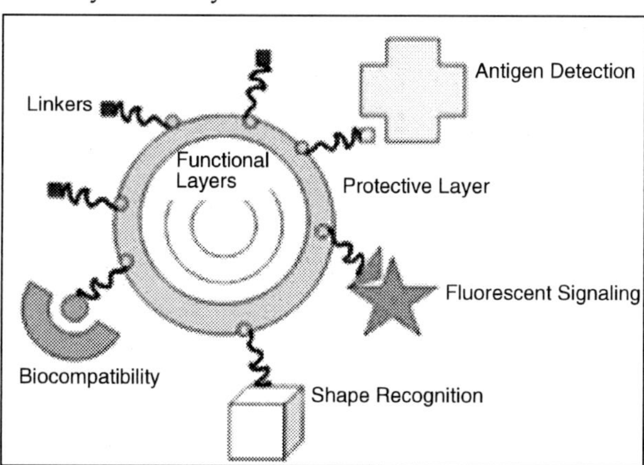

Figure: Diagram Showing how Nanotechnology Methods can Improve our Understanding of Living Cells.

Fluorescent Nanoparticles

Multi-colour labeling of both fixed and living cells with fluorescent nanoparticles conjugated with biological ligands that specifically bind against certain cellular targets enables the recording of diffusion pathways in receptor cells. Uptake of nanoparticles into the vesicular compartments around the nucleus of cells can be used to label the cells so that their pathway and fate can be followed. The nanoparticles exhibit reduced photobleaching as compared to traditional dyes and are passed on to daughter cells during cell division, therefore allowing for much longer term observation. Magnetic nanoparticles can also act as sensors for assessing how external stresses affect changes in intracellular biochemistry and gene expression.

Colloidal Gold Nanoparticles

Colloidal gold nanoparticles have been found to strongly enhance the native signals of chemical constituents in cells. Surface Enhanced Raman Spectroscopy (SERS) relies on this signal magnification and serves as a tool for ultra-sensitive monitoring of the intracellular distribution of these chemicals. Colloidal gold biofunctionalized nanomodules also have potential for use in enzyme multi-sensors. Gold nanoparticles coated with proteins have been used to detect conformation changes in the attached proteins via observation of colour changes in the solution.

DIAGNOSTICS NANOTECHNOLOGY

Improve Medical Diagnostics

Early detection of disease remains a primary goal of the medical community. Nanotechnology holds great promise for enabling the achievement of this goal. Nanoparticles in particular have exhibited tremendous potential for detecting fragments of viruses, pre-cancerous cells, disease markers, and indicators of radiation damage. Gold coatings have made it possible to use toxic cobalt nanoparticles for biomedical applications. Gold coated ferromagnetic nanoparticles tagged with HIV antibodies may be able to detect virus particles left after completion of conventional drug therapy. Metal nanoparticles in the form of dendrimers have also been functionalized with different biomolecules to detect specific proteins, antibodies, and other disease indicators. Fluorescent markers can also be attached to the dendrimers.

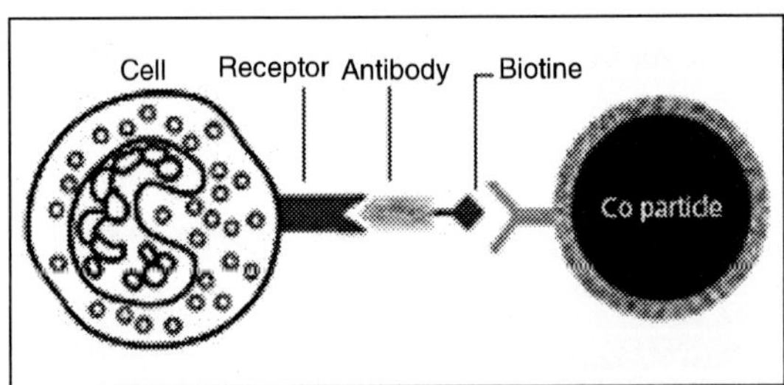

Figure: Diagram Showing how Nanoparticles can be used to Detect Diseased Cells.

Cancer Detection Using Superparamagnetic Iron Oxide

Biomolecule coated ultra small superparamagnetic iron oxide (USPIO) particles injected in the blood stream recognize target molecular markers present inside cells and induce a specific signal for detection by magnetic resonance imaging (MRI). This technology may allow for detection of individual cancer cells months or years earlier than traditional diagnostic tools, which require the presence of hundreds of cancer cells.

Biobarcode Amplification Assays

A nanoparticle-based biobarcode amplification assay (BCA) utilizes gold nanoparticles and magnetic microparticles attached to large numbers of DNA strands and antibodies for a specific disease marker. The marker binds to the nano- and microparticles forming a complex that is separated from the sample using a magnetic field. Heating the complexes releases the DNA barcodes, which emit an amplified signal due to their large numbers. This BCA technology has been applied to the detection of markers for Alzheimer's disease and is being investigated for numerous others.

TARGETED DRUG DELIVERY

Targeted Drug Delivery Methods

Targeted drug delivery systems can convey drugs more effectively and/or more conveniently, increase patient compliance, extend the product life cycle, provide product differentiation, and reduce health care costs. Drug delivery systems that rely on nanomaterials also allow for targeted delivery of compounds characterized by low oral bioavailability due to poor water solubility, permeability and/or instability and provide for longer sustained and controlled release profiles. These technologies can increase the potency of traditional small molecule drugs in addition to potentially providing a mechanism for treating previously incurable diseases.

Magnetic Nanoparticles Targeted Drug Delivery

The utilize of magnetic nanoparticles in targeted drug delivery systems is under investigation by several research groups. Therapeutic drug molecules have been immobilized on the surface of magnetic nanoparticles or nanocrystals and directed to a specific target tissue using a magnetic field gradient. The drug is released by applying a radio frequency (RF) pulse. Gold coated iron, nickel and cobalt ferromagnetic nanoparticles have been employed in this "tag and drag" approach. In hypothermal treatment, magnetic nanoparticles are directed to diseased tissue containing heat sensitive tumours. An AC magnetic field is applied such that the nanoparticles become heated, causing destruction of the cancerous cells. More effective radiation therapy for tumor treatment can also be expected using metallic nanoparticles instead of, for example, magnetite. The nanoparticles allow the application of higher dosages of radiation at the tumor while sparing normal tissue.

Other Applications for Nanomaterials

Numerous other potential applications exist in the medical and pharmaceutical field for nanomaterials from Strem. Areas currently under investigation include gene therapy, antibacterial/antimicrobial agents for burn and wound dressings, repair of damaged retinas, artificial tissues, prosthetics, enhancing signals for magnetic resonance imaging examinations, and as radio frequency controlled switching of complex biochemical processes.

DRUG DELIVERY AND NANOPARTICLES

Current years have witnessed unprecedented growth of research and applications in the area of nanoscience and nanotechnology. There is increasing optimism that nanotechnology, as applied to medicine, will bring significant advances in the diagnosis and treatment of disease. Anticipated applications in medicine include drug delivery, both in vitro and in vivo diagnostics, nutraceuticals and production of improved biocompatible materials. Engineered nanoparticles are an important tool to realise a number of these applications. It has to be recognized that not all particles used for medical purposes comply to the recently proposed and now generally accepted definition of a size ≤100 nm. However, this does not necessarily has an impact on their functionality in medical applications. The reason why these nanoparticles (NPs) are attractive for medical purposes is based on their important and unique features, such as their surface to mass ratio that is much larger than that of other particles, their quantum properties and their ability to adsorb and carry other compounds. NPs have a relatively large (functional) surface which is able to bind, adsorb and carry other compounds such as drugs, probes and proteins. However, many challenges must be overcome if the application of nanotechnology is to realise the anticipated improved understanding of the patho-physiological basis of disease, bring more sophisticated diagnostic opportunities, and yield improved therapies. Although the definition identifies nanoparticles as having dimensions below 0.1 μm or 100 nm, especially in the area of drug delivery relatively large (size >100 nm) nanoparticles may be needed for loading a sufficient amount of drug onto the particles. In addition, for drug delivery not only engineered particles may be used as carrier, but also the drug itself may be formulated at a nanoscale, and then function as its own "carrier". The composition of the engineered nanoparticles may vary. Source materials may be of biological origin like phospholipids, lipids, lactic acid, dextran, chitosan, or have more "chemical" characteristics like various polymers, carbon, silica, and metals. The interaction with cells for some of the biological components like phospholipids will be quite different compared to the non biological components such as metals like iron or cadmium. Especially in the area of engineered nanoparticles of polymer origin there is a vast area of possibilities for the chemical composition.

Although solid NPs may be used for drug targeting, when reaching the intended diseased site in the body the drug carried needs to be released. So, for drug delivery biodegradable nanoparticle formulations are needed as it is the intention to transport and release the drug in order to be effective. However, model studies to the behaviour of nanoparticles have largely been conducted with non-degradable particles. Most data concerning the biological behaviour and toxicity of particles comes from studies on inhaled nanoparticles as part of the unintended release of ultrafine or nanoparticles by combustion derived processes such as diesel exhaust particles. Research has demonstrated that exposure to these combustion derived ultrafine particles/nanoparticles is associated with a wide variety of effects including pulmonary inflammation, immune adjuvant effects and systemic effects including blood coagulation and cardiovascular effects. Since the cut-off size for both ultrafine and nanoparticles (100 nm) is the same, now both terms are used as equivalent. Based on the adverse effects of ultrafine particles as part of environmental pollution, engineered nanoparticles may be suspected of having similar adverse effects. It is the purpose of this review to use this database on combustion derived nanpoarticles (CDNP) obtained by inhalation toxicology and epidemiology and bridge the gap to engineered nanoparticles.

NANOPARTICLES, DRUG DELIVERY AND PHARMACEUTICAL DEVELOPMENT

Drug delivery and related pharmaceutical development in the context of nanomedicine should be viewed as science and technology of nanometre scale complex systems (10–1000 nm), consisting of at least two components, one of which is a pharmaceutically active ingredient, although nanoparticle formulations of the drug itself are also possible. The whole system leads to a special function related to treating, preventing or diagnosing diseases sometimes called smart-drugs or theragnostics.

The primary goals for research of nano-bio-technologies in drug delivery include:

- More specific drug targeting and delivery,
- Reduction in toxicity while maintaining therapeutic effects,
- Greater safety and biocompatibility, and
- Faster development of new safe medicines.

The main issues in the search for appropriate carriers as drug delivery systems pertain to the following topics that are basic prerequisites for design of new materials. They comprise knowledge on (i) drug incorporation and release, (ii) formulation stability and shelf life (iii) biocompatibility, (iv) biodistribution and targeting and (v) functionality. In addition, when used solely as carrier the possible adverse effects of residual material after the drug delivery should be considered as well. In this respect biodegradable nanoparticles with a limited life span as long as therapeutically needed would be optimal.

The types of chemical structures and possibilities for the preparation of nanoscale materials used as pharmaceutical carrier system (reviewed in Borm and Muller-Schulte 2006). Certainly none of the so far developed carriers fulfil all the parameters mentioned above to the full extent; the progress made in nanotechnology inter alia emerging from the progress in the polymer-chemistry, however, can provide an intriguing basis to tackle this issue in a promising way.

Table. Overview of Nanoparticles and their Applications in Life Sciences

Particle class	Materials	Application
	Chitosan	
Natural materials or derivatives	Dextrane	Drug/Gene delivery
	Gelatine	
	Alginates	
	Liposomes	
	Starch	
Dendrimers	Branched polymers	Drug delivery
Fullerenes	Carbon based	
	carriers	Photodynamics
	Polylactic acid	Drug delivery
Polymer carriers	Poly(cyano)acrylates	
	Polyethyleinemine	
	Block copolymers	
	Polycaprolactone	Drug/gene delivery
Ferrofluids	SPIONS	Imaging (MRI)
	USPIONS	
Quantum dots	Cd/Zn-selenides	Imaging
		In vitro diagnostics
Various	Silica-nanoparticles	Gene delivery
Particle class	Materials	Application
	Mixtures of above	

The aims for nanoparticle entrapment of drugs are either enhanced delivery to, or uptake by, target cells and/or a reduction in the toxicity of the free drug to non-target organs. Both situations will result in an increase of therapeutic index, the margin between the doses resulting in a therapeutic efficacy (eg, tumor cell death) and toxicity to other organ systems. For these aims, creation of long-lived and target-specific nanoparticles is needed. Chemical formulations under investigation. Most of the compounds are biodegradable polymers resulting in drug release after degradation. One of the problems in the use of particulate drug carriers including nanomaterials is the entrapment in the mononuclear phagocytic system as present in the liver and spleen. However, liver targeting of nanoparticles may be favourable when treating liver diseases like tumor metastasis or hepatitis. Surface modification with polyethylene glycol (PEG) resulted in prolonged presence in the circulation by inhibiting recognition and phagocytosis by the mononuclear phagocytic system. In addition to altering the distribution the PEG modification also reduced in vitro toxicity when gold nanorods were modified using PEG. Coating of NP may also be needed to prevent agglomeration. Several coatings can be used to prevent agglomeration and keeping the particles in colloidal suspension including various polymers like polyethylene glycol (PEG), poly(vinylpyrrolidone) (PVP) etc, natural polymers like dextran, chitosan, pullulan etc, and surfactants like sodium oleate, dodecylamine etc.

Table: Chemicals under Investigation for Drug Delivery

Albumin	Damascelli et al 2003
Cetyl alcohol/polysorbate	Koziara et al 2004
Chitosan	Dyer et al 2002; Huang et al 2004
Gelatin	Cascone et al 2002
Gold	Hainfield et al 2004; Paciotti et al 2004
Hydrogels	Gupta and Gupta 2004
Magnetic iron oxide	Gupta and Gupta 2005
Methoxy poly(ethylene glycol)/poly(ε-caprolactone)	Kim et al 2003
Polyalkylcyanoacrylate composites	Alyautdin et al 1997; Kreuter et al 2003
Poly(D,L-lactic-co-glycolic)acid (PLGA)	Panyam et al 2002; Weissenbrock et al 2004
Solid lipid formulations	Muller et al 2000; Wissing et al 2004

NP size can influence the NP distribution as was demonstrated for lipid vesicles for which a lower liver uptake was found for the smaller vesicles (200/300 nm versus 25/50 nm). Even small size differences may be of influence for the actual distribution and thus bioavailability. For liposomes with sizes >100 nm the clearance rate by the mononuclear phagocytic system increased with increasing size, while for sizes below 100 nm charge was more important. However, not all particles with sizes below 100 nm will behave similarly and composition will be important as well. Analogous to earlier findings on asbestiform and mineral fibres, the actual size and shape of nanomaterials will be of importance.

Besides degradation physical means such as heating and light may be used to provoke the therapeutic effect (cell death) or for local drug release, respectively. Thermosensitive nanoparticles may be used for selective release of the content after specific localization. An example of this principle. For doxorubicin an enhanced cytotoxicity was observed in vitro at 42 °C compared to 37 °C using copolymers of polyethylene glycol (PEG) and poly-L-lactide

(PLLA). In addition, the release of photosensitizers from nanoformulations by light, so called photodynamic therapy, was able to induce cytotoxicity as demonstrated for PLGA nanoparticles containing zinc(II) phthalocyanine and indocyanine green.

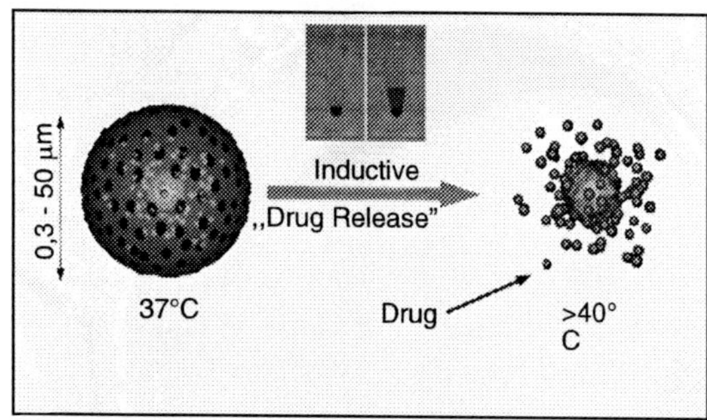

Figure: Graph Illustrating Contactless Controllable Drug Carrying System based on Thermosensitive Magnetic Nano- and Micro Particles. The Insert shows the Application of the System with Rhodamine B Encapsulated Beads that is Released after heating up to 45 °C.

Utilize of NP Formulations in Drug Delivery

One of the major challenges in drug delivery is to get the drug at the place it is needed in the body thereby avoiding potential side effects to non diseased organs. This is especially challenging in cancer treatment where the tumor may be localized as distinct metastases in various organs. The non restricted (cyto)toxicity of chemotherapeutics thus limits the full use of their therapeutic potential. Local drug delivery or drug targeting results in increased local drug concentrations and provides strategies for more specific therapy. Nanoparticles have specific particles as tools to enable these strategies. These include benefits such as their small size which allows penetration of cell membranes, binding and stabilization of proteins, and lysosomal escape after endocytosis.

The entrapment of chemotherapeutics in nanosized formulations like liposomes has been already subject of study for considerable time. Liposomes as nanosized phospholipid "fatty" structures have the advantage of being small, flexible and biocompatible thus being able to pass along the smallest arterioles and endothelial fenestrations without causing clotting. Now also other materials, including various (co-)polymers and dendrimers at the nanosize range have become available to alter the distribution of encapsulated or attached drugs.

One of the therapeutics under intensive study is paclitaxel (taxol). For paclitaxel the nanoparticle formulation resulted in enhanced cytotoxicity for tumor cells in vitro, and at the same time an increased sustainable therapeutic efficacy in an in vivo animal model. The paclitaxel was encapsulated in vitamin

E TPGS-emulsified poly (D,L-lactic-co-glycolic acid) (PLGA) nanoparticles, and this system resulted in a higher and prolonged level above the effective concentration in vivo, reflected in an increased area under the curve (AUC).

Separately from size the surface chemistry of particles is of crucial importance in particle uptake, distribution and effects. This was shown extensively with acute and chronic models of surface modified micro quartzes. Quartz which was coated with PVNO-polymer was taken op by macrophages without toxicity and showed no genotoxicity in epithelial cells or acute and chronic inflammation. On the other hand naïve quartz caused these effects to a large extent. An altered body distribution was demonstrated for two types of polymer particles. Only PMMA (about 1.4 μm and about 6.4 μm) particles but not PS (about 1.2 μm, 5.2 μm and 12.5 μm) particles could be recovered form the spleen after intraperitoneal administration. Whether a similar situation exists for nanoparticles is unknown, but studies with surface modified polystyrene particles do suggest different effects on blood coagulation, mitochondrial ROS formation and cellular oxidative burst. The coating of nanoparticles with polyethylene glycol (PEG) increases the time in circulation for the nanoparticles.

The aims for nanoparticle entrapment of drugs are either enhanced delivery to, or uptake by, target cells and/or a reduction in the toxicity of the free drug to non-target organs. For these aims, creation of long-lived and target-specific nanoparticles is needed. One of the problems is the entrapment of nanoparticles in the mononuclear phagocytic system as present in the liver and spleen. However, liver targeting of nanoparticles may be favourable when treating liver diseases like tumor metastasis or hepatitis. Also oligonucleotides for modification of gene expression were demonstrated to migrate into the liver when bound to biodegradable polyalkylcyanoacrylate nanoparticles. Surface modification with PEG resulted in prolonged presence in the circulation by inhibiting recognition and phagocytosis by the mononuclear phagocytic system.

Besides reduction of therapeutic efficacy, liver entrapment may also have an adverse effect on liver function. For cyanoacrylate and polystyrene nanoparticles (about 214 nm and about 128 nm, respectively) transient liver alterations were observed after acute and chronic intravenous administration. Inflammatory responses were characterized by secretion of acute phase protein α1-acid glycoprotein by hepatocytes. In addition, antioxidant defences of hepatocytes were depleted probably as a result of local release of oxidative species.

Although nanoformulation is aimed at enhancing drug delivery without loss of drug activity, in a study comparing insulin-chitosan nanoparticles to chitosan solution and chitosan powder formulations the insulin-chitosan nanoparticles were less effective in terms of bioavailability and lowering blood glucose level in both a rat and sheep model.

Cellular and Intracellular Targets

Intended for drug delivery not only organ or cellular targeting is of importance but also the fate of the nanoparticles within the cells. Particles generally end intracellularly in endosomes or lysosomes followed by degradation. For activity of the encapsulated drugs release into the cytosol is needed. However, for nanoparticles of about 20 nm also cellular uptake without contribution by endocytic mechanisms was demonstrated. Chemical characteristics such as surface charge may also determine the fate of nanoparticles in cells. Surface functionalization of gold nanoparticles with PEG resulted in efficient internalization in endosomes and cytosol, and localized in the nuclear region. Poly(DL-lactide-co-glycolide) nanoparticles were found to be ingested by cells by endocytosis. The escape from these endosomes into the cellular cytoplasm was suggested to be caused by a change in surface charge form negative to positive of the PLGA nanoparticles resulting in cytoplasmic delivery of the incorporated drugs. The hypothesis that the positive surface charge influenced the escape of the endosomes was supported by data obtained with negatively charged polystyrene nanoparticles which did not reach the cytosol but remained in the endosomal compartment of the smooth muscle cells used in this study.

Specific targeting to retinal pigment epithelium cells in the eye is possible. Very small quantum dots (<10 nm) have been used for specific targeting of peptide coated dots to the vasculature of lungs and tumours. In addition, polymer shells on the quantum dots might be linked to targeting molecules. For example quantum dots cores can be coated with hydrophilic polyethylene glycol (PEG) to increase the half life time. However, also uptake by lymph nodes was demonstrated in which the quantum dots could be observed up until 4 months after administration, so accumulation seems likely. PEG coating abrogated uptake by the reticuloendothelial system of liver and spleen. In contrast about 40–50 nm magnetic nanoparticles coated with PEG were quite well taken up by endocytosis.

Surface modifications of nanoparticles offer possibilities for medical applications like drug targeting in terms of cellular binding, uptake and intracellular transport. Carbohydrate binding ligands on the surface of biodegradable and biocompatible poly(D,L-lactic-co-glycolide)acid (PLGA) nanospheres were found to increase cellular binding. Such increased adherence may lead to an enhanced activity of the drug presented as or incorporated in nanoparticles. Coupling specific proteins such as antibodies to the nanoparticle surface may enable a more specific immunologically directed targeting of the particles.

The Brain–the Ultimate Target for Drug Delivery

As of several perspectives the brain is a challenging organ for drug delivery. First, the incidence of degenerative diseases in the brain will increase with the aging population. Secondly, the blood brain barrier (BBB) is well-known as the best gatekeeper in the body towards exogenous substances. Generally pharmaceuticals including most small molecules do not cross the BBB. The endothelial barrier is specifically tight at the interface with the brain astrocytes and can in normal conditions only be passed using endogeneous BBB transporters resulting in carrier mediated transport, active efflux transport and/or receptor mediated transport. However the barrier properties may be compromised intentionally or unintentionally by drug treatment allowing passage of nanoparticles. The delivery of drugs by nanocarrier was recently reviewed.

Passage of the BBB was suggested to be possible by the toxic effect of nanoparticles (about 200 nm) on cerebral endothelial cells, although for similar nanoparticles (about 300 nm) this was contradicted in another study. In addition this effect was not found for a different type of nanoparticles. Physical association of the drug to the nanoparticles was necessary for drug delivery to occur into the brain. When nanoparticles with different surface characteristics were evaluated, neutral nanoparticles and low concentrations of anionic nanoparticles were found to have no effect on BBB integrity, whereas high concentrations of anionic nanoparticles and cationic nanoparticles were toxic for the BBB. The extent of brain uptake of anionic nanoparticles at lower concentrations was superior to neutral or cationic formulations at the same concentrations. So, nanoparticle surface charges must be considered for toxicity and brain distribution profiles. Especially coating of the nanoparticles with the polysorbate (Tween) surfactants resulted in transport of drugs across the blood brain barrier. The mechanism for transport was suggested to be endocytosis via the Low Density Lipoprotein (LDL) receptor of the endothelial cells after adsorption of lipoproteins form blood plasma to the nanoparticles. Additional investigations revealed the role of apolipoprotein-E for transport of drugs across the BBB while apolipoprotein-E variants that did not recognized lipoprotein receptors failed in transporting the drug across the BBB. It was suggested that the recognition and interaction with lipoprotein receptors on brain capillary endothelial cells was responsible for the brain uptake of the drug.

Passage of the BBB may also be achieved by masking certain drug characteristics preventing or limiting binding to cellular efflux systems like p-glycoprotein, a cellular transporter associated with drug removal from cells. P-glycoprotein is one of the ATP dependent efflux transporters that has an important physiological role in limiting drug entry into the brain. In addition, p-glycoprotein also designated the multidrug resistance protein may be highly expressed in drug resistant tumor cells. Surfactant coated poly(butyl) cyanoacrylate nanoparticles have been used to deliver drugs to the CNS The effect of entrapment of a cytotoxic drug paclitaxel (PX) in cetyl alcohol/ polysorbate nanoparticles (PX NP) was evaluated in an in situ rat brain perfusion model. The results suggest that entrapment of paclitaxel in nanoparticles significantly increases the brain drug uptake and its toxicity towards p-glycoprotein expressing tumor cells (p-glycoprotein is an efflux transporter associated with drug removal from the cells). It was hypothesized that PX nanoparticles limit paclitaxel binding to p-glycoprotein and subsequent efflux from the cells, which consequently would lead to higher brain and tumor cell levels.

Other routes for reaching the brain, circumventing the BBB, may be via migration along the olfactory or trigeminal nerve endings after deposition on the olfactory mucosa in the nasal region. Translocation of ultrafine ^{13}C particles (35 nm) was detected by using this isotope measurement in the brain olfactory bulb after inhalation exposure. Also other solid NP like manganese oxide was shown to translocate to the brain by the olfactory route, based on measurements of manganese in different parts of the brain. In order to increase the specific uptake via the inhalation route nanoparticles have been functionalized by conjugation with bioactive ligands-lectins to the surface of poly (ethylene glycol)- poly (lactid acid) (PEG-PLA) nanoparticles. Wheat germ agglutinin (WGA) was used which binds to N-acetyl-D-glucosamine and sialic acid both of which are abundantly present in the nasal cavity. There was a twofold increase in the brain uptake of such functionalized NP. However, it needs to be stated that both passage of the BBB and the olfactory route only account for up to 2% nanoparticles uptake, and its efficacy with regard to drug delivery needs to make considerable increments before use.

CURRENT DATA ON THE TOXICOLOGY ENGINEERED NANOPARTICLES

Few years a number of papers have described the toxicology of newly engineered nanomaterials, including fullerenes, carbon nanotubes, quantum dots and have illustrated that apart from size and surface area, many more parameters describing the material (surface) properties have to be included. In a recent report Costigan reviewed the evidence for toxicity of NPs used in healthcare products. Her conclusions again stressed the limited availability of toxicity data of the NPs in use.

Carbon Nanotubes

Carbon nanotubes are long carbon-based tubes that can be either single- or multiwalled and have the potential to act as biopersistent fibres. Nanotubes have aspect ratios >100, with lengths of several mm and diameters of 0.7 to 1.5 nm for single-walled carbon nanotubes (SWCNT) and 2 to 50 nm for multiwalled carbon nanotubes (MWCNT). In vitro incubation of keratinocytes and bronchial epithelial cells with high doses of SWCNT results in ROS generation, lipid peroxidation, oxidative stress, mitochondrial dysfunction, and changes in cell morphology. Recent studies with carbon derived nanomaterials showed that platelet aggregation was induced by both single and multi-wall carbon nanotubes, but not by the C60-fullerenes that are used as building blocks for these CNT. MWCNT also elicit pro-inflammatory effects in keratinocytes. Several studies using intratracheal instillation of high doses of nanotubes in rodents demonstrated chronic lung inflammation, including foreign-body granuloma formation and interstitial fibrosis. In two in vivo studies SWCNTs were demonstrated to induce lung granulomas after intratracheal administration indicating that these nanotubes can not be classified as a new form of graphite on material safety data sheets. On a dose per mass basis the nanotubes were more toxic than quartz particles well known for their lung toxicity. Carbon black, carbonyl iron and graphite produced no significant adverse effects. These studies also reveal the tendency of the nonphysiologic administration route and the unrealistic high doses to lead to asphyxiation through nanotube clumping in the airways. Although it has been suggested that the granulomatous inflammation could be a biopersistent fibre effect, the high dose of the aggregated nanotubes and the presence of metal impurities (eg, Fe) could account for artificial toxicity.

Potential Fullerenes

Fullerenes are being explored as potential new antimicrobial agents in view of their potency for induction of reactive oxygen species after photoexcitation. However, this may have an impact on microbial communities if they are released into the environment via effluents. Therefore, various studies with fullerenes have been published with regard to the ecotoxicity of these important building blocks in nanomaterials. Tests with un-coated, water soluble, colloidal fullerenes (nC60) show that the 48-hour LC50 in Daphnia magna varied form 460 to 800 ppb, using standard EPA protocols. However, for sonicated C-60 fullerenes the LC50 was one order of magnitiude higher with 7.9 ppm. In largemouth brass, although no mortality was seen, lipid peroxidation was seen in the brain and glutathione depletion in the gill after exposure to 0.5 ppm nC60 for 48 hours. There are several hypotheses as to how lipid damage may have occurred in the brain, including direct redox activity by fullerenes reaching the brain via circulation or axonal translocation and dissolving into the lipid-rich brain tissue, oxygen radical production by microglia, or production of reactive fullerene metabolites by cytochrome P450 metabolism.

Natural Dendrimers

Their specific nature dendrimers are specifically suited for drug delivery purposes. Although their small size (up to 10 nm) limits extensive drug incorporation into the dendrimers, their dendritic nature and branching allows for drug loading onto the outside surfaces of the polymeric structure. Functionalization of the surface with specific antibodies may further enhance potential targeting. Apart from application in drug-delivery, dendrimers are being investigated for many other uses including bacterial cell killing, as gene transfer agents and trans-membrane transport. Little published data is available on the toxicity of this class of particles. A recent review on this topic concluded that it will only ever be possible to designate a dendrimers as "safe" when related to a specific application. The so far limited clinical experience with dendrimers makes it impossible to designate any particular chemistry intrinsically "safe" or "toxic".

Quantum Dots

Quantum dots are a heterogeneous group of nanoparticles. Quantum dot absorption, distribution, metabolism and excretion, and therefore also quantum dot toxicity, depend on multiple factors derived from both inherent physicochemical properties and environmental conditions. Quantum dots may vary in size ranges from 2.5 up to 100 nm, depending on coating thickness. Studies specifically performed to investigate quantum dot toxicity are few. In vitro studies have indicated that quantum dots may be toxic of which some toxicity could be attributed to the surface coating. Choi et al demonstrated that quantum dot toxicity was reduced after surface modification with N-acetylcysteine, while the non modified cadmium telluride quantum dots induced lipid peroxidation in the cells. Lovric, Cho et al showed "naked" quantum dots to be cytotoxic by induction of reactive oxygen species resulting in damage to plasma membranes, mitochondria and nucleus. As it is the bioactive coating which allows the use of quantum dots for specific targeting to cells and/or cell organelles, attention is warranted in using the surface molecules in terms of induction of toxic effects. However, also the quantum dot core material has an effect on the toxic potential of the quantum dots as for cadmium containing quantum dots the toxicity was suggested to be due to release of highly toxic free Cd^{2+} ions. For quantum dots composed of cadmiun/telluride cellular toxicity was found but not for cadmium selenium/ zinc sulfate quantum dots. On the other hand Hardman also reported on studies demonstrating a lack of both in vitro and in vivo toxicity. However, before there can be a responsible development of quantum dots with minimal risks more information on toxicological risks needs to be provided.

Gold Nanoparticles/nanoshells

However, such evaluations did not consider nanoformulations of gold. Metallic colloidal gold nanoparticles are widely used, can be synthesized in different forms (rods, dots), are commercially available in various size ranges and can be detected at low concentrations. Cells can take up gold nanoparticles without cytotoxic effects. For biomedical applications, they are used as potential carriers for drug delivery, imaging molecules and even genes, and for the development of novel cancer therapy products. For gold nanorods the cytotoxicity could be attributed to the presence of the stabilizer CTAB of which even residual presence after washing resulted in considerable cytotoxicity. PEG-modified gold nanorods with removing the excess CTAB did not show cytotoxicity. In an acute oral toxicity study no signs of gross toxicity or adverse effects were noted when a nanogold suspension (nanoparticle diameter ca. 50 nm) was evaluated, the single dose for acute oral LD_{50} being greater than 5000 mg/kg body weight.

Gold solutions are also used to prepare nanoshells composed of gold and copper, or gold and silver to function as contrast agents in Magnetic Resonance Imaging (RMI), and gold-silica for photothermal ablation of tumor cells. In vitro the non targeted nanoshells did not show cytotoxicity for the tumor cells, whereas after binding to the tumor cells cell death could be obtained after laser activation. Also in vivo positive results were obtained with photothermal ablation therapy in a mouse model for colon carcinoma after intraveneous administration of PEG coated gold nanoshells of approximately 130 nm.

Silica Nanoparticles

For silica nanoparticles both in vitro toxic and non toxic responses were observed. Both 15 nm and 46 nm silica nanoparticles showed similar dose dependent cytotoxicity in vitro. There was an increase in toxicity both at increasing doses and at increasing exposure time (24, 48, and 72 h). SiO_2 exposure resulted in an increased ROS levels and reduced glutathione levels indicating an increase in oxidative stress. Also Chang et al found silica nanoparticles to be toxic at high dosages as shown by a reduction in cell viability/cell proliferation and by lactate dehydrogenase (LDH) release from the cells indicating membrane damage. Cells with a long doubling time were more susceptible for the cytotoxic effects of the silica nanoparticles than cells with short doubling times. In another study only at concentrations above 0.1 mg/ml a significant reduction in cell viability was observed. In addition, an alveolar macrophage cell line (MHS) was found to be more susceptible for nanaoparticle induced cytotoxicity than a lung epithelial cell line (A549) which was suggested to be due to the phagocytic properties of the macrophage cell line. Cell death was probably not caused by apoptosis. In contrast for cationic silica nanoparticles using amino-hexyl-amino-propyltrimethoxysilane as a surface modification low or no cell toxicity was observed.

TOXICOLOGICAL HAZARDS OF NANOPARTICLES

GENERAL CONCEPTS OF NANOTECHNOLOGY

To use the potential of Nanotechnology in Nanomedicine, full attention is needed to safety and toxicological issues. For pharmaceuticals specific drug delivery formulations may be used to increase the so called therapeutic ratio or index being the margin between the dose needed for clinical efficacy and the dose inducing adverse side effects (toxicity). However, also for these specific formulations a toxicological evaluation is needed. This is particularly true for the applications of nanoparticles for drug delivery. In these applications particles are brought intentionally into the human body and environment, and some of these new applications are envisaged an important improvement of health care. Opinions started to divert when toxicologists claimed that new science, methods and protocols are needed.

However, the need for this is now underlined by several expert reports and more importantly by the following concepts:

- Nanomaterials are developed for their unique (surface) properties in comparison to bulk materials. Since surface is the contact layer with the body tissue, and a crucial determinant of particle response, these unique properties need to be investigated from a toxicological standpoint. When nanoparticles are used for their unique reactive characteristics it may be expected that these same characteristics also have an impact on the toxicity of such particles. Although current tests and procedures in drug and device evaluation may be appropriate to detect many risks associated with the use of these nanoparticles, it cannot be assumed that these assays will detect all potential risks. So, additional assays may be needed. This may differ depending on the type of particles used, *i.e.*, biological versus non-biological origin.
- Nanoparticles are attributed qualitatively different physico-chemical characteristics from micron-sized particles, which may result in changed body distribution, passage of the blood brain barrier, and triggering of blood coagulation pathways. In view of these characteristics specific emphasis should be on investigations in (pharmaco)kinetics and distribution studies of nanoparticles. What is currently lacking is a basic understanding of the biological behaviour of nanoparticles in terms of distribution in vivo both at the organ and cellular level.
- Effects of combustion derived nanoparticles in environmentally exposed populations mainly occur in diseased individuals. Typical pre-clinical screening is almost always done in healthy animals and volunteers and risks of particles may therefore be detected at a very late stage.

All would depend on the types of assays used in the preclinical evaluation, which should be considered in the light of the use of the final products. In addition, one cannot rely on the toxicological profile of the bulk material when that material is used in a nanoformulation. What is clear is that the safety evaluation and the risk benefit analysis need to be performed on a case by case basis. The use of nanoparticles as drug carrier may reduce the toxicity of the incorporated drug. In general the toxicity of the whole formulation is investigated while results of the nanoparticles itself are not described. So, discrimination between drug and nanoparticle toxicity cannot be made. So, there should be a specific emphasis on the toxicity of the "empty" non-drug loaded particles. This is especially important when slowly or non degradable particles are used for drug delivery which may show persistence and accumulation on the site of the drug delivery, eventually resulting in chronic inflammatory reactions.

Evidence for Nanoparticle Toxicity

The toxicity of nanoparticles has originated from inhalation toxicology including the PM_{10} literature (particulate matter with a size below 10 mm), where the 'NP hypothesis' has proved to be a powerful drive for research. An overview of particle terminology in relation to ambient effects. Therefore it relevant to discuss this evidence in the expectation that it will shed light on the toxicity of engineered NPs. The idea that combustion-derived NPs are an important component that drives the adverse effects of environmental particulate air pollution or PM_{10} comes from several sources:

Table. Various Denominations of Particles in Inhalation Toxicology and Drug Delivery in Relation to their Source (ambient, bulk, engineered)

Particle type	Description
PM_{10}, $PM_{2.5}$	Particle mass fraction in ambient air with a mean diameter of 10 or 2.5 μm respectively. Basis of current standards for ambient particles in Europe and USA
Coarse particles	The mass fraction of PM_{10}, which is bigger than 2.5 μm
Ultrafine particles ($PM_{0.1}$)	The fraction of PM_{10} with a size cut-off at 0.1μm. Contains primary particles and agglomerates smaller than 100 nm
PSP	Poorly soluble particles with low specific toxicity. Maybe be fine or ultrafine. Terminology used in relation to bulk synthetic particles. Examples TiO_2, carbon blacks, Amorphous silica, Iron oxides (Fe_2O_3), Zinc oxides (ZnO)
CDNP	Combustion derived nanoparticles, such as diesel exhaust particles (DEP)
DEP	Diesel exhaust particles

- Much of the mass of PM_{10} is considered to be nontoxic and so there has arisen the idea that there is a component(s) of PM_{10} that actually drives the pro-inflammatory effects and combustion-derived NP seems a likely candidate.
- Nanoparticles are the dominant particle type by number suggesting that they may be important and their small size means that they

have a large surface area per unit mass. Particle toxicology suggests that, for toxic particles generally, more particle surface equals to more toxicity.

- Substantial toxicological data and limited data from epidemiological sources support the contention that NPs in PM_{10} are important drivers of adverse effects.

The unfavourable health effects of particulate matter (PM) are measurable as exacerbations of respiratory disease and deaths as well as hospitalizations and deaths from respiratory and cardiovascular disease. Inflammation is the common factor that binds together these adverse effects and the ability of NPs to cause inflammation can be seen as an important property. It is not clear what effects of NPs have pulmonary inflammation as a prerequisite and what effects could potentially be driven by exposures below those causing inflammation. There is also the potential for pulmonary inflammation to result in changes in membrane permeability that in turn may impact the potential for particles to distribute beyond the lung. Some NPs may have the extra potential of affecting cardiovascular disease directly. Vascular function was impaired after inhalation of diesel exhaust particles. However, data to date are limited and not all studies of nanoparticles have shown significant translocation from lung to the blood. In some studies translocation has been rather minimal. Understanding clearance kinetics of inhaled ambient air nanoparticles will also be important in understanding their potential for adverse effects.

The current example in particle toxicology is that ultrafine ambient air particles have the potential of affecting cardiovascular disease both indirectly via pulmonary inflammation and directly through particle distribution. Although important, this property of redistribution has yet to be demonstrated for NPs present in real PM_{10}. It should be noted that there are several mechanisms whereby NPs could lead to inflammatory effects, as is the case for larger particles. These mechanisms are either based on the large surface area of particle core or on soluble components released by the NPs. In addition various chemicals including those of biological origin like endotoxin may be adsorbed onto the NP and released. Several toxicological studies support the contention that NPs in PM_{10} could drive inflammatory effects. There are a number of components of PM_{10} that contribute to the mass but have little toxicity – these include salts such as sulfates, chlorides and ammonium salts and nitrates, but also wind-blown or crustal dust. In fact within PM_{10} there are only few components that toxicologists would identify as likely mediators of adverse effects – *i.e.*, particle surfaces, organics, metals and endotoxin (in some PM_{10} samples). In fact, a large surface area, organics and metals are all characteristic of combustion–derived particles and so these have attracted considerable toxicological attention. However, it is difficult to untangle, in a combustion particle sample, the relative roles of surface, organics and metals, although this has been most attempted in vitro. The aggregation of multiple chemical species including biological compounds like endotoxin limits the extrapolation of the results on the toxicological effects of such particles.

Toxicological Effects of Nanoparticles

The NPs exert some very special properties that are very relevant in the further design of toxicity testing of engineered nanomaterials. An overview of most striking effects of (nano) particles that have been observed over the last decades, along with the particle type that have been tested in this response. Several effects are just quantitatively different from fine particles. In this case nanoparticles may cause the same effects as 'traditional' particles (*e.g.*, inflammation, lung cancer) but they may be more potent because of their greater surface area.

Table. Toxicity of Engineered and Combustion (nano) Particles as Illustrated by their most Unique Adverse Effects in Vivo and in Vitro

Description of finding, in vivo	Particle types
NPs cause pulmonary inflammation in the rat	All PSP
Later studies show that inflammation is mediated by surface area dose	SWCNT, MWCNT
NPs cause more lung tumours than fine particles in rat chronic studies. Effect is surface area mediated	PSP only
NPs cause progression of plague formation (ApoE -/- mice)	SWCNT, PM2.5
NPs affect immune response to common allergens	Polystyrene, CB, DEP
NsP can have access to systemic circulation upon inhalation and instillation surface coating	Specific NP, dependent on
Description of finding, in vitro	
NPs cause oxidative stress in vivo and in vitro, by inflammatory action and generation of surface radicals	PSP, NP general, CNT
NPs inhibit macrophage phagocytosis, mobility and killing	CB, TiO_2
NPs cause platelet aggregation	PM, SWCNT, fullerenes, latex-COOH
surface	
NPs exposure adversely affects cardiac function and vascular homeostasis	PM, SWCNT
NPs interfere with Ca-transport and cause increased binding of pro-inflammatory transcription factor NF-kB	CB (< 100 nm), ROFA, PM2.5
NPs can affect mitochondrial function	Ambient NP,
NPs can translocate to the brain from the nose	MnO_2, Au, carbon
NPs do affect rolling in hepatic tissue	CB

However, nanoparticles could also cause new types of effects not previously seen with larger particles (*e.g.*, mitochondrial damage, uptake through olfactory epithelium, platelet aggregation, cardiovascular effects). These effects depicted, clearly need a new way of handling their toxicology. In addition, epidemiological evidence suggests that these effects occur predominantly in subjects that have an impaired health. This finding should be considered in developing toxicological testing models.

Effects on Blood and Cardiovascular System

The ligand coated engineered nanoparticles are being explored and used as agents for molecular imaging or drug delivery tools. This has led to a considerable understanding of particle properties that can affect penetration in tissue without affecting tissue function. Cationic NPs, including gold and polystyrene have been shown to cause hemolysis and blood clotting, while usually anionic particles are quite nontoxic. This conceptual understanding maybe used to prevent potential effects of unintended NP exposure. Similarly, drug loaded nanoparticles have been used to prolong half-life or reduce side-effects and have shown which particle properties need to be modified to allow delivery, while being biocompatible.

On the other hand, one is trying to find explanations for the increased risk of patients with cardiovascular diseases upon exposure to PM and/or traffic. Several toxicological studies have demonstrated that combustion and model NPs can gain access to the blood following inhalation or instillation and can enhance experimental thrombosis but it is not clear whether this was an effect of pulmonary inflammation or particles translocated to the blood. High exposures to DEP by inhalation caused altered heart rate in hypertensive rats interpreted as a direct effect of DEP on the pacemaker activity of the heart. Inflammation in distal sites has long been associated with destabilization of atheromatous plaques and both instillation and inhalation of PM cause morphological evidence of atheromatous plaque increase and destabilization in rabbits and mice. Ultrafine carbon black instilled into the blood has been reported to induce platelet accumulation in the hepatic microvasculature of healthy mice in association with prothrombotic changes on the endothelial surface of the hepatic microvessels. Recent studies with carbon derived nanomaterials showed that platelet aggregation was induced by both single and multi-wall carbon nanotubes, but not by the C60-fullerenes that are used as building blocks for these CNT. These data show that not all nanomaterials act similar in this test, and that surface area is not the only factor playing a role here. The data also corroborate the earlier concept developed in medicine that mainly cationic species have an effect on blood clotting. Interestingly, this is the first study that allows bridging of data, since also a real life PM10 sample (SRM1648) was included in the test-series. Actually the PM sample showed a lower effect compared to the carbon nanotubes.

Uptake and Effects of Nanoparticles in the Brain

Nanoparticles can get access to the brain by two different mechanisms, *i.e.*, (1) transsynaptic transport after inhalation through the olfactory epithelium, and (2) uptake through the blood-brain barrier. The first pathway has been studied primarily with model particles such as carbon, Au and MnO_2 in experimental inhalation models in rats. The second pathway has been the result of extensive research and particle surface manipulation in drug delivery. The latter studies suggest that the physiological barrier may limit the distribution of some proteins and viral particles after transvascular delivery to the brain, suggesting that the healthy BBB contains defence mechanisms protecting it from blood borne nanoparticle exposure. When nanoparticles with different surface characteristics were evaluated, neutral nanoparticles and low concentrations of anionic nanoparticles were found to have no effect on BBB integrity, whereas high concentrations of anionic nanoparticles and cationic nanoparticles were toxic for the BBB. Nanoparticles have been shown to induce the production of reactive oxygen species and oxidative stress and this has been confirmed in the brain after inhalation of MnO_2 nanoparticles. Oxidative stress has been implicated in the pathogenesis of neurodegenerative diseases such as Parkinson's and Alzheimer's diseases. Evidence for the involvement of ambient air nanoparticles in these effects is presented by studies in biopsies from city dwellers. Alzheimer's like pathology was demonstrated in brain sections by increased markers of inflammation and AB42-accumulation in frontal cortex and hippocampus in association with the presence of nanoparticles. Also inhalation exposure of BALB/c mice to particulate matter showed activation of pro-inflammatory cytokines in the brain. Whether this is due to the fraction of combustion nanoparticles remains to be investigated.

NANOMATERIALS IN MEDICINE: NEEDS

Even though there is a considerable amount of data on the toxicity of NPs, this data is mainly based on a small panel of NPs (combustion derived NPs, TiO_2, CB) and the assumption that a lot of effects by particulate matter are driven by the ultrafine particle fraction in it. In most studies the nanoparticles were used as a model for ambient air particle toxicity. One of the more general conclusions is that indeed there is a clear tendency for very small (nano) particles to be more toxic than larger particles with the same chemical composition.

For nanoformulations used in drug delivery the focus in most papers is mainly on obtained reduction of toxicity of the incorporated drug, whereas the possible toxicity of the carrier used is not considered. Especially possible residues of such a treatment may harbor potential local and/or systemic toxic responses. Meant for medical applications certain routine assays need to be performed which will detect a number of potential hazards. However, it can be anticipated that not all hazards are at this moment known for the use of nanoparticles. In a recent report Costigan reviewed the evidence for toxicity of NPs used in healthcare products. Her conclusions again stressed the limited availability of toxicity data of the NPs in use. However, in NPs for healthcare products most if not all mechanisms of toxicity could be identified by conventional hazard identification testing as currently required to comply with the regulations for healthcare products. Costigan identified four possible mechanisms of NP toxicity, being chemical toxicity of one of the constituents with the same mode of action as the bulk chemical, toxicity due to degradation products, toxicity due to endocytosis of the NPs, and membrane lysis due to the NPs possibly via chemical toxicity. Although hazard identification is the general approach for safety evaluation of healthcare products, it is recommended to add testing driven by the anticipated application and classification by risk. Some engineered NPs, which get airborne will pose inhalation hazards, while cosmetics with NPs provide dermal exposures. For parenteral use interactions with blood components, systemic distribution and kinetics are of importance, when engineered NPs are being used as devices to target drugs to specific tissues, to increase their biological half time, or for imaging purposes. Each nanoparticle formulation should be tested on a case by case basis in the requisite ways focusing on their portal of entry. In this respect also the potential adverse (toxic) effects of empty particles should be considered.

In developing testing procedures and protocols a number of basic issues need to be considered:

- Which effects are specific for nanomaterials, and which effects are merely stronger? Nanoparticles may cause the same effects as 'traditional' particles (eg, inflammation, lung cancer) but they may be more potent because of their greater surface area. Nanoparticles could also cause new types of effects not previously seen with larger particles or bulk chemicals.
- Can we extrapolate available data and concepts? The epidemiological evidence on ultrafine particles has revealed several effects, mechanisms of action and susceptible groups upon inhalation of ultrafine particles. Whether these concepts can be used for nanoparticles released from manufactured nanomaterials is yet unknown.
- Is our current regulation robust enough to handle risks of nanomaterials? We deal with a growing set of materials of which the properties are largely unknown and for which current testing procedures and legislation might produce false negatives and/or false positives. The central question here is whether current testing and classification protocols are appropriate or sufficient. These will detect certain toxic effects as demonstrated by the studies already published. However, it can be anticipated that not all hazards will be detected and additional specific testing may be needed.

Nanotechnology also promotes convergence of technologies, and for example similar materials may be applied in the automotive and the life sciences sector. To stimulate production and marketing of safe nanomaterials exchange of data between sectors is recommended. Sharing of information on the toxicity of nanomaterials, will significantly reduce time to market for many products and producers.

- Should we use the precautionary principle in current regulatory testing? The precautionary principle (PP) is a highly debated issue in international politics and was first added in EU environmental regulation in the Maastricht treaty in Article 174 (ex Art 130r). The PP points out that scientific uncertainty is no reason for inaction if there might be strong adverse effects. It has been criticized for being too vague and too arbitrary to form a basis for rational decision-making. On the other hand the PP does not necessarily imply a complete ban of substances but may be applied in steps of decreasing uncertainty or perceived risk. It is typically applied where scientific information is insufficient, inconclusive or uncertain whilst at the same time there are indications that potential effects may occur. Approaches in risk governance range from early explorations by Swiss RE to more recent detailed reviews such by Renn and Roco and the Health Council of the Netherlands. A crucial difference between both reports is that Renn and Roco place the nanotechnology *products themselves* into the risk issue categories rather than the risk issues that accompany the use of these products. This difference may seem trivial, but is important since it is the purpose and application rather than the device itself that connect hazard to exposure, and therefore creates a risk. For example, it fails to distinguish between brain implants for enhancement purposes and those for fighting tremor in Parkinson patients. This may pave the way to one nanotechnology product compromising or allowing all the others. Furthermore, it can convey the impression that the broader societal implications and accompanying ethical issues of nanotechnologies are as new as the nanotechnology products themselves, or even are still in the future. However, most of them are not.

POTENTIAL ADVANTAGES OF NANOPARTICLE DRUG DELIVERY SYSTEMS

Nanoparticles are made from biocompatible and biodegradable materials such as polymers, either natural (*e.g.*, gelatin, albumin) or synthetic (*e.g.*, polylactides, polyalkylcyanoacrylates), or solid lipids. In the body, the drug loaded in nanoparticles is usually released from the matrix by diffusion, swelling, erosion, or degradation. Nanoparticles for the purpose of drug delivery are defined as submicron (< 1μm) colloidal particles. This definition includes monolithic nanoparticles (nanospheres) in which the drug is adsorbed, dissolved, or dispersed throughout the matrix and nanocapsules in which the drug is confined to an aqueous or oily core surrounded by a shell-like wall. Alternatively, the drug can be covalently attached to the surface or into the matrix. The following are among the important technological advantages of nanoparticles as drug carriers: high stability (*i.e.*, long shelf life); high carrier capacity (*i.e.*, many drug molecules can be incorporated in the particle matrix); feasibility of incorporation of both hydrophilic and hydrophobic substances; and feasibility of variable routes of administration, including oral administration and inhalation. These carriers can also be designed to enable controlled (sustained) drug release from the matrix. The methods for nanoparticle preparation and characterization were addressed in numerous reviews. The major data on nanoparticulate formulations of the anti-TB drugs.

Table. Drug Release and Therapeutic Efficacy of the Nanoparticle-Based Formulations of the First-Line Antituberculous Drugs-Rifampin, Isoniazid, and Pyrazinamide

Delivery System	*Animal Model*	*Administration Route*	*Plasma*	*Duration of Drug Release (d) Organs*	*Regimen Producing Sterilizing Effect Ref in Lungs and Spleen*
PLG nanoparticles	Mice	Oral	6–9	9–11	5 doses every 10 d 12
	Mice	Subcutaneous	32	36	Single injection 30
	Guinea pigs	Aerosol	4–9	up to 10 d (each drug)	5 doses every 10 d 17
	Guinea pigs	Oral	4–9	up to 10 d (each drug)	5 doses every 10 d 17
Lectin-functionalized PLG nanoparticles	Guinea pigs	Oral	7–13	up to 15 d (each drug)	3 doses fortnightly 17
	Guinea pigs	Aerosol	6–14	up to 15 d (each drug)	3 doses fortnightly 17
Solid lipid nanoparticles	Guinea pigs	Aerosol	5	7	7 doses weekly 22

- *Definition of Abbreviations:* PLG = Poly(lactide-co-glycolide);
- *The Drug-to-polymer Ratio is* 1:1 for each drug

ORAL ADMINISTRATION OF NANOPARTICLE-BASED TB DRUGS

Constancy of nanoparticles offers the possibility of oral administration. The fate of nanoparticles in the gastrointestinal tract has been investigated in a number of studies. In general, the uptake of nanoparticles occurs as follows: by transcytosis via M cells, by intracellular uptake and transport via the epithelial cells lining the intestinal mucosa, by uptake via Peyer's patches.

Pandey and colleagues demonstrated that the nanoparticles provided sustained release of the anti-TB drugs and considerably enhanced their efficacy after oral administration. Three frontline drugs, rifampin (RMP), isoniazid (INH), and pyrazinamide (PZA) were coencapsulated in poly(lactide-co-glycolide) (PLG) nanoparticles. After a single oral administration of this formulation to mice, the drugs could be detected in the circulation for 4 d (RMP) and 9 d (INH and PZA); therapeutic concentrations in the tissues were maintained for 9 to 11 d. In contrast, free (unbound) drugs were cleared from the plasma within 12 to 24 h after administration. Treatment of *M. tuberculosis*–infected mice with the nanoparticle-bound drugs (five oral doses every 10th day) resulted in complete bacterial clearance from the organs. Free drugs were able to produce bacterial clearance only after daily administration of 46 doses. Similar efficacy of the nanoparticle-bound drugs was also observed in guinea pigs.

At the same time, incorporation in microparticles was less effective: their drug-loading capacity was lower as well as the plasma half-life of the bound drugs. The behaviour of polymeric nanoparticles in the gastrointestinal tract is influenced by their bioadhesive properties; adhesion of nanoparticles to the mucosa enhances the absorption of the associated drug, thus increasing its bioavailability. Thus, lectins have been shown to improve mucoadhesion of the drug due to the biorecognition of the lectin-grafted carriers by glycosylated structures in the intestine.

The efficacy of PLG-based formulations of anti-TB drugs were further improved by covalent attachment of wheat germ agglutinin. Oral administration of wheat germ agglutinin–coated PLG nanoparticles loaded with RIF, INH, and PZA in mice produced considerably extended serum half-life: detectable RIF serum levels were observed for 6 to 7 d and INH and PZA for 13 to 14 d (vs. 4–6 d and 8–9 d for nonmodified nanoparticles). All three drugs were present in lungs, liver, and spleen for 15 d. The lectin-modified formulations produced bacterial clearance in these organs after three oral doses administered every 14 d (vs. 45 daily doses of free drugs). As suggested by the authors, the prolonged circulation of drugs encapsulated in wheat germ agglutinin–grafted nanoparticles might be attributed to the fact that lectins enhance prolonged adhesion of the particles to the intestinal surface to allow an increase in the time interval available for absorption and a localized increase in the concentration gradient between luminal and serosal sides of the membrane.

POTENTIAL FOR THE INHALATION FORM FUNCTION

The potential advantages of direct delivery of the TB drug to the lungs include the possibility of reduced systemic toxicity, as well as achieving higher drug concentration at the main site of infection. Moreover, in contrast to the oral route of administration, inhaled drugs are not subjected to first-pass metabolism. A possible obstacle to using nanocarriers for pulmonary delivery is that their mass median aerodynamic diameter, an essential parameter for the particle deposition in the lungs, is often too small.

Nevertheless, the effectiveness of pulmonary drug delivery using nanoparticles was demonstrated in a number of studies. The pharmacokinetics and antibacterial effect of the nanoparticle-bound anti-TB drugs administered via respiratory route was investigated in guinea pigs. The dose was delivered via a suitable facemask connected to the compressor–nebulizer system. A single nebulization of RMP, INH, and PZA coencapsulated in PLG nanoparticles to guinea pigs resulted in sustained therapeutic drug levels in the plasma for 6 to 8 d and in the lungs for up to 11 d. This effect was similar to that obtained after oral administration of the nanoparticulate formulation of the same drugs. In nebulization of nanoparticles to *M. tuberculosis*–infected guinea pigs at every 10th day, no tubercle bacilli could be detected in the lung after only five doses of treatment, whereas 46 daily doses of orally administered drug were required to obtain an equivalent therapeutic benefit.

Administration to infected guinea pigs of nebulized RMP, INH, and PZA coencapsulated in wheat germ agglutinin–functionalized PLG nanoparticles was even more effective: three doses administered fortnightly for 45 d were sufficient to produce a sterilizing effect in lungs and spleen.

A sterilizing effect was also achieved when the drugs were loaded in solid lipid nanoparticles. No tubercle bacilli could be detected in the lungs/spleen after seven doses of treatment of infected guinea pigs with drug-loaded solid lipid nanoparticles. It is noteworthy that the solid lipid nanoparticles display important advantages, such as the composition (physiologic compounds) and the possibility of large-scale production favoured by the feasibility to avoid organic solvents in the manufacturing process.

Intravenous Organization

In contrast to microparticles with a diameter of more than 1 μm that cannot be administered via intravascular routes, nanoparticles are small enough to allow intracapillary passage followed by an efficient cellular uptake. When administered intravenously, the nanoparticles follow the route of other foreign particulates, including intracellular pathogens. They are endocytosed by resident macrophages of the mononuclear phagocyte system and by circulating monocytes. On the other hand, in the case of infections caused by intracellularly persisting microbes (*e.g.*, *Brucella*, *Salmonella*, *Listeria*, *Mycobacterium*), macrophages become reservoirs for pathogens, thus representing one of the targets for delivery of antimicrobial agents. Preferential uptake of nanoparticles by macrophages (mainly by Kupffer cells in the liver) is achieved by the physicochemical properties of the carrier and by physiologic opportunity, thus representing an example of passive delivery. This technology improves drug delivery to macrophages, increasing the amount of the drug reaching this target site, which allows reduction of the overall therapeutic dose and decrease of the adverse effects. Accordingly, the enhanced efficacy of the nanoparticle-bound antibiotics was demonstrated in a number of experimental infections.

The potential of macrophage-targeting strategy in development of the nanoparticle-based TB drugs is supported by the *in vitro* data. Incorporation of INH and streptomycin in poly(butyl cyanoacrylate) nanoparticles not only increased the intracellular accumulation (or association) of these drugs in the cultivated human blood monocytes but also produced enhanced antimicrobial activity of these agents against intracellular *M. tuberculosis* compared with their activity in extracellular fluid. Similarly, the encapsulated ciprofloxacin and RMP produced the enhanced effect against mycobacteria in the infected macrophages. This is in contrast to the previous *in vitro* observations showing that enhanced intracellular accumulation of drugs in macrophages is rarely associated with a simultaneous increase of activity against intracellular mycobacteria (compared with activity against the extracellular bacterial population). This is presumably due to the fact that the drugs and bacteria are sequestered in different intracellular compartments. It is noteworthy that the enhanced cell uptake and activity against intracellular bacteria was demonstrated for the nanoparticulate streptomycin, an aminoglycoside agent, which, in the free form, has poor intracellular access.

Clofazimine, a riminophenazine compound, is an agent considered for treating patients with *M. avium* infection. However, use of this drug was restricted because of its poor solubility. A relatively new approach was applied to solve the problem: clofazimine was formulated as a nanosuspension consisting only of the drug and a minimum amount of surfactants (particle size, 385 nm). Intravenous injection of the nanocrystalline formulation of clofazimine resulted in a considerable reduction of bacterial loads in the liver, spleen, and lungs of mice infected with *M. avium*. This result correlated with the pharmacokinetic data: drug concentrations in these organs reached high concentrations, well in excess of the minimal inhibitory concentration for most *M. avium* strains. Interestingly, the effects of the nanocrystalline formulation of clofazimine were similar to those of the liposomal formulation used as a control in this study. This study is a vivid example of application of nanotechnology for overcoming the solubility problems of poorly soluble drugs. More details on the application of nanosuspensions in drug delivery can be found in a review on this subject. Intravenous administration of the nanoparticles has the further advantage of passive drug delivery to inflammatory sites where the endothelium becomes permeable due to pathologic processes. In this case, passive delivery of the nanoparticles would be realised by pathophysiologic opportunity. The basic principles for engineering of the nanocarriers suggest that passive accumulation in the sites with leaky vasculature would be more effective with the long-circulating (stealth) nanoparticles. These particles evade resident macrophages in the liver and have higher probability to reach other sites in the body.

Other Routes of Organization

The nanoparticle-based formulations was further demonstrated by effective subcutaneous treatment of mice infected with *M. tuberculosis*. A single subcutaneous dose of PLG nanoparticles loaded with RMP, INH, and PZA maintained therapeutic drug levels in plasma for 32 d and in lungs/spleen for 36 d. Moreover, this single subcutaneous injection produced a sterilizing effect in lungs and spleen of the infected mice (36 d post-treatment), thereby demonstrating a better chemotherapeutic efficacy, as compared with daily treatment using free drugs (35 oral doses). As suggested by the authors, the nanoparticles form a depot at the injection site that is slowly releasing drugs into the circulation.

SYSTEMIC AND TISSUE NATURE OF LIPOSOMES

Liposomes are microscopic vesicles composed of a phospholipid bilayer that are capable of encapsulating the active drug. Whether the drug is encapsulated in the core or in the bilayer of the liposome is dependent on the characteristics of the drug and the encapsulation process. In general, water-soluble drugs are encapsulated within the central aqueous core, whereas lipid-soluble drugs are incorporated directly into the lipid membrane. Liposomes can alter both the tissue distribution and the rate of clearance of the drug by making the drug take on the pharmacokinetic characteristics of the carrier. Pharmacokinetic variables of the liposomes depend on the physiochemical characteristics of the liposomes, such as size, surface charge, membrane lipid packing, steric stabilization, dose, and route of administration. The primary sites of accumulation of conventional liposomes are the tumor, liver, and spleen compared with nonliposomal formulations. The development of STEALTH liposomes was based on the discovery that incorporation of polyethylene glycol (PEG)-lipids into liposomes yields preparations with superior tumor delivery compared with conventional liposomes composed of natural phospholipids. Incorporation of PEG-lipids causes the liposome to remain in the blood circulation for extended periods of time (*i.e.*, $t_{1/2} > 40$ hours) and distribute through an organism relatively evenly with most of the dose remaining in the central compartment (*i.e.*, the blood) and only 10% to 15% of the dose being delivered to the liver. This is a significant improvement over conventional liposomes where typically 80% to 90% of the liposome deposit in the liver.

The clearance of conventional liposomes has been proposed to occur by uptake of the liposomes by the reticuloendothelial system (RES). The mononuclear phagocyte system uptake of liposomes results in their rapid removal from the blood and accumulation in tissues involved in the RES, such as the liver and spleen. Uptake by the RES usually results in irreversible sequestering of the encapsulated drug in the RES, where it can be degraded. In addition, the uptake of the liposomes by the RES may result in acute impairment of the mononuclear phagocyte system and toxicity. Sterically stabilized liposomes, such as STEALTH liposomes, prolong the duration of exposure of the encapsulated liposome in the systemic circulatio. The presence of the PEG coating on the outside of the liposome does not prevent uptake by the reticuloendothelial system, but simply reduces the rate of uptake. The exact mechanism by which steric stabilization of liposomes decreases the rate of uptake by the reticuloendothelial system is unclear.

TUMOR DELIVERY OF LIPOSOMAL AGENTS

Solid tumours have several potential barriers to drug delivery that may limit drug penetration and provided inherent mechanisms of resistance. Moreover, factors affecting drug exposure in tissue, such as alteration in the distribution of blood vessels, blood flow, capillary permeability, interstitial pressure, and lymphatic drainage, may be different in tumours and the surrounding normal tissue.

Once in the tumor, standard liposomes are localized in the extracellular fluid surrounding the tumor cell but do not enter the cell. Thus, for the liposomes to deliver the active form of the anticancer agent, such as doxorubicin, the drug must be released from the liposome into the extracellular fluid and then diffuse into the cell. As a result, the ability of the liposome to carry the anticancer agent to the tumor and release it into the extracellular fluid are equally important factors in determining the antitumor effect of liposomal-encapsulated anticancer agents. In general, the kinetics of this local release are unknown as it is difficult to differentiate between the liposomal-encapsulated and released forms of the drug in solid tissue; however, with the development of microdialysis.

Several preclinical studies have shown extensive tumor targeting and prolonged exposure of Doxil in tumours, which is consistent with the increased antitumor activity in preclinical models compared with doxorubicin and with clinical activity in patients with refractory ovarian cancer and Kaposi sarcoma. In studies comparing STEALTH liposomal cisplatin (SPI-77) and cisplatin tumor disposition in murine colon tumor xenografts, the platinum (Pt) exposure was four-fold higher and prolonged after SPI-77 compared with cisplatin administration. However, because the Pt exposure was measured in tumor extracts, it is unclear whether the Pt measured was SPI-77 (*i.e.*, liposomally encapsulated Pt), protein-bound Pt, or unbound-Pt. Moreover, although there is a four-fold higher exposure of total-Pt in tumours after SPI-77 compared with cisplatin, this has not translated into antitumor response in clinical trials. One possible explanation for the inconsistency between the high tumor exposure and low antitumor effect could be the lack of release of active unbound cisplatin from the liposome into the tumor extracellular fluid. We evaluated the exposure of unbound cisplatin in tumor extracellular fluid using microdialysis after administration of SPI-77 and compared these results to the tumor extracellular fluid exposure after cisplatin administration. The results of this study suggest that SPI-077 distributes into tumours but release significantly less Pt into tumor extracellular fluid, which results in lower formation of Pt-DNA adducts compared with cisplatin. The clinical importance of these studies is underscored by the need to select liposomal anticancer agents with high tumor penetration and delivery of the active drug to the tumour.

ALTERATION OF TOXICITY WITH LIPOSOMAL AGENTS

Liposomal formulations can also modify the toxicity profile of a drug. This effect may be due to the alteration in tissue distribution associated with liposomal formulations. Anthracyclines, such as doxorubicin, are active against many tumor types, but cardiotoxicity related to the cumulative dose may limit their use. Preclinical studies determined that liposomal anthracyclines reduced the incidence and severity of cumulative dose-related cardiomyopathy while preserving antitumor activity. There is also clinical evidence suggesting that Doxil is less cardiotoxic than conventional doxorubicin. Direct comparisons between Doxil or Caelyx and conventional doxorubicin showed comparable efficacy but significantly lower risk of cardiotoxicity with the STEALTH liposomal formulations of doxorubicin. In addition, histologic examination of cardiac biopsies from patients who received cumulative doses of Doxil from 440 to 840 mg/m^2, and had no prior exposure to anthracyclines, revealed significantly less cardiac toxicity than in matched doxorubicin controls. Administration of a drug in a liposome may also result in new toxicities. The most common adverse event associated with Doxil is hand-foot syndrome (also known as palmar-plantar erythrodysesthesia) and stomatitis, which have not been reported with conventional doxorubicin. The exact mechanisms associated with these toxicities are unknown, but are schedule and dose dependent. In general, Doxil is generally well tolerated and its side effect profile compares favourably with other chemotherapy used in the treatment of refractory ovarian cancer. Proper dosing and monitoring may further enhance tolerability while preserving efficacy; however, there is still a need to identify factors associated with hand-foot syndrome, which can be dose limiting in some patients.

OTHER LIPOSOMAL AGENTS IN DEVELOPMENT

Some other liposomal anticancer agents that are currently in development are SN-38, lurtotecan, 9NC, irinotecan, STEALH liposomal CKD-602, paclitaxel, and doxorubicin. Liposomal encapsulation of camptothecins is an attractive formulation due to the solubility issues associated with most camptothecin analogues and the potential for prolonged exposure after administration of a single dose. As compared with pegylated or coated liposomes, conventional liposomal formulations of camptothecin analogues, such as LE-SN38 and OSI-211, may result in the rapid release of the drug from the liposome in blood and thus act more as a new i.v. formulation rather than a tumor-targeting agent. However, studies evaluating encapsulated and released drug in plasma and tumor have not been reported.

Prospect generations of liposomes may contain targeting antibodies, two anticancer agents combined within a single liposome, or liposomes that are thermosensitive. Immunoliposomes combine antibody-mediated tumor recognition with liposomal delivery and are designed for target cell internalization and intracellular drug release. There are several liposomal formulations that contain fixed ratios of two anticancer agents, such as doxorubicin:vincristine, daunorubicin: cytarabine, and cisplatin: irinotecan, which are currently in preclinical development. Thermosensitive liposomes may provide a means of improving the tumor-specific delivery of anticancer agents by rapidly releasing drug from the liposome when hypothermia is applied to the tumor area.

NANOPARTICLE, MICROSPHERE AND CONJUGATE FORMULATIONS

The first protein-stabilized nanoparticle approved by the Food and Drug Administration. ABI-007 is an albumin-stabilized nanoparticle formulation of paclitaxel designed to overcome the solubility issues associated with paclitaxel that require the need for solvents such as cremophor, which have been associated with infusion-related reactions and require the need for premedication. Cremophor may also be incompatible with certain i.v. bags or tubing. The albumin-stabilized nanoparticle results in a more rapid distribution out of the vascular compartment and provides a tumor-targeting mechanism. The albumin receptor-mediated transport through the endothelial cells within blood vessels facilitates the passage of ABI-007 from the bloodstream into the underlying tumor tissue.

Comparable to liposomal agents, the dosage of ABI-007 is determined by the paclitaxel content of the formulation. The approved regimen for ABI-007 is 260 mg/m^2 i.v. over 30 minutes every 3 weeks, which is higher than the usual dose range for paclitaxel. In addition, there was a lower incidence of myelosuppression after administration of ABI-007 than previously seen with similar doses of paclitaxel. The remainder of the toxicities associated with ABI-007 were similar to high-dose paclitaxel, including sensory neuropathy and mucositis. Keratopathy, a relatively uncommon toxicity, was also associated with ABI-077. Thus, as with liposomal formulations, administration of a drug in a nanoparticle formulation can alter the pharmacokinetics, tissue and tumor distribution, and toxicity pattern. Also, similar to liposomal agents, the mechanism by which the albumin-stabilized nanoparticle is catabolized and paclitaxel is released is unclear.

Supplementary nanoparticle-formulations of paclitaxel are in clinical and preclinical development. Paclitaxel poliglumex (Xyotax), a macromolecular drug conjugate that links paclitaxel with a biodegradable polymer, poly-l-glutamic acid, has completed phase 1 studies. Paclitaxel poliglumex is a water-soluble formulation that also eliminates the need for cremophor in the formulation. Paclimer, a microsphere formulation of paclitaxel, is currently in preclinical development.

Paclimer microspheres contain paclitaxel in a polilactofate polymer microsphere and is designed to continuously deliver low-dose paclitaxel. Other conjugates of paclitaxel have been stopped in clinical development and have been associated with potential pharmacologic and pharmacokinetic problems.

Docosahexaenoic acid–paclitaxel, a novel conjugate formed by covalently linking the natural fatty acid docosahexaenoic acid to paclitaxel, was designed as a prodrug targeting intratumoral activation. At the maximum tolerated dose of docosahexaenoic acid–paclitaxel (1,100 mg/m^2), paclitaxel represented only 0.06% of the docosahexaenoic acid–paclitaxel plasma exposure.

However, the paclitaxel concentrations remained >0.01 μmol/L for an average of 6 to 7 days and the paclitaxel area under the curve was correlated with neutropenia. The results of this study suggest that most of the drug remained in the inactive prodrug conjugated form and that significant toxicity only occurred when released paclitaxel reached clinically relevant exposures. This depicts the need to perform detailed pharmacokinetic studies of conjugated and released drug in plasma and tumor.

During the past 10 years, there has been a renaissance in the field of PEG-conjugated anticancer agents. This new development has been attributed to the use of higher-molecular-weight PEGs (>20,000) and especially with the use of PEG 40,000, which has an extended $t_{1/2}$ in plasma and potential selective distribution to solid tumours. Various PEG conjugates of anticancer agents, such as doxorubicin, methotrexate, IFN, and camptothecin analogues, are currently in development.

PEG- and 20-carbonate conjugates of camptothecin analogues are especially interesting as the conjugated prodrug forms highly water soluble agents and significantly extend the duration of exposure after a single dose. Hyaluronic acid conjugates of anticancer agents are also in development. Carrier-mediated conjugates of anticancer agents also have the same pharmacologic issues (the need to evaluate the pharmacokinetics of the prodrug conjugate and released drug) as liposomal and nanoparticle formulations and the overall clinical benefit of these agents remains unclear.

ORAL LIPID BASED FORMULATIONS

Among the benefits which oral lipid-based formulations can provide are included: Improvement and reduction in the variability of GI absorption of poorly water-soluble, lipophilic drugs. Possible reduction in, or elimination of, a number of development and processing steps (salt selection or identification of a stable crystalline form of the drug, coating, tastemasking, and reduced need for containment and clean-up requirements during manufacture of highly-potent or cytotoxic drug products). Reduction or elimination of positive food effect. Relative ease of manufacture using readily available equipment.

Different types of oral lipid based formulation are like, single-component lipid solutions, self-emulsifying formulations, self-emulsifying solid dispersion formulations and melt pelletization. It has been revealed that the most frequently chosen excipients for preparing oral lipid-based formulations were dietary oils composed of medium (coconut or palm seed oil) or long-chain triglycerides (corn, olive, peanut, rapeseed, sesame, or soybean oils, including hydrogenated soybean or vegetable oils), lipid soluble solvents (polyethylene glycol 400, ethanol, propylene glycol, glycerin), and various pharmaceutically-acceptable surfactants (Cremophor; EL, RH40 or RH60; polysorbate 20 or 80; D-α-tocopherol polyethylene glycol 1000 succinate (TPGS;); Span 20; various Labrafils;, Labrasol;, and Gelucires;).

These formulations, which took the form of either bulk oral solutions or liquid-filled hard or soft gelatin capsules, were applied in instances where conventional approaches (solid wet or dry granulation, or water-miscible solution in a capsule) did not provide sufficient bioavailability, or in instances in which the drug substance itself was an oil (dronabinol, ethyl icosapentate, indometacin farnesil, teprenone, and tocopherol nicotinate). The total daily drug dose administered in these formulations, which range in complexity from simple solutions of the drug in a dietary oil up to multi-excipient, self-emulsifying drug delivery systems (SEDDS), range from less than 0.25 μg to greater than 2000 mg.

The amount of drug contained in a unit-dose capsule product ranges from 0.25 μg to 500 mg and for oral solution products, from 1 μg/ml to 100 mg/ml. The total amount of lipid excipient administered in a single dose of a capsule formulation ranges from 0.5 to 5 g, but can range from as low as 0.1 ml to as high as 20 ml for oral solution products. Some of these products tolerate room temperature storage for only brief periods of time and require long-term storage at 2-8° due to chemical and/or physical stability issues.

ROUTES OF ORGANIZATION AND THEIR BIODISTRIBUTION

The *in vivo* fate of the solid lipid nanoparticles will depend mainly on the administration route and distribution process (adsorption of biological material on the particle surface and desorption of SLN components into the biological surrounding). SLN are composed of physiological or physiologically related lipids or waxes.

Therefore, pathways for transportation and metabolism are present in the body which may contribute to a large extent to the *in vivo* fate of the carrier. Probably the most important enzymes of SLN degradation are lipases, which are present in various organs and tissues. Lipases split the ester linkage and form partial glycerides or glycerol and free fatty acids. Most lipases require activation by an oil/water interface, which opens the catalytic centre (lid opening). *In vitro* experiment indicates that solid lipid nanoparticles show different degradation velocities by the lipolytic enzyme pancreatic lipase as a function of their composition (lipid matrix, stabilizing surfactant).

Per Oral Administration

Per oral administration forms of SLN may include aqueous dispersions or SLN-loaded traditional dosage forms such as tablets, pellets or capsules. The microclimate of the stomach favours particle aggregation due to the acidity and high ionic strength.

It can be expected, that food will have a large impact on SLN performance, however no experimental data have been publish on this issue to our knowledge. The question concerning the influence of the gastric and pancreatic lipases on SLN degradation *in vivo* remains open, too. Unfortunately, only few *in vivo* studies have been performed yet.

Parenteral Administration

SLN have been administered intravenously to animals. Pharmacokinetic studies of doxorubicin incorporated into SLN showed higher blood levels in comparison to a commercial drug solution after i.v. injection in rats. Regarding distribution, SLN were found to have higher drug concentrations in lung, spleen and brain, while the solution led to more distribution into liver and kidneys.Yang *et al.* reported on the pharmacokinetics and body distribution of camptothecin after i.v. injection in mice. In comparison to a drug solution SLN was found to give much higher AUC/dose and mean residence times (MRT) especially in brain, heart and reticuloendothelial cells containing organs. The highest AUC ratio of SLN to drug solution among the tested organs was found in the brain.

Transdermal Submission

The smallest particle sizes are observed for SLN dispersions with low lipid content (up to 5%). Both the low concentration of the dispersed lipid and the low viscosity are disadvantageous for dermal administration. In most cases, the incorporation of the SLN dispersion in an ointment or gel is necessary in order to achieve a formulation which can be administered to the skin. The incorporation step implies a further reduction of the lipid content of the SLN dispersion resulting in semisolid, gel-like systems, which might be acceptable for direct application on the skin.

Applications

Solid lipid Nanoparticles possesses a better stability and ease of upgradability to production scale as compared to liposomes. This property may be very important for many modes of targeting. SLNs form the basis of colloidal drug delivery systems, which are biodegradable and capable of being stored for at least one year. They can deliver drugs to the liver *in vivo* and *in vitro* to cells which are actively phagocytic.

There are several potential applications of SLNs some of which are given below: SLNs as gene vector carrier: SLN can be used in the gene vector formulation. In one work, the gene transfer was optimized by incorporation of a diametric HIV-1 HAT peptide (TAT 2) into SLN gene vector. There are several recent reports of SLN carrying genetic/peptide materials such as DNA, plasmid DNA and other nucleic acids.

The lipid nuclic acid nanoparticles were prepared from a liquid nanophase containing water and a water miscible organic solvent where both lipid and DNA are separately dissolved by removing the organic solvent, stable and homogeneously sized lipid-nuclic acid nanoparticle (70-100 nm) were formed. It's called genospheres. It is targeted specific by insertion of an antibody-lipo polymer conjugated in the particle.

SLNs for Topical use

SLNs and NLCs have been used for topical application for various drugs such as tropolide, imidazole antifungals, anticancers, vitamin A, isotretinoin, ketoconazole, DNA, flurbiprofen and glucocorticoids. The penetration of podophyllotoxin-SLN into stratum corneum along with skin surface lead to the epidermal targeting.

By using glyceryl behenate, vitamine A-loaded nanoparticles can be prepared. The methods are useful for the improvement of penetration with sustained release. The isotretinoin-loaded lipid nanoparticles was formulated for topical delivery of drug. The soyabean lecithin and Tween 80 are used for the hot homogenization method for this. The methodology is useful because of the increase of accumulative uptake of isotretinoin in skin. Production of the flurbiprofen-loaded SLN gel for topical application after a potential advantages of delivering the drug directly to the site of action, which will produce higher tissue concentrations. Polyacrylamide, glycerol and water were used for the preparation of this type of SLN gel.

SLNs as Cosmeceuticals

The SLNs have been applied in the preparation of sunscreens and as an active carrier agent for molecular sunscreens and UV blockers. The *in vivo* study showed that skin hydration will be increased by 31% after 4 weeks by addition of 4% SLN to a conventional cream. SLN and NLCs have proved to be controlled release innovative occlusive topicals. Better localization has been achieved for vitamin A in upper layers of skin with glyceryl behenate SLNs compared to conventional formulations.

SLNs for Potential Agriculture Application

Essential oil extracted from *Artemisia arboreseens* L when incorporated in SLN, were able to reduce the rapid evaporation compared with emulsions and the systems have been used in agriculture as a suitable carrier of ecologically safe pesticides. The SLN were prepared here by using compritol 888 ATO as lipid and poloxamer 188 or Miranol Ultra C32 as surfactant.

SLNs as a Targeted Carrier for Anticancer Drug to Solid Tumours

SLNs have been reported to be useful as drug carriers to treat neoplasms· Tamoxifen, an anticancer drug incorporated in SLN to prolong release of drug after i.v. administration in breast cancer and to enhance the permeability and retention effect· Tumour targeting has been achieved with SLNs loaded with drugs like methotrexate and camptothecin.

SLNs in Breast Cancer and Lymph Node Metastases

Mitoxantrone-loaded SLN local injections were formulated to reduce the toxicity and improve the safety and bioavailability of drug. Efficacy of doxorubicin (Dox) has been reported to be enhanced by incorporation in SLNs. In the methodology the Dox was complexed with soybean-oil-based anionic polymer and dispersed together with a lipid in water to form Dox-loaded solid lipid nanoparticles. The system is enhanced its efficacy and reduced breast cancer cells.

Oral SLNs in Antitubercular Chemotherapy

Antitubercular drugs such as rifampicin, isonizide, pyrazinamide-loaded SLN systems, were able to decrease the dosing frequency and improve patient compliance. By using the emulsion solvent diffusion technique this antitubercular drug loaded solid lipid nanoparticles are prepared. The nebulization in animal by incorporating the above drug in SLN also reported for improving the bioavailability of the drug.

Stealth Nanoparticles

These provide a novel and unique drug-delivery system they evade quick clearance by the immune system. Theoretically, such nanoparticles can target specific cells. Studies with antibody labelled stealth lipobodies have shown increased delivery to the target tissue in accessible sites. Stealth SLNs have been successfully tested in animal models with marker molecules and drugs.

BIOCOMPATIBILITY AND APPLICATIONS OF CARBON NANOTUBES

The substantial progress has been achieved in nanoscience in the last decades. On the one hand the techniques for industrial production of a wide set of nanomaterials have been elaborated. On the other hand the manipulation with nanometre size objects has been demonstrated and principal schemes of nanoelectromechanical systems (NEMS) have been considered. This progress opens the door to the possibility of application of nanosystems in nanomedicine, particularly in medical nanorobots. Two main problems should be solved to select nanoobjects which are perspective for such applications. Firstly, the design of NEMS which can be used in medical nanorobots should be elaborated. Secondly, the biocompatibility of nanoobjects which are movable elements of this NEMS should be studied.

Essential characteristics of nanoobjects are a very small feature size in the range of 1–100 nanometres (nm). One nanometre spans 3–5 atoms lined up in a row. As one of the most promising nanoobjects, carbon nanotubes can be mentioned, which can occur in a variety of modifications (eg, single- or multi-walled, filled or surface modified). Carbon nanotubes are expected to find a broad field of applications in nanoelectronics (logics, data storage or wiring, as well as cold electron sources for flat panel displays and microwave amplifiers), and also as fillers for nanocomposites for materials with special properties.

The arbitrary and controlled by a nanomanipulator relative motion of walls in multi-wall carbon nanotubes allows to consider them as promising candidates for movable components of NEMS. Carbon nanotubes have a set of unique electronic and mechanical properties.

For example, the values of both the Young modulus and the bending elastic modulus of carbon nanotubes is about 1 TPA, five times greater than the steel. The walls of carbon nanotubes can be metallic or semiconducting depending on wall structure controlled by chiral indexes.

Possible nanodevices based on such motion have been proposed elsewhere: constant-force nanosprings and free sliding telescopic arms, nanogears, ultra-small switching devices, gigahertz oscillator, and perforating nanodrill. Moreover the first nanotube-based nanomotor has been realised recently by Fennimore *et al.*

Here we propose the principal schemes of the following carbon nanotube based NEMS for medical nanorobots:, nanosyringe (which can be used simultaneously as nanoprobe for study of single biological molecules and drug nanodeliver) and jet nanoengine.

Calculations show that conductance of double-wall carbon nanotubes (DWNTs) depends on the length of an overlap of the walls for telescopic DWNTs and the relative position of the walls within the unit cell of a DWNT for both telescopic and shuttle DWNTs. The measurement of individual carbon nanotube conductance has been experimentally achieved. Thus these electronic properties of DWNTs opens up the possibility to elaborate NEMS based on the relative motion of walls in DWNTs.

For example, variable nanoresistor and strain nanosensor based on DWNT have been proposed. Here we use these electronic properties to propose electromechanical nanothermometre that can be used in medical nanorobots. The chemical inertness of carbon nanotubes allows them to have a quite low toxicity (except possible inhalation toxicity). However it is necessary to take care of purification of nanotubes from metal catalyst and graphite particles that have been shown to be responsible, at least in part, for the toxicity of carbon nanotubes preparations. Conductivity and mechanical characteristics of nanotubes, as well as chemical inertness and their high aspect ratio, make them a material of choice for medical nanodevices. At present, carbon nanotubes can be produced by different methods and are already available commercially.

The description of schemes and operation of electromechanical nanodevices that can be used in nanomedicine and possible ways of these electromechanical nanodevices actuation: by a nanomanipulator, by use of electric field, or by use of magnetic field. The present-day progress in nanotechnology techniques that are necessary for assembling of NEMS based on relative motion of carbon nanotubes walls.

BIOCOMPATIBILITY OF CARBON NANOTUBES

A potential structure material for medical nanodevices is the graphene sheet which can form carbon nanotubes. Carbon fullerenes and nanotubes can be used for sensors or manipulators of medical nanorobots. They can also be used to create electrically conductive polymers and tissue engineering constructs with the capacity to provide controlled electrical stimulation.

Because of the great interest in using engineered nanomaterials for medical applications, there is some information relevant for assessing health effects reported in this area of literature. Carbon nanostructures are the only class of engineered nanomaterials that have received focused toxicological characterization.

In regard to carbon nanotubes, inhalation toxicity has been the initial concern. Huczko et al studied the effects of Carbon nanotubes containing soot on guinea pigs intratracheally instilled, by evaluating pulmonary function parameters and bronchoalveolar lavage examination. They showed that after

electrodes. For adequate operation of the nanothermometre the average displacement *d* of the shuttle in the result of diffusion during the time *t* of the nanothermometre operation should be less than the distance *L* between the electrode and the shuttle.

$$d = \sqrt{2Dt < L,}$$

where *D* is the diffusion coefficient for diffusion of the shuttle along the long wall. The expression for *D* was recently obtained.

$$D = a\exp\left(-\frac{bl}{T}\right),$$

$$a = \pi\delta\sqrt{\frac{\Delta U}{2m}},$$

$$b = \frac{\Delta U N_a}{l_c k}$$

where ΔU is the barriers to relative sliding of the walls per atom of the movable wall, *m* is the mass of carbon atom, N_a is the number of atoms in the elementary cell of the movable wall, l_c is the length of the elementary cell, and *l* is the length of the shuttle.

Thus we obtain

$$l = \frac{T}{b}\ln\frac{2at}{L^2}$$

The total length l_{tot} of nanothermometre between the electrodes is

$$l_{tot} = \frac{T}{b}in\frac{2at}{L^2} + 2L,$$

It is easy to show that this total length is minimal at $L = T/b$. Note that this result is the same for telescopic and shuttle schemes of nanothermometre and does not depend on the time t of nanothermometre operation. Using *ab initio* results for the barrier ΔU we obtain for (6,6)@(11,11) DWNT $a = 1.46 \pm 0.04 \cdot 10^{-8}$ m^2/s for the inner movable wall and $a = 1.05 \pm 0.0310^{-8}$ m^2/s for the outer movable wall, $b = 372$ K/nm for both both inner and outer movable walls. Thus we obtain that at room temperature $T = 300$ K the length *L* should exceed 1 nm, that is only several times greater than the length of the elementary translational cell of the (6,6)@(11,11) DWNT, $l_c = 0.244$ nm.

To calculate the total length of nanothermometre that operates at average during 100 years at room temperature $T = 300$ K without failure connecting with relative diffusion of walls and obtain that l_{tot} is about 40 nm or about 150 DWNT unit cells, that is $l \approx l_{tot}$ (since l_{tot} is logarithmic function of the parameter a, l_{tot} is approximately the same for the cases of inner and outer movable walls).

Note that this value of l_{tot} correspond to the range used at *ab initio* calculation of the conductance of telescopic and shuttle (6,6)@(11,11) DWNTs. In the case of telescopic scheme of the nanothermometre the length of the gap between the fixed walls should be added to the total length l_{tot}. Thus the proposed nanothermometre can be really of nanometre size and its total size can be determined by the length which is necessary to attach electrodes.

We expect the nanothermometre which can be used for accurate temperature measurement in spatially localized regions with sizes of hundred nanometres. The very small size of proposed electromechanical nanothermometre makes it to be very promising for application in medical nanorobots.

The nanothermometre can be calibrated with the use of thermocouple, and since the temperature measurement is based on conductance measurement, the same order of accuracy as in the case of thermocouple using can be achieved.

The fundamentally different scheme of nanotube based nanothermometre was proposed and realised recently. In that scheme the temperature measurement based on measurement thermal expansion of column of liquid gallium inside a nanotube. The change of Ga-column length was observed *in situ* with the use of a high-vacuum TEM or by postmeasurement TEM reading of the length of a Ga oxide layer mark visible inside nanotube. Both these techniques need in a high-vacuum TEM and therefore can not been used in medical nanorobots.

Jet Nanoengine

The relation motion of core consisting of inner walls of nanotube and shell consisting of outer walls of MWNT has been realised experimentally with the help of nanomanipulator. The simulations show the possibility of fluids flow inside carbon nanotubes. The experiment shows the internal wetting by water of carbon nanotubes with open ends. Nanopump was recently proposed, in which the core motion is used for inflow and ejection of liquid inside nanotube (that is core plays the role of a piston). We propose here that the ejection of liquid can be used for propulsion of nanorobot in this liquid ie, nanosize analog of jet engine can be made of MWNT. Several schemes of the jet nanoengine. On one side piston stroke a water is sucked through the hole into the nanoengine body made of outer wall of MWNT. On another side piston stroke a water is ejected through the nozzle made of inner wall of MWNT. Possible ways of the nanoengine actuation. Note that the gap between the plunger and the shell (the interwall distance) is 3.4 Å. Thus the proposed jet nanoengine excludes the possibility that water can penetrate inside medical nanorobot.

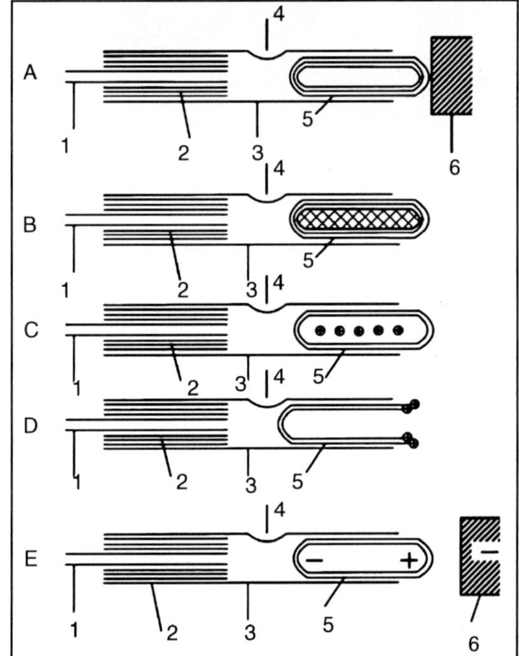

Figure: The Principal Schemes of the jet Nanoengines. The Nanoengine is Composed of the Nozzle made of the Inner Nanotube wall with small Diameter (1), the Medium walls (2) which Serve to Hold the nozzle, the body (3) Made of the outer wall with the Hole (4), and the Piston made of Another Nanotube (5). Several ways of Piston Actuation are Possible: (A) by Attached Nanomanipulator (6), (B) by a Magnetic Field Acting on the Magnetic Core of the Piston, (C) by Electric Field Acting on Encapsulated ions, (D) by Electric Field Acting on Dipole Produced by Atoms or Molecules Adsorbed on the open edge of the Walls Constituting the Piston, by electric field Acting on Dipole Induced by this Field in the Piston, (6) is the Control Electrode.

Drug Nanodeliver

Simple fluids but also large polymeric molecules such as linear polyethylene and DNA can be adsorbed inside carbon nanotubes. We believe that molecules of drug can be adsorbed inside carbon nanotubes. In this case the long outer wall of MWNT can be used as container with molecules of drug adsorbed inside it and the short core of inner walls can be used as piston for ejection of these molecules from the container.

Therefore a nanodeliver made of MWNT can be designed. Several schemes of such nanodelivers corresponding to various ways of piston actuation by electric or magnetic field. We believe that the source of electric or magnetic field can be not only outside an organism, but also medical nanorobot can be used for local actuation of some of the nanodeliveres preliminarily placed inside organism.

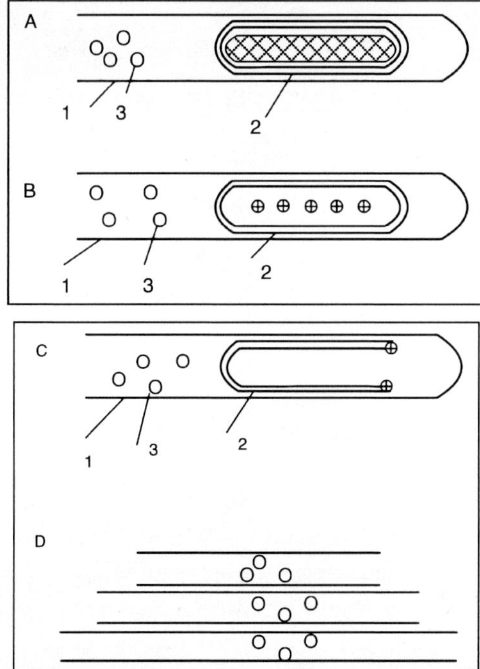

Figure: The Principal Schemes of Drug Nanodelivers. (A), (B) and (C) The Nanodeliveres Consisting of a Container (1) made of the Outer wall of the Nanotube, a Piston (2) Made of the Inner walls of the Nanotube, and a Drug (3). Several ways of Piston Actuation are Possible: by a Magnetic Field Acting on the Magnetic Core of the Piston (A), by Electric Field acting on Encapsulated ions (B) or Dipoles Produced by atoms or Molecules Adsorbed on the open edge of the walls Constituting the piston (C). (D) The Nanodeliver which Delivers a Drug as Result of its Diffusion.

Current molecular dynamics simulations show the possibility of clusters and polymeric molecules diffusion inside carbon nanotube. Therefore the delivery of large biological molecules stored inside a nanotube can occur in result of diffusion of this molecules inside nanotube and subsequent desorption of this molecule from the open end of the nanotube. Evidently that both the diffusion and desorption are random processes. So the delivery of a drug with a desirable time dependence for a one nanotube is not possible. However the charger of hundreds such nanotubes may be used as drug nanodeliver for continuous delivery of set of different molecules with the desirable time dependence. Such a nanodeliver can be transported *in vivo* at certain place by a medical nanorobot.

Nanosyringe and Nanoprobe

The relative motion of all pairs of walls of MWNT is possible. We propose that the many-walled nanotube can be used as also a telescopic nanosyringe, where outer wall of the nanotube is fixed, the motion of the medium wall relative to the outer wall can be used for penetration (that is the medium wall plays the role of the nanosyringe needle) and the motion of the core of inner walls relative to the medium wall can be used to eject molecules adsorbed inside the medium wall (that is the core of inner walls plays the role of the nanosyringe piston). In such a way the delivery of a few number of molecules of drug with nanometre scale accuracy can be possible.

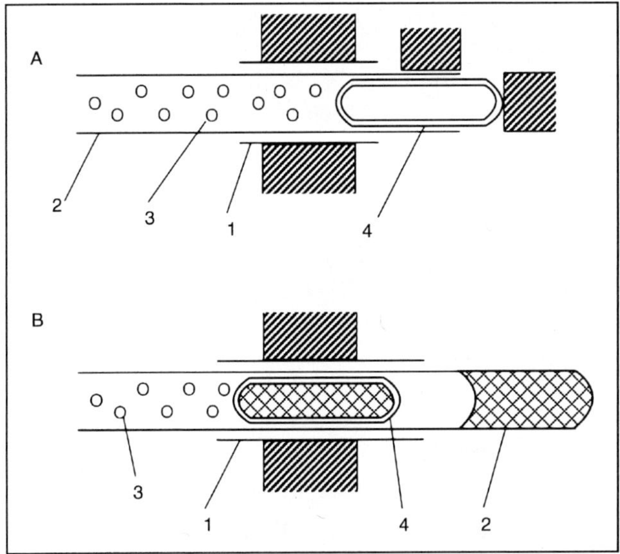

Figure: The Principal Schemes of Electromechanical Nanodevices which can Operate as a Nanosyringe or a Nanoprobe. The Nanodevice is Composed of the Fixed Outer Wall (1) of the Nanotube, the Movable Medium Wall (2) Which Serves as the Needle of the Nanosyringe or the Nanoprobe and Contains a Drug (3) and the Piston (4) Made of the Inner walls of the Nanotube. Two ways of Piston and Needle Actuation are shown: by Attached Nanomanipulator (A), by Magnetic Field Acting on the Magnetic Core of the Piston (B).

Carbon nanotubes are successfully used as the probe tips of the scanning tunnelling microscope (STM) and the atomic force microscope (AFM). Note that because of small diameter (1–10 nm) and high aspect ratio the use of carbon nanotube tip allows to study deeper features of nanoobjects and to achieve higher resolution than in the case of the use of conventional silicon tips. Both these advantages of carbon nanotubes tips were demonstrated on example of biological systems. Moreover nanotubes elastically buckle if applied force exceeds critical force 10 nN. Therefore the use of carbon nanotubes tips can prevent a damage of delicate organic and biological nanoobjects. We propose that the needle of the described above telescopic nanosyringe can serve simultaneously as a probe of STM or AFM and thus it can be used to determine the place where the molecules of drug adsorbed inside the needle should be delivered. The other usage of this device combining scanning microscope probe and telescopic nanosyringe can be also possible. Firstly telescopic carbon nanotube can be used as probe to reveal the molecule and to study its properties *in vivo* and after that this nanotube can be used to capture the molecule. Note that molecular dynamics simulation shows the insertion of long polymeric molecules inside carbon nanotubes in the result of van der Waals interaction.

ACTUATION OF NANOTUBE-BASED NEMS

The proposed nanotube-based NEMS. In principal three methods are possible: i) the direct actuation with the help of a nanomanipulator, ii) the actuation with the help of electric field, iii) the actuation with the help of magnetic field. Besides the actuation and orientation of nanotube by external electromagnetic field is possible.

The controlled by a nanomanipulator relative motion of walls in MNWNTs has been experimentally achieved. The bottleneck of using of nanomanipulator in medical nanorobots is the size of the nanomanipulator and the way how it gets a power. We believe that "artificial muscles" based on electrochemical actuation of conjugated polymers may be perspective. The calculations based on density functional theory shows the elongation of the SWNT if the electric field is directed along its axis. Such elongation has the linear dependence on the magnitude of the electric field and can be up to 13% for electric field 0.4 V/Å. This electrostrictive deformation makes carbon nanotubes high-performance actuator material. Therefore "artificial muscles" can be in one's turn based also on carbon nanotubes: bond length changes at electrochemical double-layer charging of sheets made of single-wall carbon nanotubes have been demonstrated. The actuation of jet nanoengine with the help of the nanomanipulator. The nanotube-based NEMS actuation is the usage of electric field. In this case it is necessary to charge the movable core of wall or to produce a dipole from it.

The calculations based on density functional theory shows the polarization of the SWNT if the electric field is directed along its axis. Recently it was experimentally shown that a dipole can be induced in MWNTs with the help of electric field. Moreover the rotation of the MWNT in water about the axis which is perpendicular to the MWNT axis was demonstrated in this experiment. For this purpose four electrodes placed crosswise were used. We believe that in the case of telescopically pulled out inner core of walls of a carbon nanotube, a charge can be induced on the end of this core and the relative motion of the core and the shell can be achieved with the use of nonuniform electric field produced by a control electrode. Such a way of jet nanoengine actuation.

The calculated electrostatic potentials of carbon nanotubes are sufficiently different at open and closed ends. Therefore a carbon nanotube wall with one open and another closed ends is a dipole and can be actuated with the help of nonuniform electric field. Actuation of relative motion of the inner core of walls of carbon nanotube the charge transfer in result of adsorption on the ends and inside the movable core of inner walls is suitable.

The carbon nanotube becomes a dipole in the case of dissociative adsorption of Br_2 and H_2O molecules on one open edge of the nanotube. Moreover the calculations of electrostatic potentials show distinctive polarized pattern, the direction of which is dependent on the atoms adsorbed on the open edge of the carbon nanotube.

Therefore we propose to increase the dipole moment of the movable inner core by an adsorption of electron donors and acceptors on the opposite open edges of wall constituent the movable inner core. In this case the movable inner core of the nanotube can be moved by nonuniform electric field. The jet nanoengine actuated by movable inner core with electron donors adsorbed on the open edges of the walls constituent.

The valence electron of the encapsulated potassium atom of $K@C_{60}$ is completely transferred to the C_{60} shell. Therefore it was proposed that the transfer of the valence electron of the encapsulated potassium atom to the outer SWNT takes place in the system $K@C_{60}$@SWNT. Thus the possibility that metal atoms encapsulated inside MWNTs can become ions with the transfer of valence electrons to the structure which holds the system. Evidently that the movable core of inner walls with ions encapsulated inside can be actuated with the help of electric field. The actuation of the inner wall with the encapsulated potassium atom inside relative to outer wall in the $7K^+$@(5,5)@(10,10) nanotube has been simulated with the help of molecular dynamics technique. The jet nanoengine actuated by the movable inner core with encapsulated ions.

The set of methods to fill the interior of carbon MWNTs with magnetic materials have been elaborated. In the case where a movable inner core of walls is filled with magnetic material its motion can be controlled by a magnetic field. Note that the local spin density functional calculations show the increase of the magnetic moments of iron atoms in the iron-filled carbon nanotubes in comparison with a bulk iron. The jet nanoengine with the movable inner core filled with a magnetic material.

The other way of the actuation of the relative sliding of nanotube walls with the help of the magnetic field which is nonuniform along the axis of the nanotube was proposed by Legoas et al. If the metallic wall or the the wall filled with metal moves in a nonuniform magnetic field *B* variation of magnetic flux Φ induced the current *i*

$$i = -\sigma\frac{d\Phi}{dt} = -\sigma A v\frac{dB}{dz},$$

where σ is the conductance, *A* is cross-section of the wall and υ is velocity of the wall. The magnetic energy is given by

$$U_m = -\mu B = -AiB = -\sigma A^2 vB \frac{dB}{dz},$$

where $\mu = Ai$ is the magnetic dipole moment. Then the force F_m acting on the wall is given by

$$F_m = -\frac{dU_m}{dz} = \sigma A^2 v \left[\left(\frac{dB}{dz} \right)^2 + B \frac{d^2 B}{dz^2} \right].$$

PROGRESS IN NANOTECHNOLOGY OF NANOTUBE-BASED NEMS PRODUCTION

The possibility for relative motion of carbon nanotubes walls the caps which are closing the ends of walls should be removed. Several methods of opening of carbon nanotubes by oxidation have been elaborated. These methods include oxidation by an air at temperature 750 °C, by the carbon oxide at temperature 750 °C, by the concentrated nitric acids, and by the mixture of sulfuric and nitric acids simultaneously with the sonification. MWNTs were opened also by heat treatment in NaOH at temperature between 600 °C and 750 °C. It is shown, that opened ends of walls can be closed by vacuum annealing at temperature 750 °C.

The separated walls (ie, single-wall nanotubes) should be cut into segments with desirable lengths before the assembling of NEMS. The cutting of nanotubes into segments with the length range from 100 to 300 nm has been obtained by oxidation simultaneously with sonification. The method of precise and controlled cutting of MWNTs and SWNTs has been elaborated. Walls of nanotubes covered by organo-halogen were cut by conventional transmission electron microscopy. With this reproducible method the desirable length of walls could be obtained with an error ± 3 nm.

The attachment of the nanotubes to the nanomanipulators (which is necessary at the last stage of the NEMS assembling) were achieved in the set of the experiments. The voltage was applied to the nanomanipulator arranged near raw nanotube material. The nanotube is polarized by the electric field and jumps to a nanomanipulator due to an electrostatic attraction. Then the nanotube holds by van der Waals forces. A stronger bond can be made using the electron beam to perform localized electron impact-induced dissociation and deposition of gases present in the chamber. The three-dimensional manipulation by carbon nanotube with the use of two nanomanipulators has been also demonstrated.

Because the interwall interaction energy potential relief is determined by chirality indexes of the walls, some special chirality indexes of the walls are necessary for some of nanotube-based NEMS listed above (for example, for the electromechanical nanothermometre). The direct method to determine chirality indexes of a carbon nanotube from a nanobeam electron diffraction pattern was recently demonstrated. The accuracy of this method allows to determine the chirality indexes unambiguously for single-wall nanotubes with diameters up to 4 nm.

The electrical-breakdown technique which allows to remove several walls of MWNT have been elaborated recently. The direct current passing through MWNT causes the heating of the MWNT in vacuum or heating and oxidation in air. As a result the central part of the MWNT between the electrodes (where is the peak of temperature) is destroyed.

The length of the removed part is about 100 nm. If the distance between the electrodes is less than 100 nm the walls are removed along the whole space between electrodes. It was shown that the walls of the MWNT are removed stepwise with one removed walls at each step. Therefore the desirable number of walls can be removed. Such a technique was used to destroy the outer walls near two ends of MWNT and thus fabricate MWNT-based nanomotor with movable shell of outer walls.

The possible way to produce a hole in the outer wall of a MWNT (such a hole is necessary for nanotube based jet nanoengine operation). The method of nanoscale modification of carbon nanotubes by focused electron beam irradiation has been recently developed. With the help of this method a set of knobs with diameter about 20 nanometres and controllable separation between knobs has bee fabricated on the nanotube surface. Evidently, that the matter of knobs is amorphous or graphite-like with numerous defects. In this case this matter can be easily removed by the mentioned above techniques which are used for nanotube caps removal. As the result the hole in the outer wall can be produced.

The metal atoms encapsulated inside the movable core of inner walls can become ions as a result of the transfer of valence electron to the nanotube and then to the structure which holds the fixed walls. In this case the movable inner core can be moved by electric field.

The following technique for production of such nanotubes with encapsulated ions has been proposed. At the first stage the nanotube should be filled by metallofullerenes by exposing in their vapour at high temperature. This method is demonstrated on the example of filling of the SWNTs by $Sc_2@C_{84}$. At the second stage the SWNTs filled with metallofullerenes (pea-pod nanotube) should be annealed into DWNTs encapsulated metal atoms. The annealing of pea-pod nanotubes into DWNTs by electron irradiation and by heating has been demonstrated.

The production of carbon MWNTs filled with magnetic material is perspective for control of motion of nanotube-based electromechanical nanodevices with the use of magnetic field. The following methods of such production can be listed. The suspending of nanotubes in nitric acid solution containing hydrated nickel nitrate or cobalt nitrate or iron nitrate leads to filling of nanotubes by oxides of Ni, Co of Fe, respectively. Nanotubes filled with the oxides of Ni, Co or Fe were treated under hydrogen at 400 °C. As a result the oxide has been reduced to the native metal. Nanotubes filled with Fe were produced in arc discharge in the presence of He and $Fe(CO)_5$ gases. Nanotubes filled with Ni were produced in arc discharge by doping a graphite anode with Ni and S.

REFERENCES

- Datta, S. : *Electronic Transport in Mesoscopic Systems,*Cambridge University Press, Cambridge, 1995.
- Deura, Vijay : *Nanotechnological Applications in Cancer Treatment* , Global Vision, Delhi, 2008.
- Hosono, H. : *Nanomaterials : From Research to Applications,* Elsevier, A Division of Reed Elsevier India, 2008.
- Krishna, K.: *Nanotechnology in Health and Life Sciences,* Campus Books International, Delhi, 2007.
- Lyshevski, Edward : *Nano- and Microelectromechanical Systems,* CRC Press, Delhi, 2000.
- Parthasarathy, B. K.: *Nanostructure and Nanomaterials,* Isha Books, Delhi, 2007.

Bibliography

Ahmad, Lone: *Nanoscience: Nanotechnologies and Nanophysics*, Jammu Kashmir: South Kashmir Press, 2006.

Allen, Robin: *An Introduction to Micro and Nanotechnology*, New York: Oxford University Press, 2002.

Aviram, A. : *Electronics using Hybrid-molecular and Mono-molecular Devices*, London, 2000.

Awasthi, K.P. : *Advanced Magnetic Nanostructures*, Cyber Tech Publications, Delhi, 2008.

Bainbridge, W.S. : *Societal Implications of Nanoscience and Nanotechnology*, UK: Intellect Books, 2001.

Basnet, J.: *Unique Features and Properties of Nanostructured Materials*, Delhi: Indian J Community Med, 2008.

Bhatia, A.S. : *Nano Electronics*, Deep and Deep, Delhi, 2011.

Bhatnagar, Vasudev : *Nanostructure Science and Technology*, Campus Books, Delhi, 2009.

Cerofolini, G.F.: *Nanoscale Devices*, Springer-Verlag, Berlin, 2009.

Chacko, Philip; Christabelle Noronha and Sujata Agrawa: *Small Wonder : The Making of the Nano*, Westland, Delhi, 2010.

Chrystler, H Lari : *Encyclopaedia of Nanoinformatics*, Dominant, Delhi, 2008.

Costlow, Laura and April Peter: *Nanomaterials and Nanostructures : Integral Components to Nanomachines, Nanorobots and Nanofactories Making Wonder Products for New Century*, Dominant, Delhi, 2007.

Datta, S. : *Electronic Transport in Mesoscopic Systems*,Cambridge University Press, Cambridge, 1995.

Deura, Vijay : *Nanotechnological Applications in Cancer Treatment* , Global Vision, Delhi, 2008.

Devoret, M.H. : *Single Charge Tunneling in Coulomb Blockade Phenomena in Nanostructures*, Plenum Press, New York/London, 1992.

Don Norman: *A Student Handbook of Nanobiotechnology*, Dominant, Delhi, 2008.

Drexler, K. Eric: *Nanosystems: Molecular Machinery, Manufacturing, and Computation*, New York: Wiley, 1992.

Eric, K. : *Nanosystems: Molecular Machinery, Manufacturing, and Computation*, New York: Wiley, 1992.

Fogelberg, H.: *The Material Culture of Nanotechnology*, Göteborg: Göteborg University, 2003.

Geckeler, K.E. : *Biomedical Applications of Nanostructured Materials*, Macmillan Publishers India, Delhi, 2011.

Gupta, Sushmita: *Understanding of Nano Science and Technology*, Global Vision, Delhi, 2006.

Hillebrands, B. : *Processes and Characterization of Advanced Nanostructured Materials*, Macmillan Publishers India, Delhi, 2011.

Hosono, H. : *Nanomaterials : From Research to Applications*Elsevier, A Division of Reed Elsevier India, 2008.

Ipsita Roy: *Bio-Nano-Geo Sciences : The Future Challenge*, Ane Books, Delhi, 2009.

Jana, S.: *Surface Analysis Methods for Characterizing Polymeric Biomaterials*, New Delhi: Long Life Publication, 2003.

Kapoor, Manish : *BioMEMS and Biomedical Nanotechnology*, Oxford Book Company, Delhi, 2012.

Kaushik, N. L. : *Nanotechnology and Micromachines*, Pearl Books, Delhi, 2007.

Kuhn T.: *"The Structure of Scientific Revolutions."* University of Chicago Press Chicago, 1962.

Lyshevski, Edward : *Nano- and Microelectromechanical Systems*, CRC Press, Delhi, 2000.

Marc Madou: *Fundamentals of Microfabrication*, CRC Press, Delhi, 1997.

Mathur, K.P. : *Nanotechnology Applications*, Rajat, Delhi, 2007.

Nanda, S.: *Properties of Individual Nanoparticles*, India: Indian Journal of Preventive and Medicine, 2006.

Parthasarathy, B. K.: *Nanostructure and Nanomaterials*, Isha Books, Delhi, 2007.

Phani, N. : *Principles of Nanotechnology : Materials, Tools and Process at Nanoscale*, Scitech Pub, Delhi, 2010.

Raghavan, Y.S. : *Nanostructures and Nanomaterials: Synthesis, Properties and Applications*, Arise Pub, Delhi, 2010.

Ranganathan, S.: *Handbook of Nanoparticles*, Chennai: T.T.Ranganathan Clinical research Foundation, 2008.

Reddy, K. : *Nanotechnology in Electronics*, Campus Books International, Delhi, 2007.

Siegel, R.W.: *Physics of New Materials*, Berlin, Springer 1994.

Siep L.: *"Nano-bio-ethics : Ethical Dimensions of Nanobiotechnology."* Transaction Publishers London, 2006.

Sinha, D.N.: *Nanotubes and Nanowires*, Patna: Indian Academy of Sciences, 2005.

Varghese, Thomas and Balakrishna, K.M. : *Nanotechnology : An Introduction To Synthesis Properties and Applications of Nanomaterials*, Atlantic, Delhi, 2012.

Zener, C.: *Elasticity and Anelasticity of Metals*. Chicago: University of Chicago Press, 1948.

Index

N

O

P

R

S

T

U

V